Coastal Dynamic and Evolution

Coastal Dynamic and Evolution

Editors

Giorgio Anfuso
Angela Rizzo

MDPI • Basel • Beijing • Wuhan • Barcelona • Belgrade • Manchester • Tokyo • Cluj • Tianjin

Editors
Giorgio Anfuso
University of Cádiz
Spain

Angela Rizzo
Centro Euro-Mediterraneo sui
Cambiamenti Climatici
Italy

Editorial Office
MDPI
St. Alban-Anlage 66
4052 Basel, Switzerland

This is a reprint of articles from the Special Issue published online in the open access journal *Water* (ISSN 2073-4441) (available at: https://www.mdpi.com/journal/water/special_issues/Coastal_Dynamic_Evolution).

For citation purposes, cite each article independently as indicated on the article page online and as indicated below:

LastName, A.A.; LastName, B.B.; LastName, C.C. Article Title. *Journal Name* **Year**, *Volume Number*, Page Range.

ISBN 978-3-03943-935-5 (Hbk)
ISBN 978-3-03943-936-2 (PDF)

Cover image courtesy of Angela Rizzo (Cape Trafalgar, Spain).

Contents

About the Editors

Giorgio Anfuso Ph.D. in Marine Science, has been a coastal researcher and full professor at the Faculty of Marine and Environmental Sciences at Cadiz University (Spain) since 1999. For about 25 years, he has worked on coastal processes at different time scales, from hours (by sediment transport and disturbance depth campaigns), days–months (by field surveys and monitoring programs) to decadal coastal evolution by means of aerial photos and satellite images analysis. He is also working on management issues and environmental sensitivity maps for oil spills, coastal scenery and litter characteristics and distribution, carrying out investigations in Spain, Italy, Morocco, Ireland, Colombia, Cuba, Mexico and Ecuador, among other countries. He participated in different national and international projects focused on beach morphology, evolution and response to storms' impacts. He is co-author of more than 100 peer-reviewed publications in international journals.

Angela Rizzo achieved a Ph.D. degree in Coastal Geomorphology in 2017 at the Parthenope University of Naples (Italy), where she later worked as a researcher. While obtaining her Ph.D., she collaborated with productive international research networks and was a visiting researcher at the Faculty of Marine Science at Cadiz University in Spain and at the Flood Hazard Research Centre in London (UK). Since 2018, she has been working at the Euro-Mediterranean Centre on Climate Change (REMHI Division, Italy) to support the management of international and national projects related to the assessment of climate change impacts on different land use contexts. Her research activity is focused on the assessment and mapping of coastal risks, with specific reference to the evaluation of potential impacts of expected sea-level rise. She has worked on the investigation of past and future dynamic and geomorphological evolution of different coastal sectors in Italy, Spain and Malta. Furthermore, by means of GIS spatial analysis tools, she determined the sensitivity and exposure of coastal features to forcing agents and related processes. She also collaborates in the analysis of Holocene sea-level variations by means of archaeological proxies. She participated in different national and international conferences on coastal geomorphology and vulnerability, and she is co-author of several peer-reviewed publications in international journals. Since 2014, she has been a member of the Italian Association of Geomorphology.

Preface to "Coastal Dynamic and Evolution"

Coastal areas and related landscapes have immense natural and environmental value. They represent ideal habitats for many plant and animal species, hence presenting high biodiversity and receiving extensive ecological interest. Furthermore, coastal areas and, in particular, lowland and coastal plains have always been preferred locations for human establishment, as they provide resources, assets and opportunities for commercial and economic activities and marine transport. Because of this intense and often uncontrolled urbanization, coastal areas are extremely vulnerable to direct and indirect impacts of human activities, primarily due to the high exploitation of resources and the need to protect economic interests and infrastructures. Moreover, coasts are extremely exposed to the impacts of meteorological and marine events that can cause significant socio-economic and environmental damages. In addition, coasts, with particular reference to topographically low areas that are very populated (around 10 percent of the world's population lives in areas that are less than 10 m above sea level), are currently considered hotspots of vulnerability to ongoing climate change, in terms of both increases in the frequency and intensity of extreme weather-related events and sea-level rise. For this reason, the proper management of coastal areas, which must take into account both the current dynamics and the anticipated geomorphological changes in response to global changes, is essential to ensure the short- and long-term maintenance of all ecosystem functions and services that are provided by coastal areas and related habitats.

Sustainable and integrated coastal management has now become such a high-priority objective at the international level that the most recent regulations and legislative indications foresee the adoption of specific actions aimed at reducing coastal risks and the implementation of tailored strategies to adapt to expected climate scenarios. These objectives cannot be separated from an in-depth knowledge of the dynamics that affect coastal sectors and constant monitoring of their state of conservation. For this reason, the main aim of this Special Issue was to collect studies that provide new views on coastal phenomena and processes in different coastal contexts by using different methodologies and approaches. The motivation to publish a volume on coastal dynamics stems from the desire to contribute to the dissemination of knowledge about one of the most complex and delicate environments that characterize our planet. Hopefully, the scientific collection presented herein can be of interest to different types of professionals involved in coastal studies and management so as to promote integrated research aimed at the sustainable use of resources that are provided by coastal environments.

Finally, we kindly thank all authors for their participation and contribution to the volume, as well as all reviewers, who have substantially contributed to the high quality of these published papers. We also highly appreciate the editorial support provided by the Water journal.

Giorgio Anfuso, Angela Rizzo
Editors

Editorial

Coastal Dynamic and Evolution: Case Studies from Different Sites around the World

Angela Rizzo [1,*] and Giorgio Anfuso [2]

[1]   REgional Models and geo-Hydrological Impacts (REMHI Division), Centro Euro-Mediterraneo sui Cambiamenti Climatici (CMCC), 81100 Caserta, Italy
[2]   Department of Earth Sciences (CASEM), Faculty of Marine and Environmental Sciences, University of Cadiz, 11510 Puerto Real, Cádiz, Spain; giorgio.anfuso@uca.es
*   Correspondence: angela.rizzo@cmcc.it

Received: 4 October 2020; Accepted: 9 October 2020; Published: 12 October 2020

Abstract: In recent decades, coastal areas have experienced a progressive increase in erosion and flooding processes as a consequence of the combined effect of natural factors and strong human pressures. These processes are particularly evident on low-lying areas and are expected to be exacerbated by the ongoing climate change, which will impact the littorals both in the short term, by affecting the duration and frequency of storms, and in the long term, by inducing variation in the sea-level position. In this context, this Special Issue is devoted to collecting geomorphological studies on coastal dynamic and evolution by means of multidisciplinary research methodologies and investigations, which represent a very useful set of information for supporting the integrated management of coastal zone. The volume includes 14 papers addressing three main topics (i) shoreline characterization, dynamic and evaluation; (ii) coastal hazard evaluation and impact assessment of marine events; and (iii) relevance of sediment collection and analysis for beach nourishment. Case studies from Russia, Italy, California (USA), Morocco, Spain, Indonesia, Ireland and Colombia are shown in the Special Issue, giving to the reader a wide overview of coastal settings and methodological approaches.

Keywords: coastal dynamics; coastal landscapes; coastal evolution assessment

---

1. Introduction

The present-day coastal landscape is essentially the result of the interaction among different factors, including the geological frame as well as continental and marine factors. The geological frame includes stratigraphic and tectonic assets of an area and usually determines the formation of cliffed coastlines or low coastal plains. Coastal plains show different landforms whose formation and evolution is strongly linked to continental and marine factors as sediment supplies from the land (e.g., streams and rivers) and the ocean (due to marine and aeolic processes), and the destructive action of marine energetic factors such as waves and currents. In the long term, sea level position variations acquire great importance in landscape formation and evolution [1,2]. As a result of the above-mentioned processes and factors, the coastal area shows a great variety of morphological forms ([1,3] Figure 1).

Furthermore, the coast, because of its location at the interface between land and water, represents the transitional zone between aquatic and terrestrial ecosystems and hence presents an intrinsic environmental value due to its high level of biodiversity that, at many places (e.g., at coastal areas with mangrove swamps and dune areas), supports the provision of several ecosystem services and related functions essential for human well-being too [4,5]. Last, coastal area represents a zone for recreational, cultural and industrial activities [6]. Hence, since ancient times, the coastal zone was characterized by a very high human occupation and, at present, coastal areas have become more densely populated than the hinterland and exhibit higher rates of population growth and urbanisation. Various studies estimated the population living in the coastal zone, and the most recent data show that about 10% of the world's population (ca. 600 million people) lives in low elevated coastal areas [7] and, according to Neumann et al. [8], this trend is expected to increase. Specifically, in their study [8], projections for coastal population densities under different economic and development scenarios for the year 2030 and 2060 were evaluated and then compared with the population abundance in 2000. The mentioned study also predicted that, under the "worst conditions" scenario characterized by low political governance and high global economic growth, an increase in population from 625 million (in 2000) up to 949 and 1388.2 million people could be expected in 2030 and 2060, respectively. In view of the high social, economic, and natural characteristics and related benefits, the sustainable conservation of coastal areas, as well as their integrated coastal management, are a worldwide issue [9]. To these aims, the 2030 Agenda for Sustainable Development defined, among the newly established 17 Sustainable Development Goals (SDGs), a specific one on the conservation and sustainable use of the oceans, seas and marine resources (SDG 14) [10]. Two specific targets of this goal (14.21 and 14.52) are devoted to address coastal areas and related ecosystems. Further targets under SDG 14 as well as targets under other goals, though not explicitly referred to coastal areas, are implicitly relevant for coastal areas and for the protection, conservation and management of coastal ecosystems and resources [9]. In this context, a crucial role is played by sensitivity/vulnerability and risk analysis studies, which allow for the evaluation of coastal proneness to hazardous processes such as erosion, flooding, and submersion, as well as potential coastal resilience capacity at global [11–13], regional [14–18] and local [19–24] levels. A key element in the risk evaluation procedure is represented by the assessment and damage evaluation of the natural and anthropic assets located in the areas prone to be affected by marine processes.

Figure 1. Examples of common coastal landforms. Source: authors.

2. Coastal Dynamic and Response Modalities

The strong link between forms and processes is the main characteristic of the coastal morpho-dynamic system [25]. As already introduced, the evolution of a sandy coastline is a function of highly dynamic processes acting at different spatial and temporal scales. Concerning coastal processes that occur at small scales, as in the case of storms, they generally result in rapid erosion (at a time scale of hours/days), followed by accretion that usually takes place during weeks and months, leading to negligible net change over time scales of a few months or at an annual scale [11]. Coastal changes due to the impact of hurricanes are also rapid, but recovery can take a long time, e.g., years [26]. On the contrary, if a consistent reduction of sediment supply occurs and persists for several years (i.e., at a medium-term time scale), chronic erosion processes are triggered, with consequent negative impacts on natural and anthropic assets (Figure 2). It is worth noting that, in recent times, coasts are particularly exposed to the consequences of intense erosion process mainly induced by human-related activities, both along the littoral and inland in the main watershed, which represents an additional pressure on the natural occurrence of the erosion process. Anthropic structures built along the shores of the world age back to ancient times and include different kinds of structures, such as harbours, breakwaters, and fish tanks. Nowadays, they are widely used as geo-archaeological proxies for measuring relative sea-level variations and coastal landscape evolution during the mid-late Holocene [27–31]. In more recent years, anthropic structures are mainly represented by engineering structures aimed at protecting coastal assets (both natural and anthropic) from erosion and flooding processes. Too often, these structures have generated increasing erosion processes in the adjacent zones, moving downdrift the problem without solving it and the generating causes [32–35].

Figure 2. Examples of chronic erosion along low coastal areas in Italy and Colombia. Source: Giorgio Anfuso, Angela Rizzo, Gianluigi Di Paola.

As a result of natural processes and human influences/actuations, coastal areas are facing intense erosion worldwide [11,36,37]. A qualitative study published at the beginning of 1980s [38] provided an assessment of erosion rates for sandy beaches at the global scale. It was estimated that 70% of them are eroding. More recently, Luijendijk et al. [11], based on the available optical satellite images captured since 1984, have provided a quantitative overview of the state of the world's beaches. Twenty-four percent of the world's sandy beaches are eroding at rates exceeding 0.5 m/y, and about 7% of the beaches experience erosion rates that can be classified as severe with rates up to 5 m/y. Furthermore, erosion rates exceed 5 m/y along 4% of the sandy shoreline and are greater than 10 m/y for 2% of the global sandy shoreline.

Concerning changes at large spatial and temporal scales, coastal environments shift landward or seaward as a consequence of marine transgression or regression, which cause the submersion or the emersion of coastal landforms, respectively. Conceptually, different geomorphological coastal responses can be identified as overall consequence to the combined effect of sea level changes and sedimentary processes along low-lying areas: (i) coastal submersion/rapid shoreline retreat is observed when sea level rise rates are very high and it is not balanced by sedimentary processes; (ii) coastal erosion/shoreline retreat is observed when the sea level rise rates are high and sedimentary processes allow the development and the adaptation of the coastal environments; (iii) coastal equilibrium is observed when sedimentary processes balance the sea level rates and the system is characterized by a dynamic equilibrium and (iv) and coastal progradation/shoreline advance is recorded when coastal sedimentary processes prevail on the sea level rise rates. This phenomenon is considered a marine regression.

The variation in the sea level position is not the only impact of climate change. It also affects the frequency, intensity, and persistence of climate extreme processes (such as storms, precipitations, etc.), with consequent impact on the occurrence of intense coastal storms and flood events [39,40], i.e., to favour specific and chronic erosion and flooding processes. A wide set of studies published in recent years are focused on the evaluation of the expected increasing impact in zones prone to be flooded as consequence of the sea level rise [16,20,22,41–43], storm surges [44–48] and their combined occurrence [12,49]. Very recently, global projections of extreme sea levels and resulting episodic coastal flooding over the 21th century have been assessed and mapped in Kirezci et al. [49]. The results of the mentioned paper show that the mean inundated area evaluated accounting for the future sea level is expected to increase in a range of 35–50% by the end of the century compared with the present conditions, while 0.5–0.7% of the world land area will be at risk of episodic coastal flooding by 2100 from a 1 in 100-year return period event, with an increase of 48% compared to present day.

Since the sea level-related hazard is expected to increase as a consequence of future climate scenarios, these analyses are carried out by accounting for future projections of sea positions based on both semi-empirical [50–53] and model-based methods [54,55] that provided global estimations of the expected increases in sea levels under different scenarios of an increase in temperature linked to the increase in the concentration of climate-altering gasses and their representative pathways (RCPs). Nevertheless, in order to obtain a regional estimation of the future sea-level position, the local contribution of land movements due to regional geological processes (tectonics and isostasy) have to be considered too [56,57]. Finally, a very local evaluation is obtained by adding vertical ground displacements mainly due to natural and human subsidence as a consequence of sediment compaction [58,59].

In this context, multidisciplinary research and integrated investigations aimed at the identification and evaluation of the variation of several coastal features—which include the present-day geomorphological, natural and anthropic settings such as the presence of a dune system, the main sediment composition, the recent shoreline trend and the presence of defence structures, as well as the identification of the position of ancient sea-level stands—represent a key step to define the proneness of coastal sectors to potential negative marine impacts and their capacity of coping with them. At the same time, the assessment of future coastal evolution by means of the estimation of future scenarios of sea-level positions in a climate change context can be considered as a way forward regarding the reduction of coastal risks and the definition and implementation of suitable adaptive strategies aimed at increasing the intrinsic resilience of the coastal stretches. This latter aspect is in line with the most recent international requirements and strategies for addressing climate adaptation and risk reduction challenges. At the European level, the Strategy on Adaptation to Climate Change [60], which is aimed at making Europe more resilient and minimising the effects of unavoidable climate change, has stressed the concept that coastal zones are particularly vulnerable to the impact of sea-level rise, challenging the climate resilience and adaptive capacity of coastal societies.

3. Overview of this Special Issue

This Special Issue is intended at providing a number of new geomorphological studies focused on coastal dynamics and evolution across the world. The volume includes 14 papers concerning shoreline and/or dune system morphological changes at different time scales and in a context of climate change scenarios, obtained from different kinds of operational models/instruments and field studies as well as surveys and observations by means of aerial photos and satellite images. Specifically, papers included in this SI can be grouped into three main categories:

3.1. Shoreline Characterization, Dynamics and Evaluation

Eight papers belong to this category and provide information concerning the characterization and evolution at different spatial and temporal scales of a great variety of coastal environments, including mangrove swamps (along the Caribbean coast of Colombia), and sandy and rocky coasts and dune ridges in coastal sectors located in Russia, Italy, California (USA), Morocco and Spain. A further paper included in this category is aimed at eliciting key concepts determining the aesthetic appeal of coastal dunes and forests. In this case, the example of the Curonian Spit (Lithuania) is described. The following papers are included in this category:

- Nicu et al. [61]. *Shoreline Dynamics and Evaluation of Cultural Heritage Sites on the Shores of Large Reservoirs: Kuibyshev Reservoir, Russian Federation.*
- Mammì et al. [62]. *Mathematical Reconstruction of Eroded Beach Ridges at the Ombrone River Delta.*
- Griggs et al. [63]. *Documenting a Century of Coastline Change along Central California and Associated Challenges: From the Qualitative to the Quantitative.*
- Taaouati et al. [64]. *Influence of a Reef Flat on Beach Profiles Along the Atlantic Coast of Morocco.*
- Villate Daza et al. [65]. *Mangrove Forests Evolution and Threats in the Caribbean Sea of Colombia.*
- Molina et al. [66]. *Dune Systems' Characterization and Evolution in the Andalusia Mediterranean Coast (Spain).*
- Mattei et al. [67]. *New Geomorphological and Historical Elements on Morpho-Evolutive Trends and Relative Sea-Level Changes of Naples Coast in the Last 6000 Years.*
- Urbis et al. [68]. *Key Aesthetic Appeal Concepts of Coastal Dunes and Forests on the Example of the Curonian Spit (Lithuania).*

3.2. Coastal Hazard Evaluation and Impact Assessment of Marine Events

Four study cases analyse the impact of different marine processes, such as storms (in the Tordera Delta, Spain), sea-level rise (in the Jakarta Bay, Indonesia), and tsunami (along the Ionic Sicilian coast, Italy), as well as the impact of a specific post-tropical cyclone, on the Northern coast of Ireland. The following papers are included in this category:

- Sanuy et al. [69]. *Sensitivity of Storm-Induced Hazards in a Highly Curvilinear Coastline to Changing Storm Directions. The Tordera Delta Case (NW Mediterranean).*
- Yahya Surya et al. [70]. *Impacts of Sea-Level Rise and River Discharge on the Hydrodynamics Characteristics of Jakarta Bay (Indonesia).*
- Anfuso et al. [71]. *Spatial Variability of Beach Impact From Post-Tropical Cyclone Katia (2011) on Northern Ireland's North Coast.*
- Lo Re et al. [72]. *Tsunami Propagation and Flooding in Sicilian Coastal Areas by Means of a Weakly Dispersive Boussinesq Model.*

3.3. Relevance of Sediment Collection and Analysis for Coastal Nourishment

Two papers are included in this section. The first paper is focused on the analysis of differences in sand composition and colours between natural and artificially nourished beaches in Southern Mediterranean Spain. The second paper deals with the analysis of the influence of different sieving methods on the estimation of sand size parameters to determine suitable sediments for beach nourishment, with exempla from the Southern Atlantic coast of Spain. The following papers are included in this category:

- Poullet et al. [73]. *Influence of Different Sieving Methods on Estimation of Sand Size Parameters.*
- Asensio-Montesinos et al. [74]. *The Origin of Sand and its Colour on the South-Eastern Coast of Spain: Implications for Erosion Management.*

4. Conclusions

The high variability and dynamics of coastal morphologies and landforms require the application of tailored approaches for the assessment of local changes in the short, medium, and long term. The evaluation of the present morphological changes in coastal areas and their comparison with the past conditions lies in the identification of specific indicators in terms of beach erosion rates, dune and foredune extent variation, etc. Such variations can be considered as proxies for hazardous processes, i.e., chronic erosion processes, temporary flooding and permanent inundation, which could be exacerbated by ongoing climate change.

The analysis of local dynamics, as well as the identification of the main drivers triggering these hazardous processes, represent therefore a key point for the definition and implementation of suitable management actions aimed at reducing human- and climate-induced risks on coastal zones. For ensuring the effectiveness of coastal management actions as well as the suitability of the implementation of adaptive solutions to expected changes, it is fundamental to take into account the uniqueness of each coastal area to provide site-specific and tailored solutions. In this context, mapping and zoning of the areas prone to be impacted by hazardous processes as well as the assessment of the potentially exposed natural and anthropic assets represent the most appropriate operational approaches to provide detailed information and reduce the potential impacts.

In conclusion, by proposing different case studies from different coastal areas of the world, this Special Issue contributes to increasing the dissemination of specific knowledge at the transnational level favouring the exchange of results between researchers, promoting in this way the exploitation of assessment methods and approaches.

Author Contributions: The authors led the development of the Special Issue and contributed equally to the preparation of this manuscript. Conceptualization: A.R. and G.A.; writing—original draft preparation: A.R. and G.A.; writing—review and editing: A.R. and G.A. All authors have read and agreed to the published version of the manuscript.

Funding: This research received no external funding.

Conflicts of Interest: The authors declare no conflict of interest.

References

1. Masselink, G.; Gehrels, R. *Coastal Environments and Global Change*; John Wiley & Sons: West Sussex, UK, 2014; p. 448.
2. Kelsey, H.M.; Bockheim, J.G. Coastal landscape evolution as a function of eustasy and surface uplift rate, Cascadia margin, southern Oregon. *Geol. Soc. Am. Bull.* **1994**, *106*, 840–854. [CrossRef]
3. Davidson-Arnott, R. *An Introduction to Coastal Processes and Geomorphology*; Cambridge University Press: Cambridge, UK, 2010; p. 458.
4. Reid, W.V.; Mooney, H.A.; Cropper, A.; Capistrano, D.; Carpenter, S.R.; Chopra, K.; Dasgupta, P.; Dietz, T.; Duraiappah, A.K.; Hassan, R.; et al. *Ecosystems and Human Well–Being: Synthesis*; Millennium Ecosystem Assessment: Washington, DC, USA, 2005.
5. Maes, J.; Teller, A.; Erhard, M.; Liquete, C.; Braat, L.; Berry, P.; Egoh, B.; Puydarrieux, P.; Fiorina, C.; Santos, F.; et al. *Mapping and Assessment of Ecosystems and Their Services. An Analytical Framework for Ecosystem Assessments under Action 5 of the EU Biodiversity Strategy to 2020*; Publications Office of the European Union: Luxembourg, 2013; pp. 1–58.
6. Fabbri, P. (Ed.) *Recreational Uses of Coastal Areas: A Research Project of the Commission on the Coastal Environment, International Geographical Union (Vol. 12)*; Springer Science & Business Media: Heidelberg, Germany, 2012; p. 287.
7. McGranahan, G.; Balk, D.; Anderson, B. The rising tide: Assessing the risks of climate change and human settlements in low elevation coastal zones. *Environ. Urban.* **2007**, *19*, 17–37. [CrossRef]

8. Neumann, B.; Vafeidis, A.T.; Zimmermann, J.; Nicholls, R.J. Future coastal population growth and exposure to sea-level rise and coastal flooding-a global assessment. *PLoS ONE* **2015**, *10*, e0118571. [CrossRef] [PubMed]

9. Neumann, B.; Ott, K.; Kenchington, R. Strong sustainability in coastal areas: A conceptual interpretation of SDG 14. *Sustain. Sci.* **2017**, *12*, 1019–1035. [CrossRef] [PubMed]

10. United Nations. Transforming Our World: The 2030 Agenda for Sustainable Development. UNGA Resolution A/RES/70/1. Resolution Adopted by the General Assembly on 25 September 2015. Available online: https://www.un.org/en/development/desa/population/migration/generalassembly/docs/globalcompact/A_RES_70_1_E.pdf (accessed on 10 October 2020).

11. Luijendijk, A.; Hagenaars, G.; Ranasinghe, R.; Baart, F.; Donchyts, G.; Aarninkhof, S. The state of the world's beaches. *Sci. Rep.* **2018**, *8*, 1–11. [CrossRef]

12. Kirezci, E.; Young, I.R.; Ranasinghe, R.; Muis, S.; Nicholls, R.J.; Lincke, D.; Hinkel, J. Projections of global-scale extreme sea levels and resulting episodic coastal flooding over the 21st Century. *Sci. Rep.* **2020**, *10*, 1–12. [CrossRef]

13. Kulp, S.A.; Strauss, B.H. New elevation data triple estimates of global vulnerability to sea-level rise and coastal flooding. *Nat. Commun.* **2019**, *10*, 1–12.

14. Nicholls, R.J.; Hoozemans, F.M.; Marchand, M. Increasing flood risk and wetland losses due to global sea-level rise: Regional and global analyses. *Glob. Environ. Chang.* **1999**, *9*, S69–S87. [CrossRef]

15. Molina, R.; Manno, G.; Re, C.L.; Anfuso, G.; Ciraolo, G. A Methodological Approach to Determine Sound Response Modalities to Coastal Erosion Processes in Mediterranean Andalusia (Spain). *J. Mar. Sci. Eng.* **2020**, *8*, 154. [CrossRef]

16. Aucelli, P.P.C.; Di Paola, G.; Rizzo, A.; Rosskopf, C.M. Present day and future scenarios of coastal erosion and flooding processes along the Italian Adriatic coast: The case of Molise region. *Environ. Earth Sci.* **2018**, *77*, 371. [CrossRef]

17. Rangel-Buitrago, N.; Anfuso, G. *Risk Assessment of Storms in Coastal Zones: Case Studies from Cartagena (Colombia) and Cadiz (Spain)*; Springer: Dordrecht, The Netherlands, 2015; p. 63.

18. Anfuso, G.; Martinez, J.A. Assessment of coastal vulnerability through the use of GIS tools in South Sicily (Italy). *Environ. Manag.* **2009**, *43*, 533–545. [CrossRef] [PubMed]

19. Rizzo, A.; Vandelli, V.; Buhagiar, G.; Micallef, A.S.; Soldati, M. Coastal vulnerability assessment along the North-Eastern sector of Gozo Island (Malta, Mediterranean Sea). *Water* **2020**, *12*, 1405. [CrossRef]

20. Di Paola, G.; Alberico, I.; Aucelli, P.P.C.; Matano, F.; Rizzo, A.; Vilardo, G. Coastal subsidence detected by Synthetic Aperture Radar interferometry and its effects coupled with future sea-level rise: The case of the Sele Plain (Southern Italy). *J. Flood Risk Manag.* **2018**, *11*, 191–206. [CrossRef]

21. Rizzo, A.; Aucelli, P.P.C.; Gracia, F.J.; Anfuso, G. A novelty coastal susceptibility assessment method: Application to Valdelagrana area (SW Spain). *J. Coast. Conserv.* **2018**, *22*, 973–987. [CrossRef]

22. Aucelli, P.P.C.; Di Paola, G.; Incontri, P.; Rizzo, A.; Vilardo, G.; Benassai, G.; Buonocore, B.; Pappone, G. Coastal inundation risk assessment due to subsidence and sea level rise in a Mediterranean alluvial plain (Volturno coastal plain-southern Italy). *Estuar. Coast. Shelf Sci.* **2017**, *198*, 597–609. [CrossRef]

23. Anfuso, G.; Gracia, F.J.; Battocletti, G. Determination of cliffed coastline sensitivity and associated risk for human structures: A methodological approach. *J Coast. Res.* **2013**, *29*, 1292–1296.

24. Özyurt, G.; Ergin, A. Application of sea level rise vulnerability assessment model to selected coastal areas of Turkey. *J. Coast. Res.* **2009**, *51*, 248–251.

25. Cowell, P.J.; Thom, B.G. Morphodynamics of coastal evolution. In *Coastal Evolution: Late Quaternary Shoreline Morphodynamics*; Carter, R.W.G., Woodroffe, C.D., Eds.; Cambridge University Press: Cambridge, UK, 1994; pp. 33–86.

26. Meyer-Arendt, K. Grand Isle, Louisiana: A historic US Gulf Coast Resort Adapts to Hurricanes, Subsidence and Sea Level Rise. In *Disappearing Destinations*; Jones, A., Phillips, M., Eds.; CABI: Wallingford, UK, 2011; pp. 203–217.

27. Morhange, C.; Marriner, N. Archeological and biological relative sea-level indicators. In *Handbook of Sea-Level Research*; Shennan, I., Long, A.J., Horton, B.P., Eds.; Wiley & Sons, Ltd.: West Sussex, UK, 2015; pp. 146–156.

28. Vacchi, M.; Ermolli, E.R.; Morhange, C.; Ruello, M.R.; Di Donato, V.; Di Vito, M.A.; Boetto, G. Millennial variability of rates of sea-level rise in the ancient harbour of Naples (Italy, western Mediterranean Sea). *Quat. Res.* **2020**, *93*, 284–298. [CrossRef]

29. Pappone, G.; Aucelli, P.P.; Mattei, G.; Peluso, F.; Stefanile, M.; Carola, A. A Detailed Reconstruction of the Roman Landscape and the Submerged Archaeological Structure at "Castel dell'Ovo islet" (Naples, Southern Italy). *Geosciences* **2019**, *9*, 170. [CrossRef]

30. Aucelli, P.; Cinque, A.; Mattei, G.; Pappone, G.; Rizzo, A. Studying relative sea level change and correlative adaptation of coastal structures on submerged Roman time ruins nearby Naples (southern Italy). *Quat. Int.* **2019**, *501*, 328–348. [CrossRef]

31. Mattei, G.; Troisi, S.; Aucelli, P.P.; Pappone, G.; Peluso, F.; Stefanile, M. Sensing the submerged landscape of Nisida Roman Harbour in the Gulf of Naples from integrated measurements on a USV. *Water* **2018**, *10*, 1686. [CrossRef]

32. Molina, R.; Anfuso, G.; Manno, G.; Gracia-Prieto, F.J. The Mediterranean coast of Andalusia (Spain): Medium-term evolution and impacts of coastal structures. *Sustainability* **2019**, *11*, 3539. [CrossRef]

33. Williams, A.T.; Rangel-Buitrago, N.; Pranzini, E.; Anfuso, G. The management of coastal erosion. *Ocean Coast. Manag.* **2018**, *156*, 4–20. [CrossRef]

34. Pranzini, E. Coastal erosion and shore protection: A brief historical analysis. *J. Coast. Conserv.* **2018**, *22*, 827–830. [CrossRef]

35. Manno, G.; Anfuso, G.; Messina, E.; Williams, A.T.; Suffo, M.; Liguori, V. Decadal evolution of coastline armouring along the Mediterranean Andalusia littoral (South of Spain). *Ocean Coast. Manag.* **2016**, *124*, 84–99. [CrossRef]

36. Vousdoukas, M.I.; Ranasinghe, R.; Mentaschi, L.; Plomaritis, T.A.; Athanasiou, P.; Luijendijk, A.; Feyen, L. Sandy coastlines under threat of erosion. *Nat. Clim. Chang.* **2020**, *10*, 260–263. [CrossRef]

37. Mentaschi, L.; Vousdoukas, M.I.; Pekel, J.F.; Voukouvalas, E.; Feyen, L. Global long-term observations of coastal erosion and accretion. *Sci. Rep.* **2018**, *8*, 1–11. [CrossRef]

38. Bird, E.C.F. *Coastline Changes; A Global Review*; Wiley: Chichester, UK, 1985.

39. IPCC. *Climate Change 2013: The Physical Science Basis. Contribution of Working Group I to the Fifth Assessment Report of the Intergovernmental Panel on Climate Change*; Stocker, T.F., Qin, D., Plattner, G.-K., Tignor, M., Allen, S.K., Boschung, J., Nauels, A., Xia, Y., Bex, V., Midgley, P.M., Eds.; Cambridge University Press: Cambridge, UK; New York, NY, USA, 2013; p. 1535.

40. IPCC. *Contribution of Working Group I to the Fourth Assessment Report of the Intergovernmental Panel on Climate Change, 2007*; Solomon, S., Qin, D., Manning, M., Chen, Z., Marquis, M., Averyt, K.B., Tignor, M., Miller, H.L., Eds.; Cambridge University Press: Cambridge, UK; New York, NY, USA, 2007.

41. Antonioli, F.; Falco, G.D.; Presti, V.L.; Moretti, L.; Scardino, G.; Anzidei, M.; Marsico, A. Relative Sea-Level Rise and Potential Submersion Risk for 2100 on 16 Coastal Plains of the Mediterranean Sea. *Water* **2020**, *12*, 2173. [CrossRef]

42. Antonioli, F.; Anzidei, M.; Amorosi, A.; Presti, V.L.; Mastronuzzi, G.; Deiana, G.; Marsico, A. Sea-level rise and potential drowning of the Italian coastal plains: Flooding risk scenarios for 2100. *Quat. Sci. Rev.* **2017**, *158*, 29–43. [CrossRef]

43. Nicholls, R.J.; Cazenave, A. Sea-level rise and its impact on coastal zones. *Science* **2010**, *328*, 1517–1520. [CrossRef]

44. Viavattene, C.; Jiménez, J.A.; Ferreira, O.; Priest, S.; Owen, D.; McCall, R. Selecting coastal hotspots to storm impacts at the regional scale: A Coastal Risk Assessment Framework. *Coast. Eng.* **2018**, *134*, 33–47. [CrossRef]

45. Rahmstorf, S. Rising hazard of storm-surge flooding. *Proc. Natl. Acad. Sci. USA* **2017**, *114*, 11806–11808. [CrossRef] [PubMed]

46. Castelle, B.; Marieu, V.; Bujan, S.; Splinter, K.D.; Robinet, A.; Sénéchal, N.; Ferreira, S. Impact of the winter 2013–2014 series of severe Western Europe storms on a double-barred sandy coast: Beach and dune erosion and megacusp embayments. *Geomorphology* **2015**, *238*, 135–148. [CrossRef]

47. Guisado-Pintado, E.; Jackson, D.W.T. Multi-scale variability of storm Ophelia 2017: The importance of synchronised environmental variables in coastal impact. *Sci. Total Environ.* **2018**, *630*, 287–301. [CrossRef] [PubMed]

48. Li, K.; Li, G.S. Vulnerability assessment of storm surges in the coastal area of Guangdong Province. *Nat. Hazards Earth Syst. Sci.* **2011**, *11*, 2003–2011. [CrossRef]

49. Vousdoukas, M.I.; Mentaschi, L.; Voukouvalas, E.; Verlaan, M.; Jevrejeva, S.; Jackson, L.P.; Feyen, L. Global probabilistic projections of extreme sea levels show intensification of coastal flood hazard. *Nat. Commun.* **2018**, *9*, 1–12. [CrossRef] [PubMed]

50. Kopp, R.E.; Kemp, A.C.; Bittermann, K.; Horton, B.P.; Donnelly, J.P.; Gehrels, W.R.; Rahmstorf, S. Temperature-driven global sea-level variability in the Common Era. *Proc. Natl. Acad. Sci. USA* **2016**, *113*, E1434–E1441. [CrossRef]

51. Rahmstorf, S. Modeling sea level rise. *Nat. Educ. Knowl.* **2012**, *3*, 4.

52. Rahmstorf, S. A new view on sea level rise. *Nat. Rep. Clim. Chang.* **2010**, *4*, 44–45. [CrossRef]

53. Rahmstorf, S. A semi-empirical approach to projecting future sea-level rise. *Science* **2007**, *315*, 368–370. [CrossRef]

54. Church, J.A.; Clark, P.U.; Cazenave, A.; Gregory, J.M.; Jevrejeva, S.; Levermann, A.; Merrifield, M.A.; Milne, G.A.; Nerem, R.S.; Nunn, P.D.; et al. Sea Level Change. In *Climate Change 2013: The Physical Science Basis. Contribution of Working Group I to the Fifth Assessment Report of the Intergovernmental Panel on Climate Change*; Stocker, T.F., Qin, D., Plattner, G.-K., Tignor, M., Allen, S.K., Boschung, J., Nauels, A., Xia, Y., Bex, V., Midgley, P.M., Eds.; Cambridge University Press: Cambridge, UK; New York, NY, USA, 2013.

55. Giorgi, F.; Lionello, P. Climate change projections for the Mediterranean region. *Glob. Planet. Chang.* **2008**, *63*, 90–104. [CrossRef]

56. Rovere, A.; Furlani, S.; Benjamin, J.; Fontana, A.; Antonioli, F. MEDFLOOD project: MEDiterranean sea-level change and projection for future FLOODing. *Alp. Mediterr. Quat.* **2012**, *25*, 3–6.

57. Lambeck, K.; Antonioli, F.; Anzidei, M.; Ferranti, L.; Leoni, G.; Scicchitano, G.; Silenzi, S. Sea level change along the Italian coast during the Holocene and projections for the future. *Quat. Int.* **2011**, *232*, 250–257. [CrossRef]

58. Matano, F.; Sacchi, M.; Vigliotti, M.; Ruberti, D. Subsidence trends of volturno river coastal plain (northern Campania, southern italy) inferred by sar interferometry data. *Geosciences* **2018**, *8*, 8. [CrossRef]

59. Teatini, P.; Tosi, L.; Strozzi, T. Quantitative evidence that compaction of Holocene sediments drives the present land subsidence of the Po Delta, Italy. *J. Geophys. Res. Solid Earth* **2011**, *116*. [CrossRef]

60. European Commission. *An EU Strategy on Adaptation to Climate Change*; The European Commission: Brussels, Belgium, 2013.

61. Nicu, I.C.; Usmanov, B.; Gainullin, I.; Galimova, M. Shoreline Dynamics and Evaluation of Cultural Heritage Sites on the Shores of Large Reservoirs: Kuibyshev Reservoir, Russian Federation. *Water* **2019**, *11*, 591. [CrossRef]

62. Mammì, I.; Rossi, L.; Pranzini, E. Mathematical Reconstruction of Eroded Beach Ridges at the Ombrone River Delta. *Water* **2019**, *11*, 2281. [CrossRef]

63. Griggs, G.; Davar, L.; Reguero, B.G. Documenting a Century of Coastline Change along Central California and Associated Challenges: From the Qualitative to the Quantitative. *Water* **2019**, *11*, 2648. [CrossRef]

64. Taaouati, M.; Parisi, P.; Passoni, G.; Lopez-Garcia, P.; Romero-Cozar, J.; Anfuso, G.; Vidal, J.; Muñoz-Perez, J.J. Influence of a Reef Flat on Beach Profiles along the Atlantic Coast of Morocco. *Water* **2020**, *12*, 790. [CrossRef]

65. Villate Daza, D.A.; Sánchez Moreno, H.; Portz, L.; Portantiolo Manzolli, R.; Bolívar-Anillo, H.J.; Anfuso, G. Mangrove Forests Evolution and Threats in the Caribbean Sea of Colombia. *Water* **2020**, *12*, 1113. [CrossRef]

66. Molina, R.; Manno, G.; Lo Re, C.; Anfuso, G. Dune Systems' Characterization and Evolution in the Andalusia Mediterranean Coast (Spain). *Water* **2020**, *12*, 2094. [CrossRef]

67. Mattei, G.; Aucelli, P.P.C.; Caporizzo, C.; Rizzo, A.; Pappone, G. New Geomorphological and Historical Elements on Morpho-Evolutive Trends and Relative Sea-Level Changes of Naples Coast in the Last 6000 Years. *Water* **2020**, *12*, 2651. [CrossRef]

68. Urbis, A.; Povilanskas, R.; Šimanauskienė, R.; Taminskas, J. Key Aesthetic Appeal Concepts of Coastal Dunes and Forests on the Example of the Curonian Spit (Lithuania). *Water* **2019**, *11*, 1193. [CrossRef]

69. Sanuy, M.; Jiménez, J.A. Sensitivity of Storm-Induced Hazards in a Highly Curvilinear Coastline to Changing Storm Directions. The Tordera Delta Case (NW Mediterranean). *Water* **2019**, *11*, 747. [CrossRef]

70. Yahya Surya, M.; He, Z.; Xia, Y.; Li, L. Impacts of Sea Level Rise and River Discharge on the Hydrodynamics Characteristics of Jakarta Bay (Indonesia). *Water* **2019**, *11*, 1384. [CrossRef]

71. Anfuso, G.; Loureiro, C.; Taaouati, M.; Smyth, T.; Jackson, D. Spatial Variability of Beach Impact from Post-Tropical Cyclone Katia (2011) on Northern Ireland's North Coast. *Water* **2020**, *12*, 1380. [CrossRef]

72. Lo Re, C.; Manno, G.; Ciraolo, G. Tsunami Propagation and Flooding in Sicilian Coastal Areas by Means of a Weakly Dispersive Boussinesq Model. *Water* **2020**, *12*, 1448. [CrossRef]
73. Poullet, P.; Muñoz-Perez, J.J.; Poortvliet, G.; Mera, J.; Contreras, A.; Lopez, P. Influence of Different Sieving Methods on Estimation of Sand Size Parameters. *Water* **2019**, *11*, 879. [CrossRef]
74. Asensio-Montesinos, F.; Pranzini, E.; Martínez-Martínez, J.; Cinelli, I.; Anfuso, G.; Corbí, H. The Origin of Sand and Its Colour on the South-Eastern Coast of Spain: Implications for Erosion Management. *Water* **2020**, *12*, 377. [CrossRef]

Article

Shoreline Dynamics and Evaluation of Cultural Heritage Sites on the Shores of Large Reservoirs: Kuibyshev Reservoir, Russian Federation

Ionut Cristi Nicu [1],*, Bulat Usmanov [2], Iskander Gainullin [3] and Madina Galimova [3]

[1]   High North Department, Norwegian Institute for Cultural Heritage Research (NIKU), Fram Centre, N-9296 Tromsø, Norway

[2]   Department of Landscape Ecology, Institute of Environmental Sciences, Kazan Federal University, 5 Tovarisheskaya Street, 420097 Kazan, Russia; busmanof@kpfu.ru

[3]   Khalikov Institute of Archaeology, 30 Butlerova Street, 420012 Kazan, Russia; gainullis@gmail.com (I.G.); mgalimova@yandex.ru (M.G.)

*   Correspondence: ionut.cristi.nicu@niku.no or nicucristi@gmail.com; Tel.: +47-98063607

Received: 15 February 2019; Accepted: 20 March 2019; Published: 21 March 2019

Abstract: Over the last decades, the number of artificial reservoirs around the world has considerably increased. This leads to the formation of new shorelines, which are highly dynamic regarding erosion and deposition processes. The present work aims to assess the direct human action along the largest reservoir in Europe—Kuibyshev (Russian Federation) and to analyse threatened cultural heritage sites from the coastal area, with the help of historical maps, UAV (unmanned aerial vehicle), and topographic surveys. This approach is a necessity, due to the oscillating water level, local change of climate, and to the continuous increasing of natural hazards (in this case coastal erosion) all over the world. Many studies are approaching coastal areas of the seas and oceans, yet there are fewer studies regarding the inland coastal areas of large artificial reservoirs. Out of the total number of 1289 cultural heritage sites around the Kuibyshev reservoir, only 90 sites are not affected by the dam building; the rest had completely disappeared under the reservoir's water. The scenario of increasing and decreasing water level within the reservoir has shown the fact that there must be water oscillations greater than ± 1 m in order to affect the cultural heritage sites. The results show that the coastal area is highly dynamic and that the complete destruction of the last remaining Palaeolithic site (Beganchik) from the shoreline of Kuibyshev reservoir is imminent, and immediate mitigation measures must be undertaken.

Keywords: cultural heritage; shoreline dynamics; GIS; UAV; Palaeolithic; Volga; European Russia

1. Introduction

The construction of large reservoirs along the large rivers of the world has, eventually, different effects: Local micro-climate modifications, disruption on the river flow regime [1], sediment transport [2,3], fauna [4], water chemistry [5], shore morphology [6–8], archaeology [9], fish yields [10], among other issues. They can also act as a place where different types of pollutants accumulate, and, in this way, it is easier to assess historical pollution [11]. One of the main effects is the triggering and the fast mechanic action of waves. These effects are accentuated by the global climatic changes, which are exponentially increasing every year.

Many studies deal with risk assessment [12,13], management [14,15], vulnerability [16–18], conservation strategies [19,20] and sustainability issues [21] regarding the cultural heritage of the coastal areas of seas and oceans. However, there is a lack of studies dealing with inland shorelines of large man-made reservoirs [22]. The Volga River is the largest river in Europe with a basin area of

1,360,000 km^2; it is considered the main river in Russia, and its basin represents the most significant economic region in Russia [23]. During the Soviet Union, there was a usual practice to flood large territories in order to obtain electricity and to relocate a large number of inhabitants and their houses. Unfortunately, cultural heritage sites do not enter this category; they cannot be relocated or moved.

During the Soviet period (the late 1930s), the "Great Volga Scheme" was initiated; the purpose was the construction of a chain of dams along the Volga River and one of its major tributaries—the Kama River. The reservoirs of the Volga-Kama cascade are one of the largest cascades in the world, totaling 11 reservoirs (Figure 1, Table 1). The main purpose of the dams was to produce electricity; before the 1930s, the Volga was used only for transport and fishing [24,25]. As shown in Table 1, Kuibyshev reservoir has the largest surface and the highest number of types of uses.

There have been limited studies referring to the destruction of archaeological sites around the Kuibyshev reservoir [26,27], but there are no studies referring to the entire surface of the reservoir. Therefore, this study is necessary to assess the exact number of sites impacted by the reservoir creation in 1957 and to draw attention for local authorities in their mission for future management plans [28] of the shoreline area [29]. A detailed case study was chosen to demonstrate the destructive potential of wave erosion; this was accomplished by a systematic monitoring process. The main scope of this article is (1) to track the major changes of the Volga River after the construction of the Kuibyshev reservoir with the help of GIS (2) to identify the area(s) that contain the highest concentration of archaeological sites (3) to analyse how many archaeological sites were impacted following the construction of the reservoir (4) to monitor the evolution of the only left Palaeolithic site—Beganchik from the shores of Kuibyshev reservoir, which has been specifically chosen because of its high erosion rates.

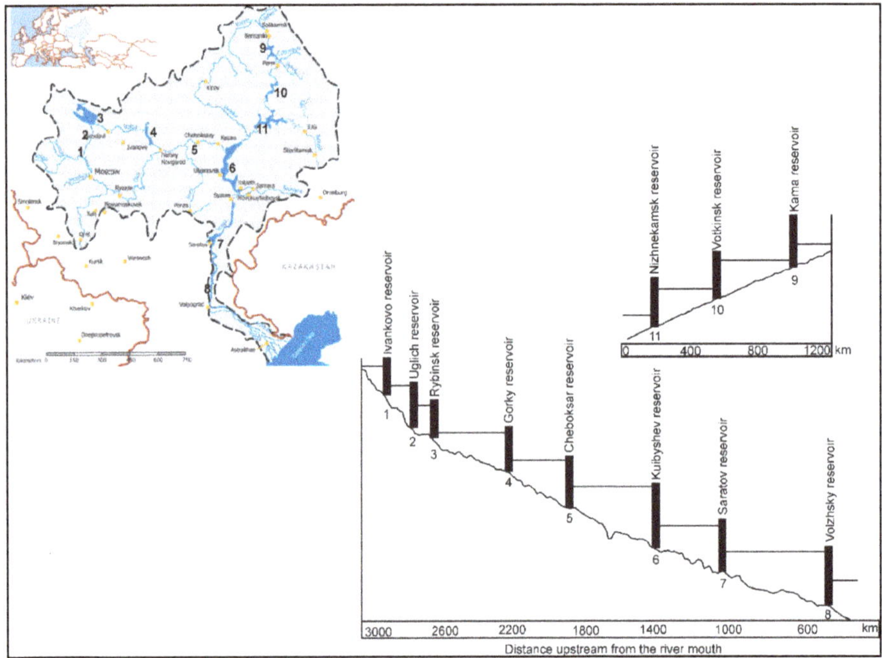

Figure 1. Detail of the reservoirs of the Volga-Kama cascade.

Table 1. The main characteristics of reservoirs from the Volga-Kama cascade [24,25] (the numbers in the first column correspond to the reservoirs from Figure 1).

No. Crt.	YOC	RA (km^2)	Volume (km^3)		IC (10^3 kW)	AO (10^9 kWh)	TOU
			Total	Useful			
1	1937	327	1.2	1	30	0.12	WNWrFPR
2	1940	249	1.2	0.8	110	0.25	PNWRF
3	1941	4550	25.4	16.6	330	1.05	PNWFFlWrRT
4	1956	1770	8.7	2.8	520	1.4	PNWFWrRT
5	1981	3780	12.6	5.4	1404	3.3	PNWRFWrT
6	1958	6500	57.3	33.9	2300	10.2	PNFIWFlWrRT
7	1968	1950	12.8	1.7	1290	5.3	PNWFlRTWr
8	1960	3165	31.4	8.2	2530	10	PNWFlFiWrRT
9	1956	1845	12.2	9.8	504	1.7	PTNFiWFRWr
10	1961	1130	9.4	3.7	1000	2.2	PNTWFiRWr
11	1978	2305	13.8	4.6	1080	2.8	PNTWFiRWr

Legend: YOC—year of commissioning; RA—reservoir area; IC—installed capacity; AO—annual output; TOU—type of use; Fi—fishery, Fl—flood control, I—irrigation, Navigation, P—power production, R—recreation, T—timber rafting, W—water supply, Wr—water releases (sanitary, irrigation).

2. Study Area

Kuibyshev reservoir is a result of the construction of the Zhiguli Hydroelectric Station, Samara region, located between Zhigulevsk city (right bank of the Volga) and Tolyatti (left bank of the Volga); the reservoir covers the territory of regions Chuvash, Tatarstan, Ulyanovsk, and Samara. Kuibyshev reservoir has a surface of 6450 km^2, a volume of water of 58 km^3, a length of approximately 510 km, a mean depth of 9.3 m; these impressive numbers make it the largest reservoir in Europe [30], with a sedimentation rate of 8 mm/year. Important changes occurred in what concerns the sedimentation rate, which has fallen to 2.7–2.9 mm/year, after the commissioning of the dam in 1957. One of the main sources of sediments is a result of the abrasion processes, collapses of a huge amount of sediments into the reservoir [25,31].

Our area of interest is located in Tatarstan region (Figure 2a), on the left bank of Kuibyshev reservoir (Figure 2b), at the junction of Kama River in the Volga, about 75 km south-east of the city of Kazan (the capital city of Tatarstan). Beganchik site is located at approximately 2.8 km North-east of Izmeri village and 1.5 km north-west of Komintern village, on an isolated hill of the terrace above the floodplain; on the left bank of the confluence of Kama and Volga rivers, at the mouth of Aktai river (Figure 2c, Figure 3a).

The geology of the area consists of Permian, Pliocene and Quaternary deposits. Quaternary sediments are dominant in Volga-Kama terraces, eight palaeohydrological phases were identified from high and low fluvial activity; the most recent active phase corresponds to the Little Ice Age [32]. There is a limited number of studies regarding the evolution of the coastal area in the Tatarstan region, Russia [25], along with the analysis of landslides [33] and gully erosion [34].

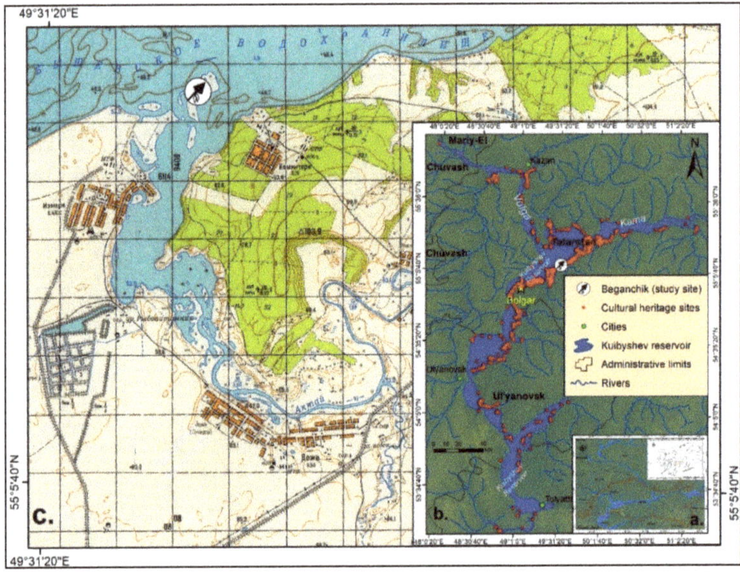

Figure 2. Geographical location of: (**a**) Kuibyshev reservoir in a regional context; (**b**) Kuibyshev reservoir and the cultural heritage sites around it; (**c**) Beganchik site on the topographic maps scale 1:50.000 (edition 1984).

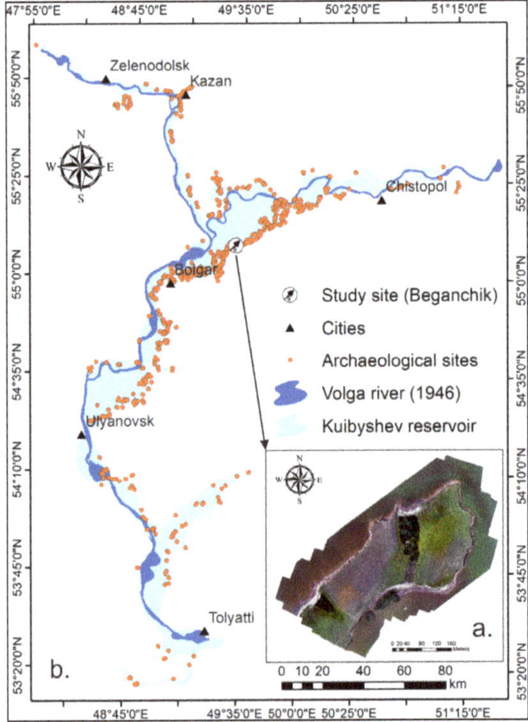

Figure 3. (**a**) Aerial image of Beganchik site from 2018; (**b**) Overlapping Volga River before and after the building of Kuibyshev reservoir, along with the inventory of archaeological sites.

3. Archaeological Background

3.1. General Overview

Ever since the Palaeolithic, large rivers and their fertile plains have been a magnet for prehistoric people to place their settlements; water is undoubtedly the most important resource that a community of people needs in order to decide where to place a settlement [35]. According to the local geographical factors and paleogeographic evolution of the landscape, the old location could be used by the next population of a different historic period. This is how multi-stratified archaeological sites were created. Around Kuibyshev reservoir, 1289 cultural heritage sites have been identified. The protection of cultural heritage assets in Russia was and is still a current issue, because of the country's complex political background; at present, cultural heritage is protected by the Federal Law 73-F3, On the Objects of Cultural Heritage (Historical and Cultural Sites) of the Peoples of the Russian Federation [36].

In this area, the oldest traces are attributed to Upper Palaeolithic-Mesolithic period. The area surrounding the Volga River has tremendous potential in regard to cultural heritage sites of Palaeolithic [37,38], Mesolithic age [39], Neolithic [40], Chalcolithic/Bronze Age [41], Early Iron Age [42], Middle Ages, etc. The only remaining Palaeolithic site which has not been impacted by the reservoir is Beganchik. Besides the Palaeolithic site, Beganchik, around the Kuibyshev reservoir there are many archaeological sites of international and national significance; among them, the Bolgar archaeological site. The Bolgar Historical and Archaeological Complex are part of the UNESCO World Heritage List since 2014; it represents the existence of the Volga-Bolgar civilisation (7–15th centuries AD), and the first capital of the Golden Horde in the 13th century [25].

3.2. Beganchik Site

Beganchik site was studied for the first time in September 1985 by M. Sh. Galimova and K. E. Istomin at the recommendation of E.P. Kazakov; the first description of the site is the islet named "the Izmeri Island". In 1981, mammoth fauna fossils were found; they were located on the towing-path in the south-western part of the islet, at the foot of the narrow, long butte, which had the shape of a peninsula with a length of about 200 m. The discovered mammoth fossils were five teeth and leg bones, together with large flint nuclei and tools in an area of 20×20 m^2. Unfortunately, by the year 2000, this peninsula was completely eroded by the Kuibyshev reservoir. In the next years, almost every autumn (until 2012) Kazakov collected stone artefacts and faunal remains at the south-western tip of the islet in conditions of low reservoir level [43].

M. Sh. Galimova in 1985–1987 and 2000 also conducted investigations of this site. In 1986, a reference point was installed at the highest point of the islet, therefore all excavations and trenches were referenced after it. An excavation area of 104 m^2 was located on the edge of the steep west coast of the islet; the cultural layer has been found at 100–130 cm depth; 1968 of artefacts were found. The specific features of the Beganchik stone industry, which was based on blade production by means of striking technique and its flint inventory, allowed M. Sh. Galimova to frame the site of initial (Upper Palaeolithic) period of the Mesolithic Ust-Kama culture. The main diagnostic tool of the Ust-Kama inventory is the arrowhead in a trapezoid shape with concave sides, which were shaped by retouching [44]. In 2000, the rescue excavations of the site were continued by M. Sh. Galimova with the participation of I. I. Gainullin. By that moment excavation territory of 1986–1987 was eroded by the reservoir. In general, the western and northern coasts of the islet were washed away by 20–25 m from the erosion ledge for 14 years (1986–2000). In the autumn of 2012, rescue investigation of the Beganchik site and Izmeri I site was conducted by the expedition of the National Centre for Archaeological Research of the Tatarstan Academy of Sciences. In the autumn of 2013, rescue investigation on the Beganchik islet was continued by a joint campaign of the "Expedition for Prehistory" of the Institute of Archaeology of the Tatarstan Academy of Sciences and "Archaeological Expedition" of the Chuvash State Institute for Humanities. The total excavated area was 20 m^2; following the excavation and surface findings, a significant collection of stone artefacts (439 items) and 80 bones of a mammoth were found [45].

4. Materials and Methods

In order to determine the shoreline dynamics, topographic map scale 1:300,000 (edition 1945), and Google Earth images from 2010, were employed. From the topographic map, the extent of the Volga River before the reservoir construction was digitised; from Google Earth images from 2010, the extent of the reservoir was digitised. They were overlapped in ArcGIS and the highest differences were observed.

The archaeological inventory, as a point feature, (Figure 3b) was provided by the Institute of Archaeology of Tatarstan Academy of Sciences. The database was compiled over a long period of time, both prior and after the filling of the reservoir; first sites were described in the early 1940s until the early 1960s, when special survey expeditions were undertaken, with the aim to find as many sites and record brief information about them. After the filling of the reservoir, more expeditions were undertaken to highlight the impact on the sites. Other sources in building the database included the descriptions of the Archaeological Maps of Tatarstan, Ulyanovsk and Samara regions. A survey has been unsystematic across the study area, sites are located to varying degrees of accuracy and the full extent of individual sites is not necessarily known. The database is still under construction, as the area is very large and only a few people within the Institute of Archaeology of Tatarstan Academy of Sciences is working to continuously update it. However, it is the most comprehensive archaeological dataset which currently exists in this area, hence will be used for this study. Density analysis of the settlements for the main historical periods was performed; this was made using the Point Density feature (using the circle as neighbourhood option) from ArcToolbox (ArcGIS).

The danger towards increasing and decreasing water level over the digital elevation model (DEM) was evaluated by making four working scenarios; the DEM used in this study is based on the Shuttle Radar Topography Mission (SRTM), with a pixel size of 30×30 m^2. First, the water level was decreased by 0.5 m and 1 m, followed by increasing the water level with 0.5 m and 1 m, respectively. Changes in reservoir level regime occur out of two main reasons: Natural seasonal changes in the flow and artificial regulations of water discharge through hydraulic structures, the difference in baric pressure, wind speed and changes in the hydraulic slope. The water level of the reservoir is controlled by a special department—RusHydro. We chose these values taking into consideration our large-scale study area and to point out the minor oscillations in water level by arbitrarily increasing/decreasing it by ±0.5 to 1 m. In order to have a better image of the changes occurred along the Volga River after the Kuibyshev reservoir was built, the entire area was divided into three sectors, as follows: Sector 1 (Figure 4b), Sector 2 (Figure 4c), and Sector 3 (Figure 4d).

For monitoring the Beganchik site, a rich cartographic background, including Soviet aerial images from 1958 and 1980, Roscosmos aerial images (2008), and UAV flights [46] from the summer of 2017 and 2018 were used. As shown in numerous studies, old maps and aerial images are an important source of information [47] that can be easily digitised, integrated into GIS [48], and used in the field of cultural heritage [49,50]. All the maps and aerial photos have been georeferenced with the help of ArcGIS. The ground control points used for the drone flights were measured with a GNSS (Global Navigation Satellite System) receiver Trimble Geoexplorer 6000 XH and Leica Zeno 20. The UAV is a drone, model DJI Phantom 4. The survey was performed with a 12-megapixel camera mounted on the quadcopter; the UAV was controlled from a smartphone using Pix4D Capture software, which allows configuring the shooting parameters. Aerial photography was performed with the following parameters, height—70 m, picture overlapping—60–80%, camera position—90 degrees, meteorological conditions—no precipitation, and wind no more than 15 m/s. The photos were processed using the algorithms built into the Agisoft Photoscan software; the resulting model was processed by a polynomial approximation exponential kernel (PAEK) method with 1 m tolerance.

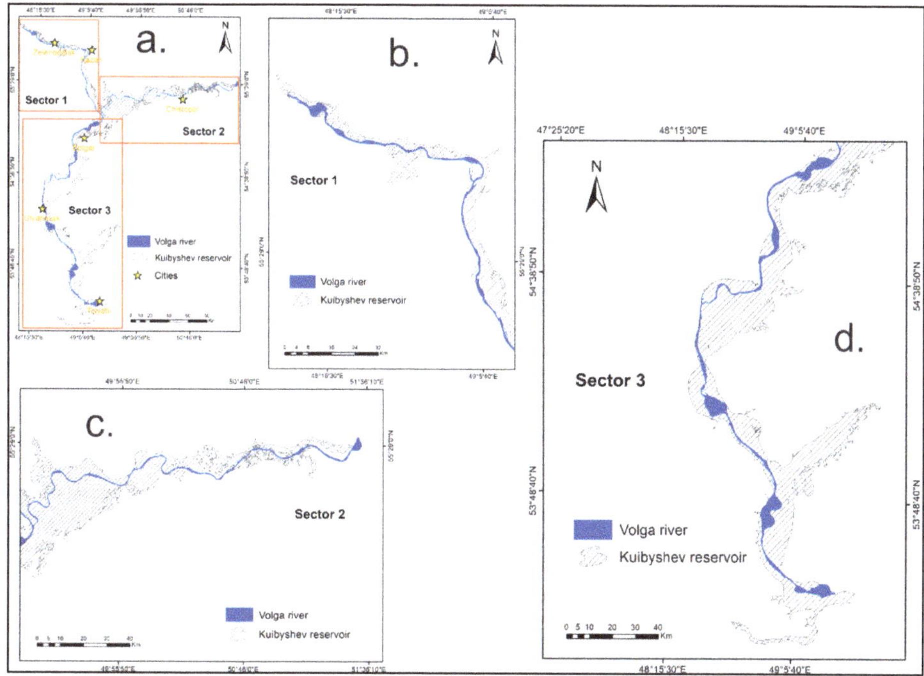

Figure 4. (**a**) General division of the Kuibyshev reservoir; (**b**) Sector 1; (**c**) Sector 2; (**d**) Sector 3.

5. Results

Each sector will be analysed according to the GIS integration of the spatial data collected from old maps and modern aerial images, followed by the analysis of the archaeological sites patterns and dynamics along the Volga River; then, the changes of Volga River will be analysed in the context of how many cultural heritage sites are directly affected by the reservoir construction. The four working scenarios will be analysed in order to evaluate the endangered sites towards increasing/decreasing water level of the reservoir. Finally, the monitoring results of the only left Palaeolithic site—Beganchik will be presented; this site has been specifically chosen because of its high erosion rates and being the only remaining Palaeolithic site around Kuibyshev reservoir.

5.1. Volga River Dynamics

From the town of Tver to Volgograd, Volga River flow velocity is affected by the 8 reservoirs. The reservoirs were built to control seasonal changes in flow; however, there are no significant changes when it comes to river discharge and the total annual discharge. In the middle Volga, the mean annual flow from 1876 to 1940 was 2876 m^3/s; after the construction of reservoirs, from 1942–1955, the mean annual flow was 2780 m^3/s [51].

Sector 1 (Figure 5a) stretches approximately in the north-western part, next to the Zvenigovo city till south-east, at the junction between Volga and Kama rivers; it has a length of approximately 145 km. The most important cities within Sector 1 are Kazan (with a population of about 1,2 mil. people) and Zelenodolsk (with a population of about 98,000 people). Out of the three sectors, Sector 1 has the fewest changes, compared with the others; the most significant changes are located about 43 km downstream from Kazan city. Initially, Volga had a width of 1.4 km, while after the building of the Kuibyshev reservoir the width of Volga reached 9.5 km. Another significant change is located between Zelenodolsk and Kazan, at the junction of Sviyaga River in the Volga; from a width of 0.6 km, Volga reached a width of 11.2 km; except this, the reservoir water has mainly covered the left side of the river. This is due to the geomorphological characteristics of the area; the right side represents the Volga uplands, while on the left side are the terraced plains of the lowland Volga River region [52].

Sector 2 (Figure 5b) stretches from east to west on a distance of approximately 150 km and represents the lower Kama River junction to Volga River; before the Kuibyshev reservoir, the area located around the junction had a significant number of villages (which were completely destroyed). Moreover, important landscape changes occurred, along with an acceleration of coastal erosion with a direct effect on cultural heritage [25]. The most important city in this sector is Chistopol (with a population of approximately 60,000 people). Along its approximately 150 km length, after the building of Kuibyshev reservoir, Sector 2 had more or less a balanced development of the right and left bank; this is because both of the sides are located within the terraced plains of the lowland Volga River region. From an average width of 0.8–1 km, the Kama River reached widths of 13–36.8 km. From these numbers, we can realise the real proportions of the consequences of the Kuibyshev reservoir being built.

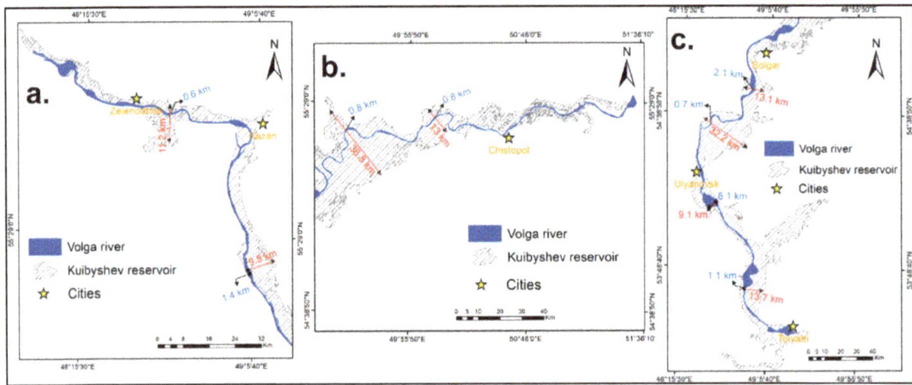

Figure 5. Results of the coastal dynamics analysis for: (**a**) Sector 1; (**b**) Sector 2; (**c**) Sector 3.

Sector 3 (Figure 5c), with a length of approximately 263 km, stretches from Kama and Volga junction until the dam of the Zhiguli Hydroelectric Station, located between the cities of Zhigulyovsk and Tolyatti. The most important cities in this sector are Tolyatti (with a population of approximately 720,000 people), Ulyanovsk (with a population of approximately 614,000 people) and Bolgar (with a population of approximately 9000 people). Bolgar is well known for the Bolgar Historical and Archaeological Complex World Heritage site. Similar in development with Sector 1, the left side of the river being more developed than the right one; again, this is to the lower altitudes of the terraced plains of the lowland Volga River region [52]. Within this sector, the width of the Volga River before Kuibyshev reservoir was ranging from 0.7–2.1 km, and from 9.1–32.2 km after the building of Kuibyshev reservoir. Having this enormous width, it is sometimes called the Kuibyshev Sea.

5.2. Archaeological Site Analysis

Following the analysis of the archaeological database provided by the Institute of Archaeology of Tatarstan Academy of Sciences, the following periods were identified: Palaeolithic/Mesolithic, Neolithic, Chalcolithic/Bronze Age, Early Iron Age, Migration Period, and Middle Ages. Large river systems, e.g., the Volga, act as a magnet when it comes to taking a decision to place a prehistoric settlement. That is why, in the close proximity of the Volga River and its tributaries, there is a high density of archaeological sites. Water represents the main resource in establishing the placement of a settlement; this is documented and well-known across the archaeologists and geo-archaeologists [53].

On the basis of the existing database, the areas with the highest concentration of archaeological settlements attributed to a certain period will be identified and highlighted accordingly. Usually, the dynamics of the settlements are influenced by different factors, like climate change [54], natural hazards [55] and threats from other populations. Having knowledge of the spatiotemporal distribution patterns of archaeological sites is a powerful tool to understand past human-environment interactions and to evaluate landscape vulnerability to natural [56] and anthropogenic changes [49].

As can be seen in Figure 6, the highest concentration of settlements for all the periods is located at the junction of the Kama and Volga Rivers, this is representing an important communication route. During the Palaeolithic/Mesolithic period (Figure 6a), the hunters-fishers-gatherers population was well adapted to the living around water bodies and forests; the highest concentration was at the junction of Kama and Volga rivers, followed by the adjacent areas of upstream and downstream of the junction. A good concentration can be observed on the Volga River, around the area where presently the city of Kazan is located; the thrive of settlements was due to the optimum climatic conditions for the Preboreal period [37,39], along with the highest levels of rivers and lakes, which was typical for the Mesolithic epoch [57]. The settlements were not very homogenous during this period. However, this can be observed during the Neolithic period (Figure 6b), when more settlements appear in the area between the today Kazan city and Kama-Volga junction. The Neolithic period is characterised by the emergence of pottery, new types of stone tools and the transition to sedentism with the help of active fishing and hunting. The majority of Neolithic sites are located on the remnants of the floodplain of the small rivers of the Kama River tributaries or on the first terrace of the Kama River [40].

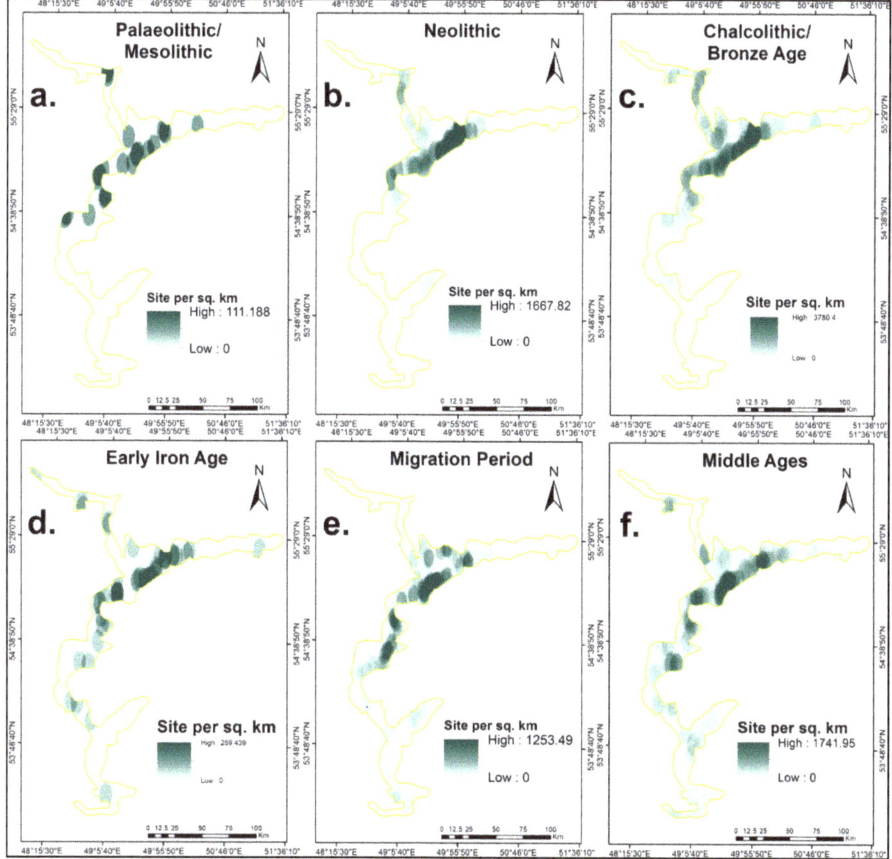

Figure 6. Location and density analysis for archaeological sites for the following periods: (**a**) Palaeolithic/Mesolithic; (**b**) Neolithic; (**c**) Chalcolithic/Bronze Age; (**d**) Early Iron Age; (**e**) Migration Period; (**f**) Middle Ages.

Following Chalcolithic/Bronze Age period (Figure 6c), it can be observed even a higher degree of homogeneity among the settlements; this is due to the fact that the lowest levels of water were recorded in the Bronze Age. As a consequence of this, even the lowest altitudes were chosen to place the settlements, which is why the analysis shows a larger continuous surface

Figure 6d illustrates the density of the Early Iron Age settlements, which started to be more fragmented. The highest concentration is at the confluence of Kama-Volga Rivers and on the territory of today Bolgar, followed by scattered low-density areas the upstream Volga, at the mouth of Sviyaga River, and the downstream Volga. The settlements appear scattered because of their higher altitudinal position (higher position throughout the Holocene), due to the associated high flood levels [57]. As can be seen in Figure 6e, the Migration period is characterised by spreading of population downstream Volga River, until today Ulyanovsk city. However, the highest concentration is still located at the Kama-Volga junction; the fact that during this period the area is very poorly populated is also indicated by [58]. Finally, the Middle Ages (Figure 6f) show the highest fragmentation of the settlements. The highest concentration remains the same (Kama-Volga junction), while the settlements are scattered downstream and the upstream Volga until today Tolyatti and Zelenodolsk, respectively. During this period, the settlements are so scattered, due to the fact that the climatic conditions were suitable for the long-term occupation of river and lake floodplains. This is the turning point when people start to settle and make semi-permanent settlements and start off using the floodplain in order to practice agriculture on a higher level [57,59].

5.3. Cultural Heritage under Erosion Threat

Since the formation of the reservoir in the middle of the 1950s, the confluence of the Kama and Volga Rivers and the left bank tributaries was flooded. As a result, many lower terraces, that were hosting archaeological sites of different periods [60], were completely flooded. The main typology of the sites is presented in Table 2; therefore, out of the total of 1289 sites, 1091 are underwater or totally impacted following the building of the Kuibyshev reservoir. According to their chronology, shown in Table 3, the only Palaeolithic/Mesolithic site that still exists, but is under high threat from coastal erosion, will be further analysed, based on the old Soviet Maps and modern surveys.

Based on the working scenarios regarding the water level increasing and decreasing to 0.5 m and 1 m, respectively it has been observed that increasing the water level, whether, with 0.5 or 1 m, a number of two extra sites will be affected (out of 1091 already underwater or impacted). If we decrease the water level by 0.5 m or 1 m respectively, the same number of sites will remain affected—1091. Having such a large surface, water level oscillations do not affect the cultural heritage sites, unless there are variations greater than ±1 m.

5.4. Beganchik Site

In order to analyse the coastal dynamics of Beganchik site, all the surveys were overlapped, and the site was divided into three sectors (Figure 7), which will be further analysed separately. Beganchik site is located at the mouth of Aktai River, on the second terrace (the first terrace being flooded by Kuibyshev reservoir) of the floodplain which formed before the Holocene [61]; the altitude is between 54–60 m a.s.l. According to the general view of the site (Figure 8a), the northern part of the site is represented by a very steep cliff (Figure 8b) which is continually eroding. Previous preliminary studies [44] have revealed that the erosion rate is about 2–3 m/year.

Table 2. Distribution of cultural heritage sites around Kuibyshev reservoir according to their typology.

Type	Building	Burial Ground	Burial Mound (s)	Complex Site	Fortified Settlement/HILLFORT	Hoard	Surface Find	Tombstone	Unfortified Settlement	Total
Number	2	103	25	4	40	17	179	4	915	1289
Affected (under water)	0	82	21	3	31	14	156	4	780	1091
Not affected	2	21	4	1	9	3	23	0	135	198

Table 3. Distribution of cultural heritage sites around Kuibyshev reservoir according to their chronology.

Age	Chalcolithic/Bronze Age	Early Iron Age	Middle Ages	Migration Period	Modern Time	Neolithic	Palaeolithic/Mesolithic	Not Identified	Total
Number	566	45	275	148	11	164	20	60	1289
Affected (under water)	490	39	223	118	6	144	19	52	1091
Not affected	76	6	52	30	5	20	1	8	198

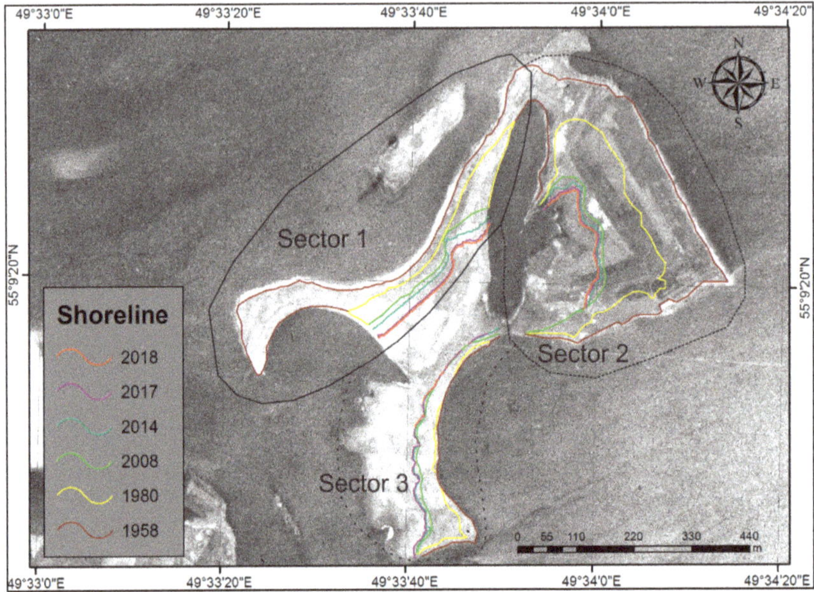

Figure 7. Shoreline limit resulted from cartographic analysis and field surveys and the division of the three analysed sectors.

Figure 8. (**a**) General view of Beganchik site (drone flight) from 2017; (**b**) Detail over the northern part of the site, fresh parts from the coast are visible in the water (August 2017); (**c**) The change of water colour, due to the clay content of the soil; (**d**) The northern part of the site, where the height of the coast is decreasing.

5.4.1. Sector 1

Sector 1 was not actively eroded between 1958 and 1980 because it was protected by another island (60–90 m north-west, Figure 7), as indicated by the relatively low values of the shoreline retreat (Table 4); in this way, the site was protected from the mechanical action of waves (Figure 8c). Later on, it can be seen that after the island disappeared, the yearly erosion has considerably increased, along with the specific land loss and volume. The direction of the Kama River flow is from north, north-east; being located at the "shelter" of Sector 2 from the speed and currents of the Kama River, this section was in some way protected. However, this sector became likely to be eroded, due to the high erosion rates of Sector 2 and having an elongated shape; this is highlighted of the specific land loss for 1958–1980.

Table 4. Detailed morphometric indicators from different observation periods for Sector 1.

Observation Period	Years	Shoreline Retreat		Eroded Area		Specific Land Loss	Specific Volume Loss
		m	m/year	ha	ha/year	$n * 10^{-3}$ ha/km * year	thousands m^3/km * year
1958–1980	22	32.84	1.5	2.77	0.13	253.13	12.65
1980–2008	28	33.88	1.21	1.26	0.04	133.48	6.67
2008–2014	6	11.57	1.93	0.39	0.06	191.87	9.59
2014–2017	3	17.03	5.68	0.50	0.17	469.53	23.48
2017–2018	1	3.23	3.23	0.09	0.09	276.55	13.83

Note: * defines multiplication.

Very high values of the specific land loss, in comparison with other sectors, is due to the height of the coast; which, in some parts can reach 5 m in height. Following the analysis, Sector 1 can be characterised as an extremely dangerous one.

5.4.2. Sector 2

Unlike Sector 1, Sector 2 was and still is under the direct exposure of the Kama River flow and currents. As can be seen in Table 5, the specific land loss is at extremely high rates (Figure 8d). This sector is the most exposed and threatened by erosion. The shoreline retreat is generally stable, varying within 2 m. According to the specific land loss indicator, it can be observed that the destruction occurred, especially within the first two periods, which is typical for the initial stage of lowland reservoir development. During this period, the extremities of this sector are cut off, after which the erosion process stabilises. Between 1958–2008, approximately 70% of the eastern part of the site was eroded. Following that, part of the river's current's strength was redistributed along the north-western part, which explains the sudden decrease in land loss. The height of the coast does not exceed 2 m, therefore, Sector 2 can be classified as moderately dangerous.

Table 5. Detailed morphometric indicators from different observation periods for Sector 2.

Observation Period	Years	Shoreline Retreat		Eroded Area		Specific Land Loss	Specific Volume Loss
		m	m/year	ha	ha/year	$n * 10^{-3}$ ha/km * year	thousands m^3/km * year
1958–1980	22	44.85	2.04	4.68	0.21	279.75	4.2
1980–2008	28	64.81	2.31	3.51	0.13	288.84	4.33
2008–2014	6	9.61	1.6	0.27	0.04	89.91	1.57
2014–2017	3	4.76	1.59	0.13	0.04	82.24	1.44
2017–2018	1	2.8	2.8	0.08	0.08	189.95	3.32

Note: * defines multiplication.

5.4.3. Sector 3

Sector 3 is located in the close proximity to Aktai River mouth, where is protected from the mechanical action of waves and Kama River strong currents. This portion of the Beganchik site shoreline is the most stable. As it can be seen from Table 6, there have been no significant changes regarding this part of the coast from 1958 to 2018; except the period 1980–2014, when the specific land loss is higher when compared to other periods, but considerably lower when compared with the other two sectors. The most intensive processes of coastal transformation in the study area were observed in Sectors 1 and 2, open to the destructive effect of the currents and the mechanic action of waves. The erosion intensity may vary from year to year, depending on the water level oscillations in the reservoir. In order to have a more detailed situation on the Beganchik site erosion rates, continuous annual observations are needed.

Table 6. Detailed morphometric indicators from different observation periods for Sector 3.

Observation Period	Years	Shoreline Retreat		Eroded Area		Specific Land Loss	Specific Volume Loss
		m	m/year	ha	ha/year	$n * 10^{-3}$ ha/km * year	thousands m^3/km * year
1958–1980	22	5.59	0.25	0.06	-	16.55	0.25
1980–2008	28	13.85	0.49	0.22	0.01	47.49	0.71
2008–2014	6	7.37	1.23	0.05	0.01	46.24	0.69
2014–2017	3	1.53	0.51	0.01	-	14.17	0.21
2017–2018	1	0.31	0.31	0.002	0.002	9.11	0.14

Note: * defines multiplication.

Particular attention should be paid to Sector 1, in which the most important part of the site is located. If the erosion rates remain stable, the site will be completely impacted in about two or three decades. This imposes urgent mitigation measures from local authorities, along with the sustainable management of cultural heritage sites.

6. Discussions

Reservoir construction had a significant impact on the flow regime because the current velocity decreased. The currents are very complex, as river flows are under the direct effect of convective flows and wind effects formed in the reservoirs. These are characteristic for Kuibyshev reservoir, where wind effect and bottom relief have a high influence on hydrological conditions [51]. Volga River frames itself into the future increase of global river flow as a consequence of climate change; predictions have shown an increase of 4–8% during 2071–2100 [62]. To be more specific, future trends in the area show an increase in precipitation, temperature and in the use and levels of waters in rivers [63]. The cyclic oscillations have occurred in the Volga River basin in the last half-century; this has influenced the water level in the reservoir, and therefore the erosion rates of the shoreline. The numbers in the Tables 4 and 5 are related to the cyclic oscillations that brought two high-water periods (1951–1962, 1977–1995) and two low-water periods (1963–1976, 1996–present) [51]. For Sectors 1 and 2 high-water levels are associated with low erosion, while the low-water level is associated with higher erosion rates. The research presented in this paper continues our endeavour to monitor the endangered cultural heritage sites from the shoreline of the Kuibyshev reservoir [25–27,60]. Combining old maps with new data collected from field surveys shows high efficiency in establishing the erosion rates of archaeological sites located on shorelines of big reservoirs. When comparing erosion rates with the previous study [25], the average shoreline retreat is close (~3–4 m/year).

7. Conclusions

In this study, the main changes along the largest reservoir in Europe—Kuibyshev (Russian Federation) were analysed, in strong connection to cultural heritage sites. Following the analysis, Sector 2 has been identified as one with the highest values of width oscillation, from 0.8–1 km to 13–36.8 km. Cultural heritage sites located in the close proximity to big rivers and/or big reservoirs are especially subjected to erosion from water, water level oscillations, and the mechanical action of waves. A diachronic analysis of the archaeological sites located along the Volga River and its main tributaries has highlighted the fact that the most inhabited area was located at the junction of Kama River into the Volga. As highlighted in our analysis, 85% of the cultural heritage around Kuibyshev reservoir is impacted. However, a more thorough process of monitoring and evaluating the present state of cultural heritage is needed. This has to be done with the cooperation of local authorities and stakeholders. The survey of the only left Palaeolithic site—Beganchik, has shown a fast degradation with no mitigation measures from the local authorities. Beganchik site remains promising for regular rescue archaeological excavations, despite the loss of more than a half of its surface, due to coastal erosion over the last 30 years. Working on scenarios regarding the management of archaeological sites around Kuibyshev reservoir represents one of our future goals.

Author Contributions: Conceptualisation, I.C.N. and B.U.; methodology, I.C.N., B.U. and I.G.; software, I.C.N., B.U. and I.G.; formal analysis, I.C.N. and B.U.; investigation, I.C.N., B.U. and I.G.; resources, I.C.N., B.U., I.G. and M.G.; data curation, I.C.N., B.U., I.G. and M.G.; writing—original draft preparation, I.C.N. and B.U.; writing—review and editing, I.C.N., B.U., I.G. and M.G.; visualisation, I.C.N. and B.U.; supervision, I.C.N. and B.U.; review corrections, I.C.N. and B.U.

Funding: This work is performed according to the Russian Government Program of Competitive Growth of Kazan Federal University.

Acknowledgments: The archaeological database was kindly provided by the Institute of Archaeology of Tatarstan Academy of Sciences (through Leonid Vyazov). James S. Williamson (Memorial University of Newfoundland, Canada) is kindly acknowledged for the English language editing of the manuscript. The authors are grateful for the constructive comments of two anonymous reviewers.

Conflicts of Interest: The authors declare no conflict of interest. The funders had no role in the design of the study; in the collection, analyses, or interpretation of data; in the writing of the manuscript, or in the decision to publish the results.

References

1. Khon, V.C.; Mokhov, I.I. The hydrological regime of large river basins in Northern Eurasia in the XX–XXI centuries. *Water Resour.* **2012**, *39*, 1–10. [CrossRef]
2. El-Younsy, A.R.; Essa, M.A.; Wasel, S.O. Sedimentological and geoenvironmental evaluation of the coastal area between Al-Khowkhah and Al-Mokha, southeastern Red Sea, Republic of Yemen. *Environ. Earth. Sci.* **2017**, *76*, 50. [CrossRef]
3. Romanescu, G.; Mihu-Pintilie, A.; Ciurte, D.L.; Stoleriu, C.C.; Cojoc, G.M.; Tirnovan, A. Allocation of flood control capacity for a multireservoir system. Case study of the Bistrita River (Romania). *Carpath. J. Earth. Environ.* **2019**, *14*, 223–234. [CrossRef]
4. Avakyan, A.B.; Podol'skii, S.A. Impact of reservoirs on the fauna. *Water Resour.* **2002**, *29*, 123. [CrossRef]
5. Berka, R. *Inland Capture Fisheries of the USSR*; FAO Fisheries Technical Paper 311; FAO (Food and Agriculture Organization): Rome, Italy, 1989; p. 143.
6. Vilmundardóttir, O.K.; Magnússon, B.; Gísladóttir, G.; Thorsteinsson, T. Shoreline erosion and aeolian deposition along a recently formed hydro-electric reservoir, Blöndulón, Iceland. *Geomorphology* **2010**, *114*, 542–555. [CrossRef]
7. Su, X.; Nilsson, C.; Pilotto, F.; Liu, S.; Shi, S.; Zeng, B. Soil erosion and deposition in the new shorelines of the Three Gorges Reservoir. *Sci. Total Environ.* **2017**, *599–600*, 1485–1492. [CrossRef]
8. Jayakumar, K. Analysis of shoreline changes along the coast of Tiruvallur District, Tamil Nadu, India. *J. Geogr. Cartogr.* **2018**, *1*, 1–9. [CrossRef]
9. Svyatko, S.V.; Reimer, P.J.; Schulting, R. Modern freshwater reservoir offsets in the Eurasian steppe: Implications for archaeology. *Radiocarbon* **2017**, *59*, 1597–1607. [CrossRef]
10. Górski, K.; van den Bosch, L.V.; van den Wolfshaar, K.E.; Middelkoop, H.; Nagelkerke, L.A.J.; Filippov, O.V.; Zolotarev, D.V.; Yakovlev, S.V.; Minin, A.E.; Winter, H.V.; et al. Post-damming flow regime development in a large lowland river (Volga, Russian Federation): Implications for floodplain inundation and fisheries. *River Res. Appl.* **2012**, *28*, 1121–1134. [CrossRef]
11. Majerová, L.; Bábek, O.; Navrátil, T.; Nováková, T.; Štojdl, J.; Elznicová, J.; Hron, K.; Matys Grygar, T. Dam reservoirs as an efficient trap for historical pollution: The passage of Hg and Pb through the Ohře River, Czech Republic. *Environ. Earth. Sci.* **2018**, *77*, 574. [CrossRef]
12. Pranzini, E. Coastal erosion and shore protection: A brief historical analysis. *J. Coast. Conserv.* **2017**, *22*, 827–830. [CrossRef]

13. Pourkerman, M.; Marriner, N.; Morhange, C.; Djamali, M.; Amjadi, S.; Lahijani, H.; Beni, A.N.; Vacchi, M.; Tofighian, H.; Shah-Hoessein, M. Tracking shoreline erosion of "at risk" coastal archaeology: The example of ancient Siraf (Iran, Persian Gulf). *Appl. Geogr.* **2018**, *101*, 45–55. [CrossRef]

14. Chapman, H.P.; Fletcher, W.G.; Thomas, G. Quantifying the effects of erosion on the archaeology of intertidal environments. A new approach and its implications for their management. *Conserv. Manag. Arch.* **2001**, *4*, 233–240. [CrossRef]

15. Jazwa, C.S.; Johnson, K.N. Erosion of coastal archaeological sites on Santa Rosa Island, California. *West. N. Am. Nat.* **2018**, *78*, 302–327. [CrossRef]

16. Pantusa, D.; D'Alessandro, F.; Riefolo, L.; Principato, F.; Tomasicchio, G.R. Application of a Coastal Vulnerability Index. A Case Study along the Apulian Coastline, Italy. *Water* **2018**, *10*, 1218. [CrossRef]

17. Westley, K. Refining Broad-Scale Vulnerability Assessment of Coastal Archaeological Resources, Lough Foyle, Northern Ireland. *J. Isl. Coast. Archaeol.* **2018**. [CrossRef]

18. O'Rourke, M.J.E. Archaeological site vulnerability modelling: The influence of high impact storm events on models of shoreline erosion in the western Canadian Arctic. *Open Archaeol.* **2017**, *3*, 1–17. [CrossRef]

19. Flatman, J. Conserving marine cultural heritage: Threats, risks and future priorities. *Conserv. Manag. Arch.* **2009**, *11*, 5–8. [CrossRef]

20. Dillenia, I.; Troa, R.; Triarso, E. In situ preservation of marine archaeological remains based on geodynamic conditions, Raja Ampat, Indonesia. *Conserv. Manag. Arch.* **2016**, *18*, 364–371. [CrossRef]

21. Howard, P.; Pinder, D. Cultural heritage and sustainability in the coastal zone: Experiences in south-west England. *J. Cult. Herit.* **2003**, *4*, 57–68. [CrossRef]

22. Usmanov, B.; Nicu, I.C.; Gainullin, I.; Khomyakov, P. Monitoring and assessing the destruction of archaeological sites from Kuibyshev reservoir coastline, Tatarstan Republic, Russian Federation. A case study. *J. Coast. Conserv.* **2018**, *22*, 417–429. [CrossRef]

23. Demin, A.P. The efficiency of water resources management in Volga basin. *Water Resour.* **2005**, *32*, 594–604. [CrossRef]

24. Micklin, P.P. Environmental costs of the Volga-Kama cascade of power stations. *Water Resour. Bull.* **1974**, *10*, 565–572. [CrossRef]

25. Avakyan, A.B. Volga-Kama cascade reservoirs and their optimal use. *Lakes Reserv. Res. Manag.* **1998**, *3*, 113–121. [CrossRef]

26. Gainullin, I.I.; Sitdikov, A.G.; Usmanov, B.M. Destructive abrasion processes study in archaeological sites placement (Kuibyshev and Nizhnekamsk reservoirs, Russia). In Proceedings of the SGEM 2014 Scientific SubConference on Anthropology, Archaeology, History and Philosophy, Albena, Bulgaria, 3–9 September 2014; Volume 3, pp. 339–346. [CrossRef]

27. Gainullin, I.I.; Khomyakov, P.V.; Sitdikov, A.G.; Usmanov, B.M. Study of anthropogenic and natural impacts on archaeological sites of the Volga Bulgaria period (Republic of Tatarstan) using remote sensing data. *Proc. SPIE 9688* **2016**. [CrossRef]

28. Nicu, I.C. Natural risk assessment and mitigation of cultural heritage sites in North-eastern Romania (Valea Oii river basin). *Area* **2019**, *51*, 142–154. [CrossRef]

29. Döring, M.; Ratter, B.M.W. Coastal landscapes: The relevance of researching coastscapes for managing coastal change in North Frisia. *Area* **2018**, *50*, 169–176. [CrossRef]

30. Ratushnyak, A.A. The role of aquatic macrophytes in hydroecosystems of Kuibyshev Reservoir (Republic of Tatarstan, Russia). *Am.-Eurasian J. Agric. Environ. Sci.* **2008**, *4*, 1–8.

31. Zakonnov, V.V.; Ivanov, D.V.; Zakonnova, A.V.; Kochetkova, M.Y.; Malanin, V.P.; Khaidarov, A.A. Spatial and temporal transformations of bottom sediments in the Middle Volga reservoirs. *Water Resour.* **2007**, *34*, 540–548. [CrossRef]

32. Panin, A.; Matlakhova, E. Fluvial chronology in the East European Plain over the last 20 ka and its palaeohydrological implications. *Catena* **2015**, *130*, 46–61. [CrossRef]

33. Koposov, E.V.; Sobol, I.S.; Ezhkov, A.N. Prognozirovanie abrazionnoy i opolznevoy opasnosti poberezhiy Volzhskikh vodokhranilishch (Peculiar Predicting of Formation of Abrasion and Landslide Hazard Shores of the Volga Water Reservoirs). *Vestnik MGSU* **2013**, *6*, 170–176. (In Russian) [CrossRef]

34. Golosov, V.; Yermolaev, O.; Rysin, I.; Vanmaercke, M.; Medvedeva, R.; Zaytseva, M. Mapping and spatial-temporal assessment of gully density in the Middle Volga region, Russia. *Earth Surf. Process. Landf.* **2018**, *43*, 2818–2834. [CrossRef]

35. Biswas, A.K. Impacts of large dams: Issues, opportunities and constraints. In *Impacts of Large Dams: A Global Assessment*; Tortajada, C., Altinbilek, D., Biswas, A.K., Eds.; Springer-Verlag: Berlin Heidelberg, Germany, 2012; pp. 1–18.

36. Petrov, N. Cultural heritage management in Russia. In *Cultural Heritage Management. A Global Perspective*; Mauch, M.P., Smith, G.S., Eds.; University Press of Florida: Gainesville, FL, USA, 2010; pp. 153–161.

37. Leonova, N.B. The Upper Palaeolithic of the Russian steppe zone. *J. World Prehist.* **1994**, *8*, 169–210. [CrossRef]

38. Galimova, M.S. Final Palaeolithic—Early Mesolithic cultures with trapezia in the Volga and Dnieper basins: The question of origin. *Archaeol. Baltica* **2005**, *7*, 136–148.

39. Zhilin, M.G. Early Mesolithic hunting and fishing activities in Central Russia: A review of the faunal and artefactual evidence from wetland sites. *J. Wetl. Archaeol.* **2014**, *14*, 91–105. [CrossRef]

40. Lychagina, E.L.; Vybornov, A.A. Chronology of Kama Neolithic culture. *Doc. Praehist.* **2017**, 152–161. [CrossRef]

41. Demkina, T.S.; Borisov, A.V.; Demkin, V.A.; Khomutova, T.E.; Kuznetsova, T.V.; El'tsov, M.V.; Udal'tsov, S.N. Paleoecological Crisis in the Steppe of the Lower Volga Region in the Middle of the Bronze Age (III-II centuries BC). *Eurasian Soil Sci.* **2017**, *50*, 791–804. [CrossRef]

42. Patrushev, V.S. Textile ceramics of the Early Iron Age in the Ardino settlement. *Perm Univ. Herald-Hist.* **2017**, *36*, 63–73. [CrossRef]

43. Kazakov, E.P. *Pamyatniki epohi kamnya v Zakame (Arheologicheskiy ocherk). Ch. 1. Stone Age Sites in the Trans-Kama Region (Archaeological Essay). Part 1*; Institute of History named after Sh. Marjani: Kazan, Russia, 2011. (In Russian)

44. Galimova, M.S. *Pamyatniki pozdnego paleolita i mezolita v uste reki Kamyi (Upper Palaeolithic and Mesolithic Sites at the Mouth of the Kama River)*; «Yanus-K» Publ.: Moskow-Kazan, Russia, 2001. (In Russian)

45. Galimova, M.S. *Otchet ob arheologicheskih issledovaniyah stoyanki Beganchik, raspolozhennoy v zone abrazionnoy deyatelnosti Kuybyishevskogo vodohranilischa v Spasskom rayone Respubliki Tatarstan v 2013 godu (Report on Archaeological Investigation the Site of Beganchik Located in the Area of the Abrasion Activity of the Kuybyshev Reservoir in the Spassky District of the Republic of Tatarstan in 2013)*; The Archive of the Institute of Archaeology of Russian Academy of Sciences: Kazan, Russia, 2017. (In Russian)

46. Callow, J.N.; May, S.M.; Leopold, M. Drone photogrammetry and KMeans point cloud filtering to create high resolution topographic and inundation models of coastal sediment archives. *Earth Surf. Process. Landf.* **2018**, *43*, 2603–2615. [CrossRef]

47. Kinda, A. Nature of old maps: As primary source materials for historical geography. *Earth Sci.* **2018**, *7*, 260–267. [CrossRef]

48. Gainullin, I.I.; Kasimov, A.; Khomyakov, P.V.; Usmanov, B.M. An integrated approach for Medieval hillforts study (Republic of Tatarstan, Russia). In Proceedings of the 3rd International Association of SGEM Conference, Albena, Bulgaria, 30 September 2016; Volume 3, pp. 247–254. [CrossRef]

49. Nicu, I.C. Cultural heritage assessment and vulnerability using Analytic Hierarchy Process and Geographic Information Systems (Valea Oii catchment, North-eastern Romania). An approach to historical maps. *Int. J. Disaster Risk Reduct.* **2016**, *20*, 103–111. [CrossRef]

50. Nicu, I.C. Tracking natural and anthropic risks from historical maps as a tool for cultural heritage assessment: A case study. *Environ. Earth. Sci.* **2017**, *76*, 330. [CrossRef]

51. Litvinov, A.S.; Mineeva, N.M.; Papchenkov, V.G.; Korneva, L.G.; Lazareva, V.I.; Shcherbina, G.K.; Gerasimov, Y.V.; Dvinskikh, S.A.; Noskov, V.M.; Kitaev, A.B.; et al. Volga River Basin. In *Rivers of Europe*; Tockner, K., Uehlinger, U., Robinson, C.T., Eds.; Academic Press: Cambridge, MA, USA, 2009; pp. 23–57.

52. Melnikova, G.L. Peculiarities of Shallows in Regulated Reservoirs. In *IIASA Professional Paper*; IIASA (International Institute for Applied System Analysis): Laxenburg, Austria, 1977; p. 77.

53. Nicu, I.C. *Hydrogeomorphic Risk Analysis Affecting Chalcolithic Archaeological Sites from Valea Oii (Bahlui) Watershed, Northeastern Romania*; An Interdisciplinary Approach, 1st ed.; Springer: Cham, Switzerland, 2016.

54. Capuzzo, G.; Zanon, M.; Dal Corso, M.; Kirleis, W.; Barceló, J.A. Highly diverse Bronze Age population dynamics in Central-Southern Europe and their response to regional climatic patterns. *PLoS ONE* **2018**, *13*. [CrossRef] [PubMed]

55. Nicu, I.C. Application of analytic hierarchy process, frequency ratio, and statistical index to landslide susceptibility: An approach to endangered cultural heritage. *Environ. Earth Sci.* **2018**, *77*, 79. [CrossRef]

56. Hosner, D.; Wagner, M.; Tarasov, P.E.; Chen, X.; Leipe, C. Spatiotemporal distribution patterns of archaeological sites in China during the Neolithic and Bronze Age: An overview. *Holocene* **2016**, *26*, 1576–1593. [CrossRef]

57. Panin, A.V.; Nefedov, V.S. Analysis of variations in the regime of rivers and lakes in the Upper Volga and Upper Zapadnaya Dvina based on archaeological-geomorphological data. *Water Resour.* **2010**, *37*, 16. [CrossRef]

58. Ivanov, V.A. Southern CIS-Urals in the Great Migration Period—Archaeological and geographical context. *Povolzhskaya Arkheol.* **2017**, *4*, 8–23. [CrossRef]

59. Demkin, V.A.; Yakimov, A.S.; Alekseev, A.O.; Kashirskaya, N.N.; El'tsov, M.V. Paleosol and paleoenvironmental conditions in the Lower Volga steppes during the Golden Horde period (13th–14th centuries AD). *Eurasian Soil Sci.* **2006**, *39*, 115–126. [CrossRef]

60. Gainullin, I.I.; Khomyakov, P.V.; Sitdikov, A.G.; Usmanov, B.M. Qualitative assessment of the medieval fortifications condition with the use of remote sensing data (Republic of Tatarstan). *Proc. SPIE 10444* **2017**. [CrossRef]

61. Yermolaev, O.P.; Igonin, M.E.; Bubnov, A.Y.; Pavlova, S.V. *Landshaftyi Respubliki Tatarstan. Regionalnyiy Landshaftno-Ekologicheskiy Analiz (Landscapes of the Republic of Tatarstan. Regional Landscape-Ecological Analysis)*; «Slovo» Publ.: Kazan, Russia, 2007. (In Russian)

62. Falloon, P.D.; Betts, R.A. The impact of climate change on global river flow in HadGEM I simulations. *Atmos. Sci. Lett.* **2006**, *7*, 62–68. [CrossRef]

63. Kouzmina, J.V.; Treshkin, S.E. Climate changes in the basin of the Lower Volga and their influence on the ecosystem. *Arid Ecosyst.* **2014**, *4*, 142–157. [CrossRef]

Article

Mathematical Reconstruction of Eroded Beach Ridges at the Ombrone River Delta

Irene Mammì [1,2,*], **Lorenzo Rossi** [1,2] and **Enzo Pranzini** [1,3]

[1] GNRAC—Gruppo Nazionale per la Ricerca in Ambiente Costiero, Corso Europa, 26-16132 Genova, Italy; lrossi@geocoste.com (L.R.); enzo.pranzini@unifi.it (E.P.)
[2] GeoCoste s.n.c., Via Corsi, 19-50141 Florence, Italy
[3] Department of Earth Science, University of Florence, Via Micheli, 4-50121 Florence, Italy
* Correspondence: irenemam3@gmail.com

Received: 10 September 2019; Accepted: 28 October 2019; Published: 31 October 2019

Abstract: Several remotely sensed images, acquired by different sensors on satellite, airplane, and drone, were used to trace the beach ridges pattern present on the delta of the River Ombrone. A more detailed map of these morphologies, than those present in the literature, was obtained, especially at the delta apex, where beach ridges elevation in minor. Beach ridges crests, highlighted through image enhancement using ENVI 4.5 and a DTM based on LiDAR data, were then processed with ArcGIS 9.3 software. Starting from this map, a method to reconstruct beach ridges segments deleted by the transformations of the territory is proposed in this paper. The best crest-lines fitting functions were calculated through interpolation of their points with Curve Expert software, and further extrapolated to reconstruct the ridges morphology where human activity, riverbed migration, or coastal erosion eliminated them. This allowed to reconstruct the ridges pattern also offshore the present delta apex, where the shoreline retreated approximately 900 m in the last 150 years. Results can be further used to implement conceptual and numerical models of delta evolution.

Keywords: beach ridges; mathematical reconstruction; curve fitting; Ombrone River Delta

---

1. Introduction

Deltas are dynamic depositional landforms, sensitive to changes in both the terrestrial and marine environment [1,2], constantly reshaped by river input, tides, and waves [3,4]. Wave-dominated deltas are characterized by a cuspate morphology, a steep offshore profile, and the presence of more or less curved beach ridges parallel to past shoreline positions [5]. This allowed to reconstruct Holocene delta formation phases for many rivers, and to date, the various morphologies when archaeological remnants were present or if radiometric dating were executed [6,7].

Fluctuations in fluvial sediment supply were also assessed for several deltas, being the most important factor controlling the spatial and temporal variation in beach ridges development [8,9]. Spits are frequently present near the river mouth, sometimes overpassed by the coastal progradation and detectable in the delta area, others eroded by waves. Beach ridges frequently evolve in coastal dunes which reflect a huge amount of sediments that are redistributed by waves and moved inland by winds forming elongated and parallel morphologies behind the coastline [10]. A fast accreting coast makes the formation of several low crests, whereas a slowly prograding or a stable one allows the growth of few high dunes [11,12]. Beach ridges morphology and related pattern could be used as markers of morphodynamic variations: they can give information on past wave regime, climate conditions, sediment supply, and sea level change [13–16].

In historical times, fast delta progradation is frequently induced by fluvial input increase, ascribed to the population development on the catchment area accompanied by intensive deforestation [17,18].

Just short (months) and medium time (years) changes in sediment input can induce delta apex reshaping, resulting in different delta lobes morphology and beach ridges patterns. Fast progradation periods, due to higher sediment input by the river, sees a sharp cusp and the rotation of the river mouth towards the dominant wave front direction, as described for the Ombrone and Arno River deltas [19]; on the contrary, a reduced fluvial input determines a flattening of the cusp up to a rectification of the coastline and downdrift shift of the river mouth [20,21].

Conceptual models of cuspate delta evolution have been proposed by ref. [19,22], whereas other authors ref. [2,21,23,24] used numerical models to explain delta asymmetry and wave reshaping of the river mouth during erosion and accretion phases. For deltas characterized by alternating phases of growth and erosion, not all the beach ridges at the delta apex can be identified, since their most offshore part has been beheaded during the erosion phases [12]; in addition, some segments can be leveled by human activity, especially agriculture, or cut by the river course migration.

A full reconstruction of the delta evolution needs to know the delta shape at its extreme expansions, even if later cut by erosion; this is of outmost importance when conceptual or numerical models based on wings surface are to be developed.

The possibility to observe the secular trends of beach ridges on the Ombrone delta plain represents a valuable opportunity to obtain useful information on its complex geomorphological evolution, where one of the main drawbacks is represented by the loss of geographical information due to natural and anthropic transformations of the territory over time and the lack, or inaccuracy, of ancient cartography. The problem was solved through a mathematical reconstruction of the missing portions of the beach ridges inspired form by the observation of their geometry and how their planar forms remind of specific mathematical functions.

On Earth, a number of geomorphic elements have a regular shape, including small-scale landforms such as cirques, drumlins, sand dunes, alluvial fans, dolines, polygonal soils and cracks, columnar jointing, and scoria cones, to large-scale landforms such as composite and shield volcanoes [25]. Geomorphometry is the science of quantitative land-surface analysis and a variety of landforms are fitted by mathematical functions, often by means of regression analysis [26].

Just to remain in coastal environment, zeta bays in equilibrium with approaching refracted waves are fitted by a logarithmic spiral as proposed by Silvester and Hsu [27], or the equilibrium nearshore profile is mathematically described by the Dean equation [28].

Following a similar approach, a method to reconstruct beach ridges, where they have been eroded or razed by human activity, is hereafter proposed using fitting equations.

Results can be further used for a detailed model on the delta evolution reconstruction, which is not the aim of this study.

2. Study Area

The Ombrone River delta is in Italy, located on the Southern Tuscany coast (Italy), between Castiglione della Pescaia and Monti dell'Uccellina, in the Maremma Regional Park (Figure 1). The dominant waves direction is from Souh-West [24], and for the present research, is assumed to be a constant wave climate over the historical times [29,30]. The regional longshore transport direction is from south to north [31,32]; subsidence in the alluvial plain was measured in 3 mm/year [33] and this is the only significant vertical movement in the considered time lapse. Sea level rise is about 1 m since the roman period [34].

The ancient gulf of Grosseto started to fill with the material transported by the Ombrone River during the Holocene transgression, but the most internal beach ridge is very likely associated to a sand bar closing a lagoon. When the lagoon was almost silted, the river could directly empty into the sea and started the delta formation. This is dated to the early Roman period [35] and explained with the intense deforestation carried out within the watershed by the Etruscan, first, and by the Romans, later.

During the last twenty centuries, the delta apex grew for approximately 6.5 km at the apex, with a very fast expansion in the 14th–18th century [36], when an intense deforestation occurred. Some authors also see in this delta expansion, the effect of the Little Ice Age [35,37]. Two main erosional phases have been identified, the first after the fall of the Roman Empire, and the second with the Black Death epidemic; in both cases a strong population reduction induced the abandonment of many cultivated areas and the growth of the forest [36].

Furthermore, at the end of the 19th century, these tendencies drastically reversed with the erosion of the coastline at the river mouth, caused by agriculture abandonment, riverbed quarrying, dams and weirs construction; whereas wetland reclamations, through river course diversion, started some centuries before depriving the river load mostly in its finer fraction. Results of these works were poor since sediment compaction was restoring the original wetlands [38].

Since the end of the 19th century, the delta apex started to retreat, and it is now approximately 900 m back from its most prominent position (depicted in a 1881 topographic map). Before shore protection works were carried out in 2013–2014 on the southern lobe (submerged groins) and milder in 2016 on the northern one (short wood permeable groins), the erosion rate was approximately 10 m/year [17].

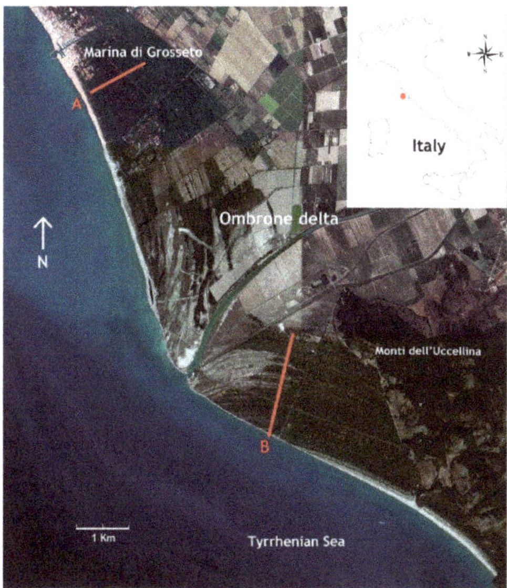

Figure 1. The Ombrone River delta and the position of the two topographic profiles drawn in Figure 3 (Google Earth image).

Delta morphology is slightly different in the two lobes [6,36,37]. The northern side hosts inter-dune swales and, near the river course, some permanent ponds, possibly old channels to connect salterns to the sea. Dunes progressively merge to the north in few parallel ridges; the inner one is attributed to the Etruscan period, the next five to the Roman time, followed by seven Medieval and the last four belonging to the Modern Age [39].

Agriculture activity and urban development at Marina di Grosseto deleted some segments of these dunes. The southern side is almost uninhabited, dunes are closer, and their continuity is more preserved, but at the apex, where they are far lower, seasonal ponds form and in the wetter period most of the apex is flooded. The Monti dell'Uccellina, with their seaward projection, limit sediment dispersion and, even if in the past centuries the prevalent longshore transport was to the north, the delta could symmetrically grow. (Smaller volume but shorter base in the southern side). The River Ombrone delta, hosts many beheaded ridges; but the well-preserved morphology of those remained allows to test a methodology to mathematically reconstruct the lost segments.

3. Material and Methods

For this study, a detailed beach ridge map was obtained through the combination of several data sets, produced by active and passive remote sensors, and processed with various methodologies. A LiDAR (light detection and ranging) survey acquired in 2008 by the Ministry of the Environment, with 1×1 m ground resolution, was used to derive a 3D model of the beach ridges area. Where agriculture flattened the ridges, ad hoc topographic surveys were carried out, and a higher vertical resolution DTM (digital terrain model) was produced from low altitude drone imagery using the structure from motion (SfM) technique.

Satellite images used span from Landsat TM and Landsat ETM+ (from 1986 to 2002); Quickbird (2004); Landsat 8 (2015) and Sentinel 2a (2017); in addition, some Google Earth images were used. Pixel size goes from 2.44 m (Quickbird) to 30 m (Landsat TM).

Remotely sensed images helped in tracing ridge crests where ridges were flattened, and soil moistures helps to discriminate sand from silty/clayey soils present in the interdune swales. This was obtained with specific image processing, such as normalized difference of vegetation index (NDVI), tasseled cup transformation (brightness, greenness, and wetness) [40], principal components analysis (PCA), and RGB (red-green-blue) color composites of the results of the different elaborations. Recent topographic surveys, regional cartography, and old aerial photos acquired in 1944 by RAF (Royal Air Force) were also used. Images were then wrapped on the Lidar 3D model to assist their photo interpretation.

All these data were processed using ENVI 4.5 (Harris Geospatial Solutions Inc., Broomfield, CO, USA) and ArcGIS 9.3 (ESRI) softwares in order to produce a detailed beach dunes/beach ridges map of the delta area in the WGS84 UTM32 reference system (Figure 2) and leading information was derived from LiDAR data, whose resolution was maintained also after the following image overlapping.

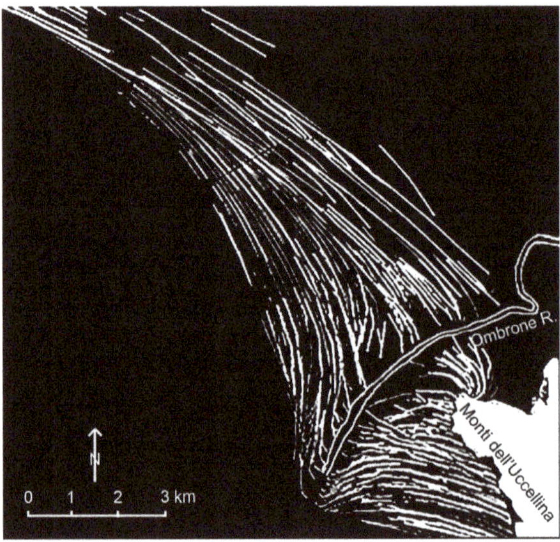

Figure 2. Ombrone River delta, beach ridges pattern mapped through the combination of different data sets and elaboration methodologies.

From LiDAR-DTM, several cross-shore profiles were extracted, and minor beach ridges sequence and evolution phases were recognized and relatively dated, starting from the main delta evolution periods described in the literature [36,37].

From the position, geometry, and height of some ridges it was possible to pair them across the gap formed by the river or by agricultural flattering. In Figure 3, profiles crossing the dunes/ridges system on the two delta wings are drawn and ridges associated and numbered (see Figure 1 for profiles location). In the northern profile, all the morphologies are compressed, both because the profile is located far from the delta apex, and because sediment is distributed along a wider area (the northern limit of the cell—Punta delle Rocchette—is 24 km from the river mouth, whereas the southern, one—Monti dell'Uccellina—is 7 km only).

Profiles show the differences in height and distance between ridges, and how some of the highest/largest ones on the delta wings and associated to apex erosion or stability periods, comprise more than one crest, due to an overlapping of more dunes.

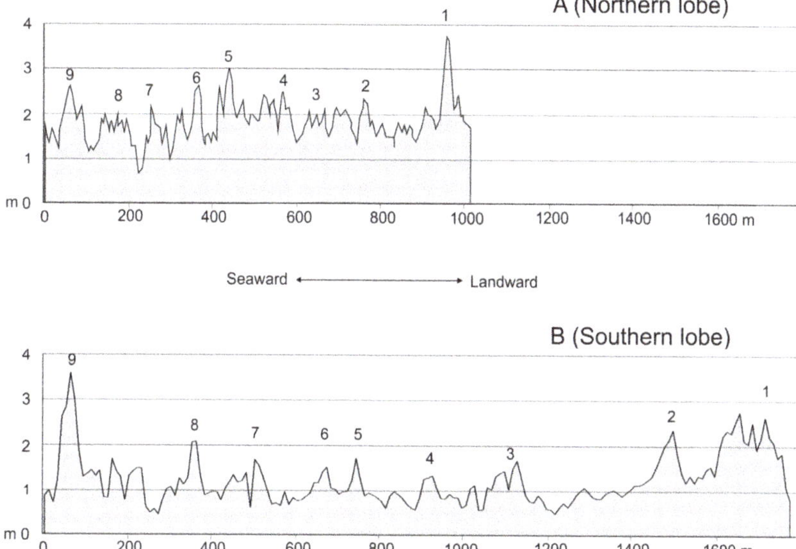

Figure 3. North (**A**) and south (**B**) profiles from the early 15th (1) century until the limit of the progradation in middle 19th century (9). Due to the difference in deposition rates, horizontal scale cannot be considered a time scale.

Sixteen beach ridges from the early 15th to the middle-19th century have been selected representing different periods of the delta evolution, both when accreting and when eroding. In the first case, beach ridges plant geometry is more curved at the apex with ridges sharply converging (Figure 4); on the contrary, soon after a period of delta erosion their planform is more rectilinear since beach retreats at the apex but progradation goes on at the wings (actually it is a form of beach rotation) [12].

Along each beach ridge, several x,y points were digitized where each ridge crest was more evident; one point every 50 m on average. Along this distance, ridges are almost rectilinear and more points do not increase the accuracy. They were further rotated and shifted in a Cartesian coordinate system with the y-axis coincident with the direction of the terminal river course.

In order to mathematically reconstruct the missing parts of the beach ridges, 33 families of linear and not linear regression models were tested, both for progradation and for erosion beach ridge shapes. The calculation was done using a specific curve fitting software (CurveExpert Pro 1.4, Hyams Development, Chattanooga, TN, USA). The software provides a ranking of the best fitting functions and they are also displayed in the graphing window with statistical results, such as standard error and correlation coefficient. Values x,y for points of interpolated and extrapolated curves are also available and useful for the reconstruction of the missing parts of the delated beach ridges.

4. Results

Finding equations that well fit the ridges allows to reconstruct their original plan form in the stretches where they have been erased by human activities or by coastal erosion.

Among all the 33 regression models tested, two distinct families of functions, respectively Weibull (Equation (1)) and Morgan–Morgan–Finney (MMF) (Equation (2)), were found out with lowest values of standard error and highest correlation coefficient. The former function better approximates more rectilinear beach ridges of older coastlines or of erosional phases (more flat lines); whereas the latter function better fits curved beach ridges of fast accreting stages (Table 1).

To a lesser extent, Hoerl (Equation (3)) and 3th degree polynomial (Equation (4)) were the following two better equations in the fitting ranking.

$$\text{Weibull}: \ y = a - be^{-cx^d}. \tag{1}$$

$$\text{MMF}: \ y = \frac{ab + cx^d}{b + x^d}. \tag{2}$$

$$\text{Hoerl}: \ y = a\,b^x x^c. \tag{3}$$

$$\text{Polynomial}: \ 3\text{th} - \text{degree} \ y = a + bx + cx^2 + dx^3. \tag{4}$$

Obviously, the robustness of the method depends on the extension of interpolated beach ridges too, and is correlated to the segment form: linear or arcuate. For very short segments, results exponentially lose significance.

Table 1. Relation between beach ridge and best four fitting mathematical functions. ER = apex erosion, PR = apex progradation. In bold, the best standard error for each ridge.

Year (ca.)	Evolution Type	Standard Error [m] for Four of the Best Different Fitting Models			
		MMF	Weibull.	Hoerl	3rd Degree Polynomial
1500	ER	5.1	**3.2**	4.0	3.3
1550	PR	**14.4**	16.6	16.4	22.2
1600	ER	16.6	**12.6**	14.5	19.2
1615	PR	**7.0**	11.4	15.8	27.5
1625	ER	14.3	**12.8**	16.8	23.9
1650	ER	12.5	8.2	**7.2**	16.3
1660	PR	**5.5**	9.3	12.3	18.9
1675	ER	16.6	**10.0**	11.9	12.6
1700	ER	15.2	**7.1**	9.4	8.4
1725	ER	13.4	**8.5**	10.2	18.2
1750	PR	**9.3**	12.1	21.2	16.2
1775	ER	18.8	**14.7**	19.6	32.7
1800	ER	16.4	11.5	13.1	**9.5**
1820	ER	16.5	**5.6**	18.7	21.1
1830	PR	**6.4**	14.9	15.5	23.8
1850	ER	29.7	13.2	20.7	**11.8**

In the table are reported precise year, just to have a temporal reference to be attributed to every mapped beach ridge, but there may be indeed an uncertainty of several years. The correlation coefficients range are from 0.999 to 0.997.

In Figures 4 and 5, the real 1625 (apex erosion phase) and 1750 (apex accretion phase) beach ridges on the south lobe of the delta are drawn together with the best 4 fitting models produced through extrapolation, near the river, of the first 4 km of each ridge. The graphs show Weibull (Equation (1)) better approximates more rectilinear erosional beach ridges and MMF (Equation (2)) curved beach ridges of progradation phases.

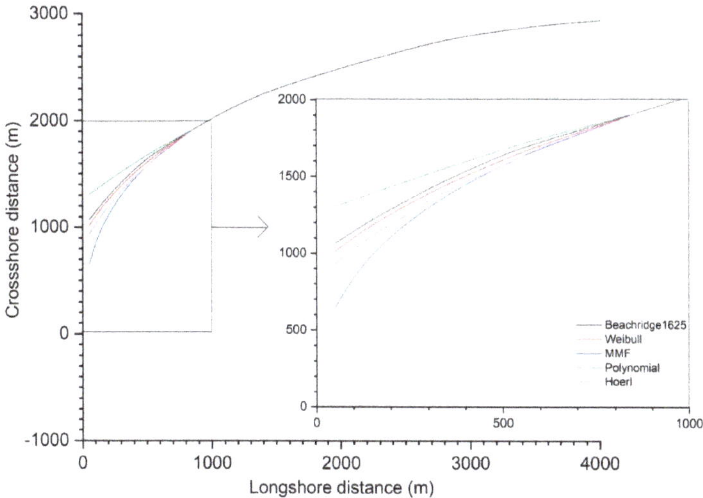

Figure 4. Real beach ridge 1625 ca., correlated to an erosional period, with the four best fitting mathematical functions.

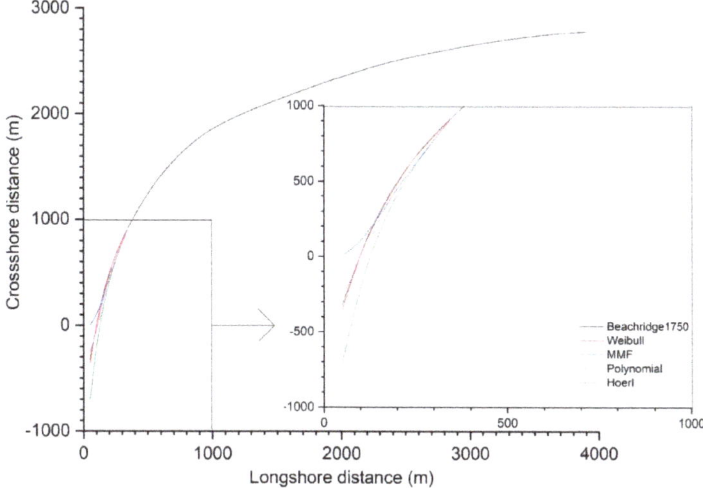

Figure 5. Real beach ridge 1750 ca., correlated to a progradation period, with the four best fitting mathematical functions.

Starting from all the beach ridges mapped in Figure 2, the main evolution phases of the Ombrone River delta, for the considered period, are reported in Figure 6. The considered period is from the early 15th to the middle 19th century, after which reliable topographic surveys are available.

Beach ridges deleted near the river because of watercourse shifting or truncated by a subsequent erosive event have been mathematically reconstructed according to the methodology proposed.

Beach ridges beheaded at the river mouth are represented in cyan and those related to the cuspate progradation in red. In blue and purple, their reconstructed morphologies with mathematical functions.

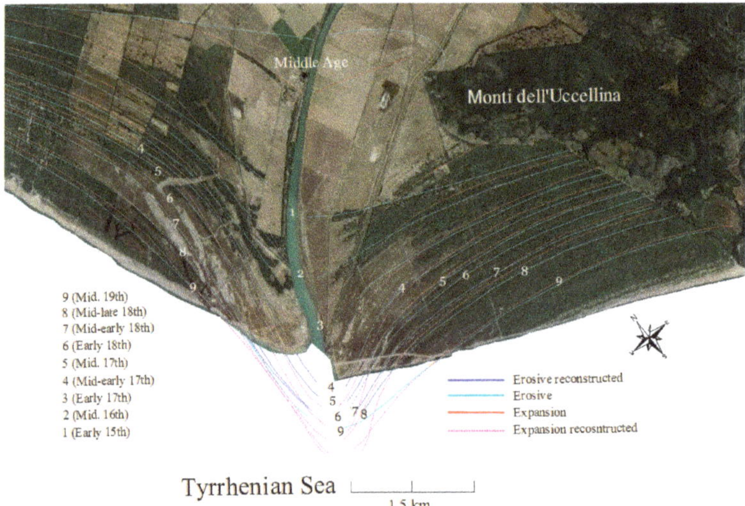

Figure 6. Beach ridges reconstruction for main historical phases from the 15th to the 19th century.

5. Discussion

Quantitative approach in Geomorphology became established in the mid-20th century [41], but equations fitting coastal landforms arrived a few decades later, like those aforementioned, by Dean [28] and by Silvester and Hsu [27]; both, although developed on actual morphologies, are used to forecast the shape a landform (nearshore profile and zeta-bays) will reach at equilibrium. What is done in this paper is the reconstruction of ancient landforms in those parts which have been later eroded.

In a period in which approximately 31% of the world beaches are eroding [42], and considering that erosion generally starts at the river mouth and gradually expand laterally [22], many present cuspate river deltas lost their apex. Further, agricultural works and urbanization frequently flattened, or at least hid, original morphologies. This makes it impossible to fully reconstruct the natural delta morphology and study its historical evolution.

The proposed method needs high-resolution data, which was obtained from active and passive remote sensing systems. In this way, a detailed reconstruction of beach ridges morphologies was done, and several accretion and erosion phases mapped. Thirty-three different fitting curves were generated though interpolation of control points on the beach ridges/dune crests and their accuracy in describing those morphologies was assessed.

The further extrapolation of the last seaward points allowed to reconstruct the ridge segments which were eroded during the last 150 years, thus obtaining information on the past delta apex shape. On the other sides, interpolation made it possible to identify reaches flattened by agriculture or by other anthropogenic activities. For ridges formed during delta expansions phases (more curved at the tip), the MMF model proved to be more accurate than all the other tested; whereas ridges formed during delta tip erosion, more flattens, were better reconstructed by the Weibull equation.

6. Conclusions

The mathematical interpolation and extrapolation of the functions fitting the ridge crests at the Ombrone River delta allowed to reconstruct their entire morphologies, not only in the part eroded during the last 150 years, but also where similar processes worked in the Middle and Modern Age.

On these bases, it will be possible to develop more accurate models of delta evolution in order to better explain the different morphology characterizing the updrift and the downdrift delta sides, and to forecast future shoreline displacement in this coast rich in natural and economical elements.

The hoped application of this method to other deltas, using different equation coefficients, could help to solve similar problems and improve the method itself.

Author Contributions: Conceptualization, I.M., L.R., E.P.; methodology, I.M., L.R., E.P.; validation, L.R.; formal analysis, I.M.; investigation, I.M.; resources, E.P.; data curation, L.R.; writing—original draft preparation, I.M. and L.R.; writing—review and editing, L.R. and E.P.; supervision, E.P.

Funding: This work was supported by the Project Interreg Marittimo-IT-FR Maritime "Management des Risques de l'Erosion cotière et actions de Gouvernance Transfrontalière" (MAREGOT).

Conflicts of Interest: The authors declare no conflict of interest.

References

1. Syvitski, J.P.; Milliman, J.D. Geology, geography and humans battle for dominance over the delivery of sediment to the coastal ocean. *J. Geol.* **2007**, *115*, 1–19. [CrossRef]
2. Nienhuis, J.H.; Ashton, A.D.; Roos, P.C.; Hulscher, S.J.M.H.; Giosan, L. Wave reworking of abandoned deltas. *Geophys. Res. Lett.* **2013**, *40*, 5899–5903. [CrossRef]
3. Wright, L.D.; Coleman, J.M. Variations in morphology of major river deltas as functions of ocean wave and river discharge regimes. *AAPG Bull.* **1973**, *57*, 370–398.
4. Galloway, W.E. Process framework for describing the morphologic and stratigraphic evolution of deltaic depositional systems. In *Deltas, Models for Exploration*; Broussard, M.L., Ed.; Houston Geological Society: Houston, TX, USA, 1975; pp. 87–98.
5. Van Maren, D.S.; Hoekstra, P. Cyclic development of a wave dominated delta. In Proceedings of the Coastal Sediments '03, Clearwater Beach, FL, USA, 18–23 May 2013; p. 12.
6. Bellotti, P.; Caputo, C.; Davoli, L.; Evangelista, S.; Valeri, P. Lineamenti morfologici e sedimentologici della piana deltizia del Fiume Ombrone (Toscana Meridionale). *Boll. Soc. Geol. It.* **1999**, *118*, 141–148.
7. Otvos, E.G. Beach ridges–definition and significance. *Geomorphology* **2000**, *32*, 83–103. [CrossRef]
8. Anthony, E.J. Beach ridge progradation in response to sediment supply examples from West Africa. *Marine Geol.* **1995**, *129*, 175–186. [CrossRef]
9. Anthony, E.J. Wave influence in the construction, shaping and destruction of river deltas: A review. *Marine Geol.* **2015**, *361*, 53–78. [CrossRef]
10. Vespremeanu-Stroe, A.; Preoteasa, L.; Zainescu, F.; Rotaru, S.; Croitoru, L.; Timar-Gabor, A. Formation of Danube delta beach ridge plains and signatures in morphology. *Quat. Int.* **2016**, *415*, 268–285. [CrossRef]
11. Psuty, N.P. Spatial variation in coastal foredune development. In *Coastal Dunes: Geomorphology, Ecology and Management*; Balkema: Rotterdam, The Netherlands, 1992; pp. 3–13.
12. Pranzini, E. Airborne LIDAR survey applied to the analysis of the historical evolution of the Arno River delta (Italy). *J. Coast. Res.* **2007**, 400–409.
13. Curray, J.R.; Emmel, F.J.; Crampton, P.J.S. Holocene history of a strand plain, Lagoonal Coast, Nayarit, Mexico. In *Lagunas Costeras*; Costonares, A.A., Phelger, F.B., Eds.; Un Simposio; Universidad Nacional Autonoma de Mexico: Mexico City, Mexico, 1967; p. 686.
14. Stapor, F.W. Beach ridge and beach ridge coast. In *Encyclopedia of Beaches and Coastal Environments*; Schwartz, M.L., Ed.; Hutchinson Ross: Stroudsburg, PA, USA, 1982; pp. 160–161.
15. Fairbridge, R.W.; Hillaire-Marcel, C. An 8000-yr palaeoclimatic record of the double-hale 45-yr solar cycle. *Nature* **1977**, *268*, 413–416. [CrossRef]
16. Taylor, M.; Stone, G.W. Beach-bridges: A review. *J. Coast. Res.* **1996**, *12*, 612–621.
17. Pranzini, E. Bilancio sedimentario ed evoluzione storica delle spiagge. *Il Quaternario* **1994**, *7*, 197–202.
18. Syvitski, J.P.M.; Vörösmarty, C.J.; Kettner, A.J.; Green, P. Impact of humans on the flux of terrestrial sediment to the global coastal ocean. *Science* **2005**, *308*, 376–380. [CrossRef] [PubMed]
19. Pranzini, E. Updrift river mouth migration on cuspate deltas: Two examples from the coast of Tuscany (Italy). *Geomorphology* **2001**, *38*, 125–132. [CrossRef]

20. Komar, P.D. Computer model of delta growth due to sediment input rivers and longshore transport. *Geol. Soc. Am. Bull.* **1973**, *84*, 2217–2219. [CrossRef]
21. Nienhuis, J.H.; Ashton, A.D.; Giosan, L. Littoral steering of deltaic channels. *Earth Planet. Sci. Lett.* **2016**, *453*, 204–214. [CrossRef]
22. Pranzini, E. A model for cuspate delta erosion. In Proceedings of the 6th Sympoium on Coastal and Ocean Management/ASCE, Charleston, SC, USA, 11–14 July 1989; pp. 4345–4357.
23. Ashton, A.D.; Giosan, L. Wave-angle control of delta evolution. *Hydrol. Land Surf. Stud.* **2011**, *38*, 4051–4056. [CrossRef]
24. Aminti, P.; Pranzini, E. Variations in longshore sediment transport rates as a consequence of beach erosion in a cuspate delta. *Proc. EUROCOAST Assoc.* **1990**, 130–134.
25. Pike, R.J.; Evans, I.S.; Hengl, T. Geomorphometry: A brief guide. In *Developments in Soil Science*; Hengl, T., Reuter, H.I., Eds.; Geomorphometry—Concepts, Software, Applications; Elsevier: Amsterdam, The Netherlands, 2009; Volume 33, pp. 3–30.
26. King, C.A.M. Geometrical forms in geomorphology. *Int. J. Math. Educ. Sci. Technol.* **2006**, *2*, 153–169. [CrossRef]
27. Silvester, R.; Hsu, J.R.C. *Coastal Stabilization. Innovative Concepts*; PRT Prentice Hall: Upper Saddle River, NJ, USA, 1993; p. 578.
28. Dean, R.G. Equilibrium beach profiles: U.S. Atlantic and Gulf coasts. In *Ocean Engineering Report, 12*; Department of Civil Engineering, University of Delaware: Newark, NJ, USA, 1977.
29. Li, L.; Walstra, D.R.; Storms, J. The impact of wave-induced longshore transport on a delta-shoreface system. *J. Sediment. Res.* **2015**, *85*, 6–20.
30. Meini, L.; Mucci, G.; Vittorini, S. Ricerche meteo marine sul litorale toscano: Centoventi anni di osservazioni metereologiche a Livorno (1857–1976). *Boll. Soc. Geogl. It.* **1979**, *VIII*, 449–474.
31. Aiello, E.; Bartolini, C.; Caputo, C.; D'Alessandro, L.; Fanucci, F.; Fierro, G.; Gnaccolini, M.; La Monica, G.B.; Lupia Palmieri, E.; Piccazzo, M.; et al. Il trasporto litoraneo lungo la costa toscana tra la foce del Fiume Magra ed i Monti dell'Uccellina. *Boll. Soc. Geol. It.* **1976**, *94*, 1519–1571.
32. Aminti, P.L. Ricostruzione del clima ondoso della Toscana Meridionale sulla base di misure anemometriche. *Bolettino Degli Ingegneri* **1983**, *4*, 34–56.
33. Salvioni, G. I movimenti del suolo nell'Italia centro-settentrionale. *Bollettino di Geodesia e Scienze Affini* **1957**, *16*, 325–366.
34. Antonioli, F.; Silenzi, S. Variazioni relative del livello del mare e vulnerabilità delle pianure costiere italiane. *Quaderni della Società Geologica Italiana* **2007**, *2*, 1–29.
35. Citter, C.; Arnoldus-Huyzendveld, A. Archeologia urbana a Grosseto, la città nel contesto geografico della bassa valle dell'Ombrone. Origine e sviluppo di una città medievale nella "Toscana delle città deboli". In *Le Ricerche 1997–2005*; Edizioni all'Insegna del Giglio s.a.s: Fiorentino, Italy, 2007; pp. 41–60.
36. Innocenti, L.; Pranzini, E. Geomorphological evolution and sedimentology of the Ombrone River Delta, Italy. *J. Coast. Res.* **1993**, *9*, 481–493.
37. Bellotti, P.; Caputo, C.; Davoli, L.; Evangelista, S.; Garzanti, E.; Pugliese, F.; Valeri, P. Morpho-sedimentary characteristics and holocene evolution of the emergent part of the Ombrone River delta (southern Tuscany). *Geomorphology* **2004**, *61*, 71–90. [CrossRef]
38. Barsanti, D.; Rombai, L. *La Guerra delle Acque in Toscana*; Edizioni Medicea: Firenze, Italy, 1986; p. 169.
39. Mori, A. L' evoluzione della costa grossetana dal Pliocene ad oggi. In *Annuario 1932–1935 del Regio Liceo di Grosseto*; La Maremmana: Grosseto, Italy, 1935; p. 21.
40. Crist, E.P.; Cicone, R.C. A physically based transformation of thematic mapper data. *IEEE Trans. Geosci. Remote Sens.* **1984**, *GE-22*, 256–263. [CrossRef]

41. Pitty, A.F. *The Nature of Geomorphology*; Methuen & Co.: London, UK, 1982; p. 161.
42. Luijendijk, A.; Hagenaars, G.; Ranasinghe, R.; Baart, F.; Donchyts, G.; Aarninkhof, S. The state of the world's beaches. *Nat. Sci. Rep.* **2018**, *8*, 6641. [CrossRef]

Article

Documenting a Century of Coastline Change along Central California and Associated Challenges: From the Qualitative to the Quantitative

Gary Griggs [1,*], **Lida Davar** [1,2] **and Borja G. Reguero** [3]

[1] Department of Earth and Planetary Sciences, University of California Santa Cruz, Santa Cruz, CA 95064, USA

[2] Department of Environmental Sciences, University of Tehran, P.O. Box 14155-6453, Tehran, Iran; ldavar@ucsc.edu

[3] Institute of Marine Sciences, University of California Santa Cruz, Santa Cruz, CA 95064, USA; breguero@ucsc.edu

* Correspondence: griggs@ucsc.edu

Received: 21 November 2019; Accepted: 13 December 2019; Published: 15 December 2019

Abstract: Wave erosion has moved coastal cliffs and bluffs landward over the centuries. Now climate change-induced sea-level rise (SLR) and the changes in wave action are accelerating coastline retreat around the world. Documenting the erosion of cliffed coasts and projecting the rate of coastline retreat under future SLR scenarios are more challenging than historical and future shoreline change studies along low-lying sandy beaches. The objective of this research was to study coastal erosion of the West Cliff Drive area in Santa Cruz along the Central California Coast and identify the challenges in coastline change analysis. We investigated the geological history, geomorphic differences, and documented cliff retreat to assess coastal erosion qualitatively. We also conducted a quantitative assessment of cliff retreat through extracting and analyzing the coastline position at three different times (1953, 1975, and 2018). The results showed that the total retreat of the West Cliff Drive coastline over 65 years ranges from 0.3 to 32 m, and the maximum cliff retreat rate was 0.5 m/year. Geometric errors, the complex profiles of coastal cliffs, and irregularities in the processes of coastal erosion, including the undercutting of the base of the cliff and formation of caves, were some of the identified challenges in documenting historical coastline retreat. These can each increase the uncertainty of calculated retreat rates. Reducing the uncertainties in retreat rates is an essential initial step in projecting cliff and bluff retreat under future SLR more accurately and in developing a practical adaptive management plan to cope with the impacts of coastline change along this highly populated edge.

Keywords: coastline; cliff and bluff retreat; erosion rate; uncertainty; sea-level rise; adaptive management

---

1. Introduction

Many diverse natural forces and processes interact along the shoreline, making the coastline one of the world's most dynamic environments [1–3]. Waves, tides, wind, storms, rain, and runoff combine to build up, wear down, and continually reshape the interface of land and sea [3–5]. Through the 20th century, however, global sea-level rise, due in a large part to human-induced climate change [6–8], contributed to increase both cliff and beach erosion [9,10]. Coastline (cliff and bluff) erosion (covered in this study) is different from shoreline (beach) erosion and is defined as the actual landward retreat of a cliff or bluff. While a number of older references indicate that cliffs occur along about 80% of the world's coasts [10–12], more recent work using a GIS-based global mapping analysis and a detailed literature review suggest that coastal cliffs likely exist on about 52% of the global shoreline [13]. Cliff retreat is distinct from beach erosion in that it is not recoverable, at least within our lifetime, by any natural processes [10]. The terms cliff and bluff are often used interchangeably [10,14], but in this study cliff refers to coastal landforms that consist of harder and more resistant rocks that stand higher and steeper than bluffs, which are generally composed of weaker materials and stand at gentler slopes (Figure 1) [10].

Figure 1. (**a**,**b**): Coastal cliffs, (**c**,**d**): Coastal bluffs. Photos: © 2002-2015, California Coastal Records Project [15].

The world's coastlines will respond to global climate changes and the associated adjustment to oceanographic forcing [16]. For cliffed coasts with limited beach development, there appears to be a relationship between long-term cliff retreat and the rate of sea-level rise [17]. Satellite altimetry has shown an average rise in global mean sea level (GMSL) of ~3.4 mm/year since 1993 [10,18,19] and this rate is increasing by about 0.08 mm annually, which implies that global mean sea level could rise at least 65 ± 12 cm by 2100 compared with 2005 [20], enough to cause significant problems for coastal cities around the planet [21]. However, more recent studies along California's coastline indicate the possibility of significantly higher sea levels by 2100, with levels at specific future dates highly dependent on future global greenhouse gas emissions [22]. Future sea-level rise will increase the frequency at which waves will attack the base of coastal cliffs and bluffs [23–28], and as a result, coastal erosion will almost certainly be accelerated during the 21st century [23,27,29]. In addition, changes in regional meteorological and climate patterns, including the frequency and intensity of El Niño events, coupled with rising sea level, are predicted to result in increasing extremes in sea level [30] and wave power [31]. Waves riding on these higher water levels will cause increased coastal erosion and shoreline damage, more than that expected from sea-level rise alone [30]. Many major coastal cities were developed in areas vulnerable to shoreline and/or coastal erosion [32]. With coastal populations and associated economic assets continuing to increase [33], cliff, bluff top, and shoreline development will be increasingly threatened by erosion and retreat [34]. This has led to an increased need for accurate information on rates and trends of coastal recession [35] in order to respond and adapt to expected future shoreline changes.

In this study, we focused on California's coast, which is experiencing well-documented sea-level rise [22,36] and related coastal impacts including coastal erosion [10,30,37]. California's coast reflects a complex geological history and the interplay of tectonic or mountain building processes, geology, climate, and the sea, and has always been identified with change [1,38]. At the close of the last ice age 18,000 years ago, the coastline stood several to as far as 50 km offshore to the west [4]. As the climate warmed, seawater expanded and ice melted. In response, sea level rose about 130 m and advanced inland, moving the cliffs, bluffs, and beaches eastward. About 8000 years ago, the rate of sea-level rise slowed from an average of about 11mm/year over the previous 10,000 years to less than a millimeter per year. Over the past century or so, however, due primarily to anthropogenic global warming, the global rate of sea-level rise has accelerated to about 3.4 mm/year (13.4 inches/century) leading to an increase in rates of shoreline and coastline retreat.

The great majority (72% or about 1272 km) of California's 1760 km coastline consists of actively eroding sea cliffs and bluffs [4], and of the 1272 km cliffed coast—including the 5.8 km (3.6 mile) long section of Santa Cruz coast covered in this paper—about 1040 km consists of low to moderate relief cliffs and bluffs ranging in height from about 10 to 100 m, which are typically eroded into uplifted marine terraces (Figure 2). Some cliff rocks are so hard and resistant, however, that photographs taken of the coast 75 years ago look identical to those of today. Elsewhere, however, coastal bluff materials are so soft and weak that the coast is being eroded at average rates of 2 m or more each year. These changes are easily recognized when comparing historic ground photographs.

Figure 2. Typical morphology of much of California coast with cliffs eroded into an uplifted marine terrace (photo by Gary Griggs, 2006).

The coast of California is dominated by uplifted marine terraces fronted by low cliffs, but also includes steep coastal mountains and areas of coastal lowlands, estuaries, and dunes [30]. The two sea-level rise related hazards of greatest concern to any oceanfront development along the California coast, whether public or private, are (1) coastal cliff and bluff erosion (Figure 3a) and (2) more frequent flooding of low-lying areas by storm waves and high tides (Figure 3b), followed in time by permanent inundation [39]. California's coastline is approaching a crisis point, which has resulted from a combination of natural processes and cycles, combined with human intervention and population growth [10]. California's population and economic centers are concentrated along its coast [40]. Although California's 19 coastal counties (including San Francisco Bay) make up only 22% of the state's land mass, they account for 68% of its population [41], 80% of its wages, and 80% of its GDP [42]. In addition, California's coastal population is expected to continue to grow significantly over the coming decades [43], which will only compound the erosion and flooding problems at the edge. A recent study [27], showed that for California, the world's 5th largest economy, over $150 billion in property, equating to more than 6% of the state's GDP and 600,000 people, could be impacted by coastal flooding by 2100.

Figure 3. (**a**) The erosion of a coastal cliff has removed an ocean front street about 10 km from the project site. (**b**) Flooding of a coastal parking lot during a period of elevated sea level and large waves approximately 20 km from the study area.

2. Study Area

The West Cliff Drive coastline, which has been selected as a case study in this paper, extends 5.8 km along the western edge of the city of Santa Cruz between Point Santa Cruz and Natural Bridges State Beach on the central California coast (Figure 4). This section of the coastline is somewhat unique in California in having a public street (West Cliff Drive) extending the entire 5.8 km (3.6 mile) length, along with a pedestrian/bicycle path. This allows unobstructed views of this dramatic coast without the presence of homes or other development on the top of bluff, which is more typical of many of the state's coastal communities. As a result, this road has been a popular area for residents and visitors alike for well over a century.

Figure 4. West Cliff Drive study area.

2.1. Geologic Setting

The striking features of California's diverse landscape, the San Andreas Fault (which lies just 25 km east of the project area) and its associated earthquakes, the rugged coastal mountains, and the uplifted marine terraces [36] and coastal cliffs that characterize much of the coastline, all have their origins in millions of years of large-scale tectonic processes that continue today [4]. The rocks exposed along the coastline and in the sea cliffs provide evidence of this complex geological history and the changes the landscape has undergone. Coastal cliffs along the state's coast may consist of granitic, volcanic, metamorphic, or sedimentary rocks.

The West Cliff Drive coastline consists of near vertical cliffs varying in height from about 6 to 12 m (20 to 40 feet), which form the outer edge of the lowest uplifted marine terrace along this coast (Figure 5a). The lower bedrock portion of the seacliff consists of two different geologic units: The older, harder, and more resistant Santa Cruz Mudstone of Miocene age (~5–7 million years old; Figure 5a) and the younger and weaker Purisima Formation of Pliocene age (~3–5 million years old) (Figure 6). The mudstone extends approximately 2100 m along the western section of the coastal area studied, while the overlying Purisima Formation makes up the approximately 3700 m long eastern section. The Purisima consists of interbedded mudstones, siltstones, and sandstones that are pervasively jointed. It is the orientation of the joints sets that exerts a major control on the erosion of the bedrock and the shape or morphology of the coastline (Figure 5b). The uppermost 2 to 4 m of the cliffs consist of much younger (~100,000 year old) poorly consolidated, marine and non-marine terrace deposits, primarily sand, gravel, and cobbles (Figure 6). The same processes that formed the coastal landscape continue to act on it today, although at a nearly imperceptible rate. More noticeable, however, is the rate of coastline erosion, as waves attack and undercut the emergent land and force the nearly vertical bluffs to recede inland.

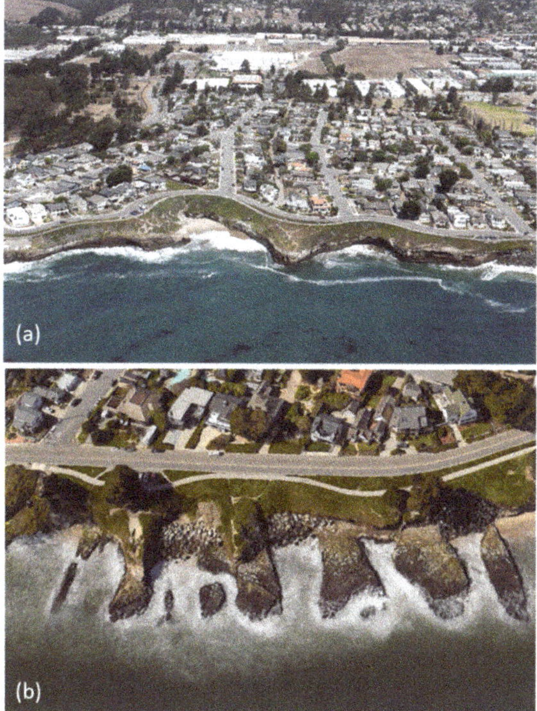

Figure 5. (a) Unprotected section of West Cliff Drive study area showing low eroding cliffs eroded into the Santa Cruz Mudstone [15], (b) oriented embayments eroded along parallel joints in the Santa Cruz Mudstone (Google Earth, 2018).

Figure 6. This section of coastal bluff consists of three different geologic units: The Santa Cruz Mudstone, the Purisima Formation, and the overlying unconsolidated terrace deposits, which erode differentially and make selecting a bluff edge subjective.

2.2. Oceanographic Conditions

The central California coast experiences a mixed semi-diurnal tide with a maximum range of about 2.5 m (8.2 feet), ranging from +2.0 to −0.50 m (+6.6 to −1.6 feet). During years with strong El Niño events, however, water levels may be elevated as much as 30 cm above predicted water levels for days [30,44], which brings large waves closer to the cliffs, and when coincident with high tides, often produce failure of the bedrock, as well as erosion of the overlying and much more erodible terrace deposits. Current wave conditions along West Cliff Drive were defined using historical data from the Global Ocean Waves database [45] that covers the time period between 1948 and 2008. The offshore wave data was propagated to the shore using the SWAN wave propagation model based on the models developed for California with nearshore bathymetry information [46]. The wave propagation results were used to reconstruct hourly time series of wave parameters (significant wave heights, Hs; and mean periods), as described in Camus et al. [47], and calculate hourly wave energy at the 10 m depth contour along West Cliff Drive. Prevailing winter storm waves approach this stretch of coastline dominantly from the northwest and west and undergo little loss of energy through refraction as they approach the coastline along West Cliff Drive. Significant wave heights of 1 to 2 m occur frequently in the winter months (December through March). During major storms, however, wave heights may reach 4 m or more. This is a high-energy coastline with occasional severe wave attack (Figure 7).

Figure 7. Storm waves at high tide overtopping West Cliff Drive bluff (photos by Gary Griggs).

During the periods of largest waves and highest tides, waves are attacking essentially all of the cliffs along this entire section of coast. About 600 m of the 5800 m long stretch of sea cliffs is buffered by small pocket beaches, which come and go seasonally. These vary in length from about 30 to 300 m with maximum widths of 25–50 m in the summer months. The beaches undergo strong seasonal fluctuations in size in response to changing wave conditions, with sand levels dropping 2 m or more from summer to winter months (Figure 8a,b).

Figure 8. (a) Summer and (b) winter beach sand levels along West Cliff Drive (photos by Gary Griggs).

The mean and 25% and 75% percentiles of annual wave energy, and corresponding directions were then calculated using the hourly time series. There is significant variation in a high percentile of wave heights (95% percentile of significant wave height, Hs95) and wave energy intensity and direction (Figure 9). Annual mean wave energy decreases from west to east and rotates slightly anticlockwise (Figure 9a). Changes in wave energy (Figure 9b) are more marked than the differences in high-values of wave heights (Figure 9c). This is because annual wave energy not only represents conditions of a single storm, but, in addition, high wave conditions accrue proportionally more energy than calmer sea states [31]. Therefore, wave power not only shows a significant spatial variation across West Cliff as a result of wave propagation, but may also serve as an indicator of erosion potential of wave action on the cliffs over cumulative periods of time.

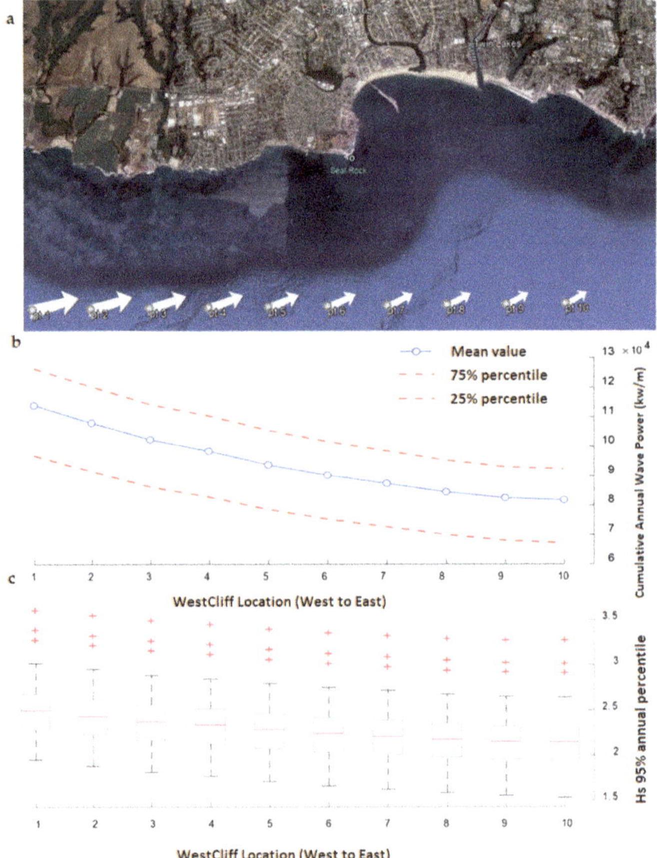

Figure 9. Spatial variation of wave climate along West Cliff. (**a**) Locations of data points with time series of wave parameters and mean direction and intensity of annual Wave Power. (**b**) Eastward variation of cumulative annual wave power. (**c**) Eastward variation of the 95% percentile of significant wave heights (Hs). The boxes represent the range between the 20% to 75% values, where the mean is indicated in red. The crosses represent values exceeding that range.

3. Materials and Methods

In this research we used both qualitative and quantitative assessments to analyze coastal cliff and bluff retreat along the West Cliff Drive coast as a case study to identify the challenges of determining historical erosion rates.

3.1. Qualitative Coastline Change Assessment

We conducted a qualitative assessment by reviewing literature and relevant documents on California coastline erosion, in addition to investigating the geological history, cliff geomorphic differences, sea-level rise, and related impacts on coastal erosion and history of coastline changes throughout the study area.

3.2. Quantitative Coastline Change Analysis

- Materials

We selected two historical sets of aerial photograph from 1953 and 1975 (scale = 1:10,000), and satellite imagery (Google Earth) from 2018 (Figure 10). This gave us a significant span of time to achieve useful results for extracting coastlines and comparing them to detect coastline change, identifying cliff and bluff retreat, as well as measuring the erosion rates along this coast. We also used a hill-shaded digital surface model (USGS, 2018), as well as historical ground and modern photos of the area to digitize and adjust the selected reference line in each segment as accurately as possible. GIS tools were used to perform coastline change detection.

Figure 10. Aerial photographs (**a**) 1953, (**b**) 1975, and (**c**) satellite imagery (2018).

- Methods

Cliff and bluff retreat were analyzed through three main stages:

1. Defining the reference coastline: For digitizing the coastlines and being able to compare them in order to measure coastal retreat, we defined the reference lines that were detectable on both the aerial photographs and satellite imagery and that also could represent the cliff or bluff retreat over the studied time span. Due to the diverse cliff and bluff profiles (Figure 11), three different lines were defined and extracted from the vertical images: a. Cliff/bluff edge; b. the base of bluff; and c. the base of cliff (Figure 12). Depending on data resolution and along different segments of the coast, we selected the line that was the clearest and most detectable for extraction on both the aerial photographs and satellite imagery to use in the analysis.

2. Geometric correction and extracting reference lines: In order to georeference the aerial photographs and perform the nonsystematic geometric correction to reduce the distortion of both the aerial and satellite imagery, we used 19 ground control points (GCPs) that were identified on the two historical aerial photographs and the satellite imagery and collected their coordinates during a field survey. To remove the spatial error (coastline positioning error) as much as possible, we also performed a geometric correction on 31 different coastal segments (Figure 13) using common features which were distinguishable on both aerial photographs and satellite imagery. We then digitized the identified reference line of each segment independently.

3. Coastline change assessment: Analyzing extracted coastlines position led us to historical cliff or bluff retreat rate measurements along West Cliff Drive and allowed us to evaluate the challenges of widely used methods in coastline change studies.

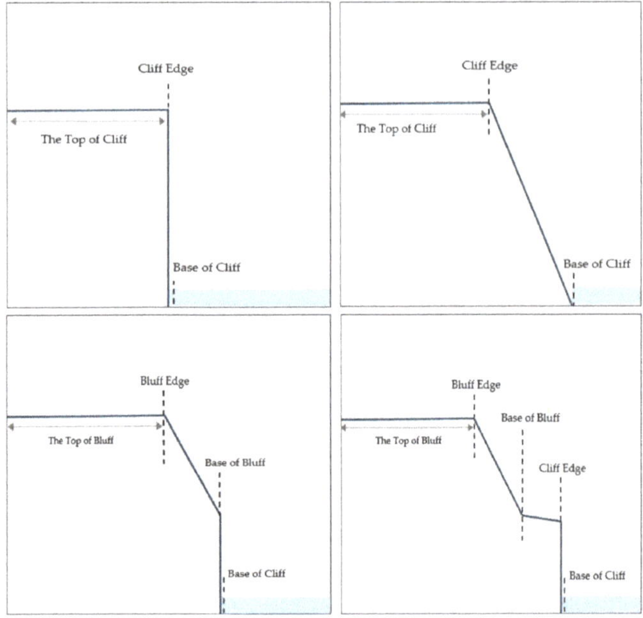

Figure 11. The various profiles and related reference lines (coastline) along cliffed coasts.

Figure 12. The defined reference lines along a cliffed coast. (**a,b**) profile view (**c**) vertical view. (In this segment of coast, the detectable reference line on both aerial photographs and satellite imagery was "the base of bluff").

Figure 13. Map of the 31 identified coastal segments, armored coastline, and unprotected areas along the West Cliff Drive coast—(**a**) West side (**b**) East side.

4. Results

4.1. Qualitative Coastline Change Assessment

- Geology

Erosion along the West Cliff Drive coastline typically occurs through a combination of wave impact, weathering and abrasion of the bedrock, rainfall and terrestrial runoff to a lesser degree, and also relatively infrequent seismic shaking during large earthquakes. Bedrock erosion along weaker stratigraphic layers or joint sets leads to focused erosion and the frequent formation of undercuts, arches, caves, and embayments that made this area an early attraction for residents and visitors. The Santa Cruz Mudstone is more resistant to wave attack than the Purisima Formation and retreats at slower rates overall. As described above, it is primarily the well-developed joint in the latter formation that focuses erosion and typically leads to the failure of large joint-bounded blocks or the collapse of arches and caves. The unconsolidated sandy terrace deposits are much less resistant to wave attack, and it is during periods of large storm waves coincident with very high tides that waves overtop the bedrock and attack and erode the weaker terrace deposits. It is this erosion that has historically threatened and undermined sidewalks, West Cliff Drive, and also a historic lighthouse. Coastal erosion or cliff and bluff retreat in the study area, as along most of the California coast, is an episodic process with most of the major cliff failures occurring during the simultaneous arrival of large storm waves and elevated sea levels.

- Coastal Protection and Erosion

Efforts to stabilize or protect this stretch of shoreline from wave erosion began in 1926 and have continued intermittently to the present. Concrete retaining walls along the upper bluff and rip-rap revetments at the base of the cliff have been the dominant type of armor emplaced, although broken slabs of old streets and sidewalks and stacked bags of concrete were also used in the early years in attempts to halt or slow cliff retreat. Today of the 5800 m of coastline studied, about 2600 m or 44.8% has been armored (Figure 13), with 91% of this armor consisting of rock revetments (Figure 14). While this has served to reduce and halt erosion, it has completely changed the natural condition and appearance of this stretch of coast. These large revetments have also covered large areas of sandy beach that are now removed from public use. The local government agency (City of Santa Cruz Department of Public Works) received emergency permits to install much of the rip-rap, but some of these permits were never finalized with the permitting agency (California Coastal Commission). The Coastal Commission has recently required that the city evaluate the coastal erosion and protection issues, public use, and economics of this 5800 m stretch of coastline and develop a long-term management plan for the future.

Figure 14. The placement of rip-rap over a 60 year period has reduced or eliminated bluff retreat along about 50% of West Cliff Drive, thus complicating any measurements of natural rates of bluff retreat.

- Coastal Change from Historical Ground Photographs

As soon as cameras became widely available, residents, visitors, and commercial photographers began to take pictures of the Santa Cruz coastline. The earliest dated photographs we have discovered of this coast were taken 143 years ago (1876). Certain areas, the picturesque arches, sea stacks, and distinct rock formations along West Cliff Drive, for example, were photographed frequently and memorialized in hand-colored postcards and family albums. Over the subsequent years, as winter storms have periodically battered the bluffs and beaches, and sea level has gradually risen, the coastline has slowly retreated. Some areas have changed dramatically (Figures 15a–c and 16a,b), and others have changed surprisingly little. The natural bridges, arches, and sea stacks that owe their origins to wave attack of the weaker sandstones and mudstones have been destroyed by the same forces that created them, with many fascinating and revealing photographs taken of these natural and unnatural features along the way. While it is very difficult to get any quantitative measurements of cliff or bluff retreat from old ground photographs, they do provide a clear record of the extent of change or retreat that has taken place since the time the original photograph was taken. In many cases, and for most people, a then and now set of photographs provides a more understandable record of coastal change than a rate of retreat in cm/year [48].

Figure 15. (a–c) Progressive erosion of an arch in the study area over a period of 50 years to ultimately form a sea-stack. Dates of photos: (**a**) Before 1888, (**b**) ~1890, (**c**) 2005.

Figure 16. (a–c) Promontory in foreground and Bird Rock arch (in background) in 1909, 2005, and 2019.

4.2. Quantitative Coastline Change Assessment

- Documenting Coastal Erosion from Vertical Aerial Photographs and Maps

The first aerial photographs were taken of the Santa Cruz coast in 1928 and were in stereo. This is quite amazing as Charles Lindberg had just made the first solo flight across the Atlantic Ocean the year before. These photographs were taken in order to study the route of a potential highway between San Francisco and Santa Cruz and aerial photographs were the easiest way to accomplish that. These images provide us with a photographic record extending back 90 years and can be used to determine qualitative changes; because of the only moderate resolution and lack of features from which to take measurements from, they are of limited value in quantitative assessment of historic cliff erosion. Vertical stereo photos were then taken in subsequent years along the Santa Cruz coast, which became quite regular beginning in the 1940s and extending to the present. Aerial photographs were taken more often in later years as various state and federal agencies became interested in documenting the landscape including forest cover, agriculture land use, coastal conditions, and development, highway and railway routes, among other purposes. Until relatively recently, historical aerial photographs were the most common sources for documenting or measuring rates of coastal change such as coastal bluff and cliff retreat. This required first determining the scale of the photograph, and then finding locations where the position of the cliff edge could be measured from some fixed feature (a road, building, etc.) over time. There are multiple challenges involved in this approach, including: a. The scale and resolution of the aerial photographs; b. the sharpness or ease of recognition of the cliff edge; c. the presence of vegetation obscuring the cliff edge making measurements difficult or unreliable; d. the lack of a reference point on older photographs to measure from; and e. the time period covered by the photographs.

The older aerial photographs are usually not of as high resolution as recent photos and useful reference features present in more recent images may well not have existed at the time the older photographs were taken. Even though we have aerial photographs that extend back 90 years for the Santa Cruz coast, using conventional methods (such as an optical comparator or loupe, for example) for measurement is hindered by photograph resolution and appropriate features from which to take repeated measurements.

- Documenting Coastal Erosion from Satellite Imagery and DEMs

In recent decades, the availability of satellite imagery (Google Earth, for example) and Lidar (Light Detection and Ranging) derived DEMs also provided high-resolution data sets for documenting coastal cliff or bluff erosion. With the inclusion of a series of historical satellite photographs in Google Earth, adjusted to precisely the same scale on the website, typically extending back into the early 1990s, or in some cases back to the 1980s, and a built-in measuring tool, a user can determine the distance from some landmark or feature to the cliff edge relatively easy on multiple images all of the same scale. The same issues that can affect the reliability of cliff erosion measurements from historic hard copy aerial photographs still exist, however, a landmark or feature that can be recognized on all images, the resolution and scale of the images, and the clarity or ease of recognizing the cliff edge. In addition, and this is an issue along the section of Santa Cruz coast covered in this study, there are many coastal bluffs and cliffs where the feature designated as the bluff or cliff edge is somewhat subjective because of the varying geomorphology, which is related to the differences in geologic materials (Figures 6 and 17). This makes determining erosion rates difficult.

Figure 17. A section of coastal bluff consisting of the overlying weaker terrace deposits that are mostly vegetated, and the underlying Santa Cruz Mudstone that is being undercut. As in Figure 6, selecting the edge of the bluff for comparative measurements is difficult and somewhat subjective.

In addition, applying the DEMs, as well as topographic maps to extract the indicator lines such as the cliff edge are other approaches to conduct coastal change analysis. The lack of high-resolution elevation data, and the differences in resolution, and scale of available elevation data, are the main constraints to conducting a time-series analysis over a relatively long-period study and would increase the uncertainties of cliff retreat rate retreat estimations.

Further complicating the measurement of changes in cliff or bluff edges is the armoring of this coast, which has gone on for about 60 years (Figures 13 and 14), and depending upon when the rip-rap was placed, this essentially brings erosion to a near halt for a number of years. As of 2019, approximately 45% of the entire West Cliff Drive area had been armored.

- Documenting Coastal Erosion along West Cliff Drive as a Case Study

In addition to investigating the evolution of coastline change over time, we used both aerial photographs and satellite imagery, as well as the hill-shaded DSM (USGS, 2018) to assess coastline retreat along West Cliff Drive coast. The coastline was divided into 31 individual segments that were evaluated independently (Figure 18). The results (Table 1 and Figures 19 and 20) showed the maximum coastline retreat over the studied time span occurred in coastal segment number 16 (Figure 21), where the retreat rate over 65 years ranged from 0.3 to 32 m (Figure 19), and the maximum retreat rate was 0.5 m/year. (Figure 20).

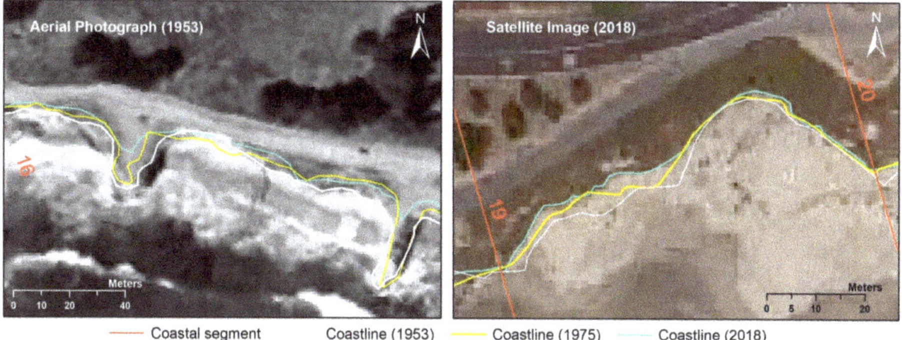

Figure 18. Two examples of digitized coastlines in segments 16 and 19.

Table 1. Coastline retreat along the West Cliff Drive Coast.

	Time span		
	1953–1975	**1975–2018**	**1953–2018**
The range of coastline retreat (m)	~0.3–19	~0.3–28	~0.3–32
The maximum retreat rate (m/year.)	~0.9	~0.6	~0.5

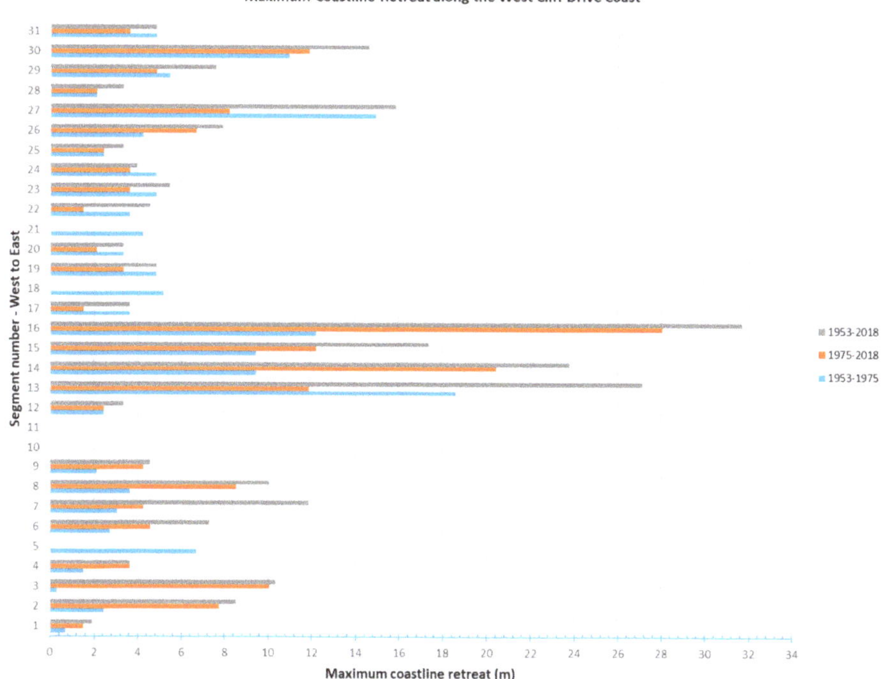

Figure 19. Maximum coastline retreat along the 31 coastal segments (zero value indicates that the change of the segment was undetectable).

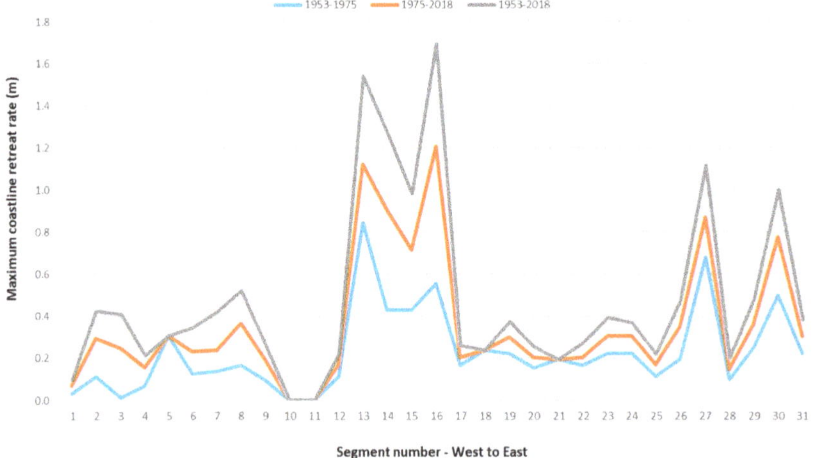

Figure 20. Maximum coastline retreat rate along the 31 coastal segments (zero value shows the coastline of the segment was undetectable).

Figure 21. Oblique aerial photographs from coastal segment number 16 (**a**) 1972 [15], (**b**) 2018 [USGS,2018].

- A Review of Cliffed Coast Retreat Studies

A number of coastal researchers have endeavored to document historical coastal cliff retreat and project future retreat along the coast of California including the Santa Cruz coastline over the years from aerial photographs, satellite imagery, and DEM, with all of the inherent challenges involved. Hapke and Reid [37] completed the most comprehensive assessment. They evaluated cliff retreat using map and photographic data for more than 350 km of the California coast over a period of approximately 70 years, as part of the US Geological Survey's Assessment of Coastal Change Program. They compared one historical cliff edge digitized from old maps dating from 1920–1930, with a recent cliff edge interpreted from LIDAR topographic surveys from either 1998 or 2002. Long-term (~70 year) rates of the retreat were calculated using differences in the locations of the two different cliff edges. The average rate of coastal cliff retreat over this time period for the sections of California coast studied was 0.3 +/− 0.2 m/year. The average amount of total cliff retreat over the 70 year period was 17.7 m. Due to the regional scale of the area studied, however, the shoreline projections were not always accurate, which affected the erosion rate determinations in specific areas. The book "Living with the Changing California Coast" [1], includes cliff and bluff erosion rates where they were published or available for a number of locations along the state's coast. Moore et al. [49] utilized aerial photos corrected through softcopy photogrammetry for a detailed study of cliff erosion rates for both Santa Cruz and San Diego counties as part of a national FEMA (Federal Emergency Management Agency) assessment of coastal erosion hazards. Unfortunately, that study did not include the West Cliff Drive area. Griggs and Johnson [50] reported on cliff erosion rates along the Santa Cruz County coastline, including a few measurements along West Cliff. Their rates were based on comparative measurements from aerial photographs taken in 1940 and 1960, but were limited by the photograph issues discussed above. Young et al. [51] detected 30 individual cliff edge failures and maximum landward retreats from 0.8 to 10 m along the 7.1 km of unprotected coastal cliffs near Point Loma in San Diego over a 5.5-year period (2003–2009). In a recently published study, decadal-scale coastal cliff retreat in southern and central California [52], cliff erosion was detected along 44% of the 595 km of shoreline evaluated, while the remaining cliffs were relatively stable.

Revell et al. [53] evaluated potential future erosion hazards along the coast of California by 2100 under a 1.4 m sea-level rise scenario. For cliff-backed shorelines future potential erosion is projected to average 33 m, with a maximum potential erosion distance of up to 400 m. Young et al. [54] studied cliff and shoreline retreat considering sea-level rise in southern California, and based on their model's results, mean and maximum scenario cliff retreat over 100 years ranged from 4–87 and 21–179 m, respectively. Barnard et al. in a study on coastal vulnerability [16], demonstrated that El Niño events result in wave directional shifts, elevated wave energy, and severe coastal erosion for the Central Pacific and California. Limber et al. [29] applied a multimodel ensemble to project time-averaged sea cliff position of the 475 km long coastline of Southern California over multidecadal time scales and large (> 50 km) spatial scales. Results showed that future retreat rates could increase relative to mean historical rates by more than twofold for the higher SLR scenarios, causing an average total land loss of 19–41 m by 2100.

5. Discussion

One of the most important types of information needed prior to initiating or approving coastal human and natural communities' protection plans, as well as any development along the coast, whether private or public, are the long-term rate at which the cliffs or bluffs are eroding. The longer the period of record covered by the aerial photographs or other data sources, the more representative will be the erosion rate calculated. Additionally, in recent decades, the challenges of a continuing rise in sea level at an accelerated rate, and changes in wave climate, which will affect the long-term cliff erosion or retreat rates, have increased the demand for historical erosion rate data that can be used to project future cliff and bluff positions in highly developed areas. Knowing where the edge of the coastal bluff or cliff is likely to be over time is important for future planning. However, most coastal change studies conclude that there are always different levels of uncertainty in erosion rate measurements and coastline retreat projection results. In this study, some of the identified challenges that could increase the result uncertainties were:

(a)   Extracting a constant and continuous coastline on both the aerial photographs and satellite imagery. The combination of a complex cliffed coast profile (Figure 22), as well as diverse geology, led to several lines which could be considered as the erosion (cliff retreat) indicator. We extracted the line that was the most detectable on 31 individual studied coastal segments independently.

(b)   In some areas the existing armor, as well as vegetation coverage made it impossible to extract the coastline retreat indicator, the cliff edge, for example.

(c)   The low-resolution of historical aerial photographs made it difficult to recognize and extract the identified indicator line along several coastal segments.

d)   Considering the data availability, we employed a nonsystematic geometric correction, however it did not completely remove the geometric errors along the coastal segments and, as a result, we did not measure the retreat of such areas.

(e)   There were two other types of coastal erosion that were not detectable on vertical aerial and satellite images including undercuts and sea caves (Figure 23).

(f)   The scale of the study could directly affect the accuracy of coastline change analysis. In this study, examining quantitative coastline retreat along a relatively short section of coast enabled us to use the ground photographs of the study area to adjust the modern coastline we had extracted from satellite imagery.

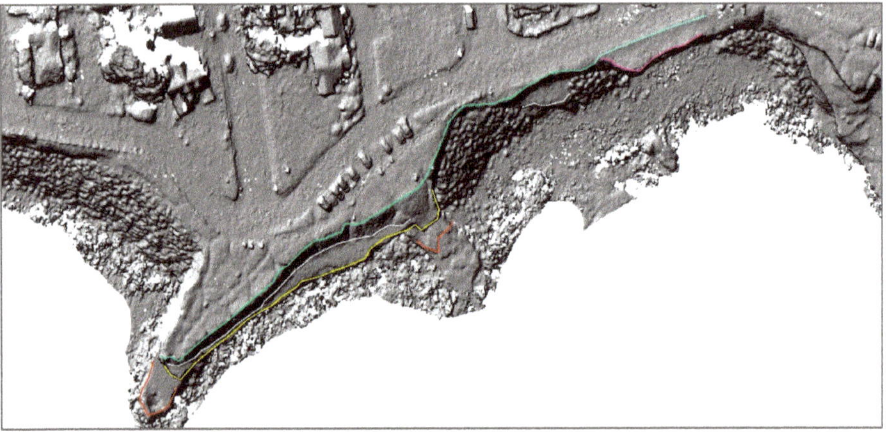

Figure 22. A hill-shaded Digital Surface Model—DSM (USGS-2018) shows the complex profile of cliffed coasts, as well as various lines which could be used in cliff and bluff retreat analysis.

Appropriately addressing the identified challenges in coastal cliff erosion studies (both qualitative and quantitative assessments) could reduce the uncertainty of the historical erosion rate measurements (cliff or bluff retreat) and, as a result, would improve the future bluff or cliff retreat projections. Erosion models include substantial uncertainty, not only derived from the definition of future sea levels and waves but also from the estimates of historical coastline retreat rates. Cliff erosion brings even more complexity and uncertainty given the interaction of the coastal geology with sea levels and waves, which produce different coastal sections with collapsing and eroding modes. Therefore, coastline change models should be contrasted with historical rates from remote sensing and historical imagery as a ground truth and expected erosion potential. While this approach may not provide quantitative definition of the future coastline, it is adequate to identify the historical impacts, delineate erosion hotspots, and establish priorities for management, today and in the foreseeable future, both from regular oceanographic conditions and episodic cycles such as El Niño.

Figure 23. Seacave (**a**) and undercuts (**b**) along West Cliff Drive.

6. Conclusions

As sea level continues to rise at an accelerated rate, the intensive development and infrastructure along California's coastline is under an increasing threat. Whether construction on coastal bluffs or cliffs, or along low-lying shoreline areas, higher sea levels combined with storm waves and high tides will lead to increased rates of cliff and bluff retreat and more frequent coastal flooding. Planning and adapting to a new but uncertain coastline position is and will continue to be a major challenge for many coastal communities. However, there are challenges and uncertainties in both accurately documenting historic cliff and bluff retreat and in projecting these values into the future. There are significant obstacles for developing and implementing future sea-level rise adaptation strategies, managed retreat for example, along the coast, in particular cliffed coastal regions. Reducing or avoiding the problems and concerns identified in this study in determining erosion rates as much as possible through using the most reliable data sets and applying appropriate approaches is an essential process in developing a roadmap for the future management of the area.

Coastal cliff retreat is the product of a complex interaction between the (1) intrinsic properties of the cliff or bluff materials (lithology or rock type, internal rock weaknesses such as joint patterns, and stratigraphic variations, for example) that combine to resist erosion, and (2) the extrinsic processes (rock weathering, rainfall, wave energy, tidal range, storm frequency and intensity, and sea-level rise, for example) that work to weaken the cliff materials and produce failure or erosion. While the historic coastline data from ground and aerial photographs, maps, and satellite imagery can be used with caution and experience to provide the most accurate measurements possible of past changes, with the increase in sea-level rise rates and other aspects of climate change, conditions are shifting. Implementing a detailed coastal monitoring program to document and track the present location and condition of cliff and bluff edges, delineating armored and unarmored sections of coastline, documenting rates of sea level change, and identifying erosion hazard zones will, over time, provide a more robust foundation for future decision making.

Author Contributions: G.G., designed research; G.G., L.D., and B.G.R., performed research; G.G. and L.D., field survey and data collection; G.G., L.D., B.G.R., writing, review, and editing.

Funding: This research received no external funding.

Acknowledgments: We would like to express our appreciation to Patrick Barnard for providing the scientific advice, as well as logistic support, and Alexander Snyder for his help with our field survey to collect the GCPs coordinate. Daniel Hoover for facilitating the field survey, Mark Fox and Jeff Nolan for their valuable contribution to a brainstorming exercise on indicator coastlines.

Conflicts of Interest: The authors declare no conflict of interest.

References

1. Griggs, G.B.; Patsch, K.; Savoy, L.E. *Living with the Changing California Coast*; University of California Press: Berkeley, CA, USA, 2005; p. 340.
2. Bush, D.M.; Young, R. Coastal features and processes. In *Geological Society of America*; Young, R., Norby, L., Eds.; Geological Monitoring: Boulder, CO, USA, 2009; pp. 47–67.
3. Short, A.D. Coastal Processes and Beaches. *Nat. Educ. Knowl.* **2012**, *3*, 15.
4. Griggs, G.B. *Introduction to California's Beaches and Coast*; University of California Press: Berkeley, CA, USA, 2010.
5. Paeniu, L.; Iese, V.; Des Combes, H.J.; De Ramon, N.A.; Korovulavula, I.; Koroi, A.; Sharma, P.; Hobgood, N.; Chung, K.; Devi, A. *Coastal Protection: Best Practices from the Pacific. Pacific Centre for Environment and Sustainable Development; (PaCE-SD)*; The University of the South Pacific: Suva, Fiji, 2015.
6. IPCC. Summary for Policymakers. In *Climate Change 2013: The Physical Science Basis. Contribution of Working Group I to the Fifth Assessment Report of the Intergovernmental Panel on Climate Change*; Stocker, T.F., Qin, D., Plattner, G.-K., Tignor, M., Allen, S.K., Boschung, J., Nauels, A., Xia, Y., Bex, V., Midgley, P.M., Eds.; Cambridge University Press: Cambridge, UK; New York, NY, USA, 2013.
7. NOAA. Is Sea Level Rising? National Ocean Service Website. Available online: https://oceanservice.noaa.gov/facts/sealevel.html (accessed on 30 July 2019).
8. Church, J.A.; White, N.J. Sea-Level Rise from the Late 19th to the Early 21st Century. *Surv. Geophys* **2011**, *32*, 585–602. [CrossRef]
9. Nicholls, R.J.; Wong, P.P.; Burkett, V.R.; Codignotto, J.; Hay, J.; McLean, R.; Ragoonaden, S.; Woodroffe, C.D. Coastal systems and low-lying areas in Parry. In *Climate Change 2007: Impacts, Adaptation and Vulnerability. Contribution of Working Group II to the Fourth Assessment Report of the Intergovernmental Panel on Climate Change*; Canziani, M.L., Palutikof, O.F., van der Linden, J.P., Hanson, C.E., Eds.; Cambridge University Press: Cambridge, UK, 2007; pp. 315–356.
10. Griggs, G. *Coast in Crisis, A Global Challenge*; University of California Press: Berkeley, CA, USA, 2017; p. 243.
11. Emery, K.O.; Kuhn, G.G. Sea cliffs: Their processes, profiles, and classification. *Geol. Soc. Am. Bull.* **1982**, *93*, 644–654. [CrossRef]
12. Naylor, L.A.; Stephenson, W.J.; Trenhaile, A.S. Rock coast geomorphology: Recent advances and future research directions. *Geomorphology* **2010**, *114*, 3–11. [CrossRef]
13. Young, A.P.; Carilli, J.E. Global distribution of coastal cliffs. *Earth Surf. Process. Landf.* **2018**. [CrossRef]
14. US Geological Survey (USGS). *Formation, Evolution, and Stability of coastal Cliff—Status and Trends*; Montly, A., Hampton, G., Griggs, G.B., Eds.; USGS: Reston, VA, USA, 2004. Available online: http://pubs.usgs.gov/pp/pp1693 (accessed on 15 December 2019).
15. Adelman, K.; Adelman, G. California Coastal Records Project. Available online: http://www.californiacoastline.org (accessed on 30 June 2019).
16. Barnard, P.L.; Short, A.D.; Harley, M.D.; Splinter, K.D.; Vitousek, S.; Turner, I.L.; Allan, J.; Banno, M.; Bryan, K.R.; Doria, A.; et al. Coastal vulnerability across the Pacific dominated by El Niño/Southern Oscillation. *Nat. Geosci.* **2015**, *8*, 801–807. [CrossRef]
17. Walkden, M.; Dickson, M. *The Response of Soft Rock Shore Profiles to Increased Sea-Level Rise*; Working Paper; Tyndall Centre for Climate Change Research: Norwich, UK, 2006; Volume 105, p. 22.
18. Nerem, R.S.; Chambers, D.P.; Choe, C.; Mitchum, G.T. Estimating Mean Sea Level Change from the TOPEX and Jason Altimeter Missions. *Mar. Geod.* **2010**, *33*, 435–446. [CrossRef]
19. IPCC. *Climate Change 2014: Synthesis Report. Contribution of Working Groups I, II and III to the Fifth Assessment Report of the Intergovernmental Panel on Climate Change*; Pachauri, R.K., Meyer, L.A., Eds.; IPCC: Geneva, Switzerland, 2014; p. 151.
20. Nerem, R.S.; Beckley, B.D.; Fasullo, J.T.; Hamlington, B.D.; Masters, D.; Mitchum, G.T. Climate-change–driven accelerated sea-level rise detected in the altimeter era. *Proc. Natl. Acad. Sci. USA* **2018**, *115*, 2022–2025. [CrossRef]
21. NASA. New Study Finds Sea Level Rise Accelerating. By Katie Weeman and Patrick Lynch. 2018. Available online: https://climate.nasa.gov/news/2680/new-study-finds-sea-level-rise-accelerating/ (accessed on 1 September 2019).

22. Griggs, G.; Cayan, D.; Tebaldi, C.; Fricker, H.A.; Arvai, J.; DeConto, R.; Kopp, R.E.; Whiteman, E.A. *California Ocean Science Protection Council Advisory Team Working Group. Rising Seas in California: An Update on Sea-Level Rise Science*; California Ocean Sciences Trust: Oakland, CA, USA, 2017; p. 71.

23. Stephens, S.A.; Bell, R.G.; Lawrence, J. Applying Principles of Uncertainty within Coastal Hazard Assessments to Better Support Coastal Adaptation. *J. Mar. Sci. Eng* **2017**, *5*, 40. [CrossRef]

24. U.S. Global Change Research Program. *Climate Change Impacts in the United States: The Third National Climate Assessment*; Melillo, J., Richmond, T., Gary, Y., Eds.; U.S. Global Change Research Program: Washington, DC, USA, 2014; p. 841. [CrossRef]

25. Nicholls, R.J. Planning for the impacts of sea level rise. *Oceanography* **2011**, *24*, 144–157. [CrossRef]

26. Vitousek, S.; Barnard, P.L.; Fletcher, C.H.; Frazer, N.; Erikson, L.; Storlazzi, C.D. Doubling of coastal flooding frequency within decades due to sea-level rise. *Sci. Rep.* **2017**, *7*, 1–9. [CrossRef] [PubMed]

27. Barnard, P.; Erikson, H.; Foxgrover, L.; Amy, A.; Finzi, H.; Limber, J.; Patrick, C.; O'Neill, A.; van Ormondt, M.; Vitousek, S.; et al. Dynamic flood modeling essential to assess the coastal impacts of climate change. *Sci. Rep.* **2019**, *9*. [CrossRef] [PubMed]

28. Limber, P.W.; Barnard, P.L.; Vitousek, S.; Erikson, L.H. A Model Ensemble for Projecting Multidecadal Coastal Cliff Retreat during the 21st Century. *JGR: Earth Surf.* **2018**, *123*, 1566–1589. [CrossRef]

29. Orton, P.S.; Vinogradov, N.; Georgas, A.; Blumberg, N.; Lin, V.; Gornitz, C.; Little, K. Horton, J.R. New York City Panel on Climate Change 2015 Report: Dynamic coastal flood modeling. *Ann. N. Y. Acad. Sci.* **2015**, *1336*, 56–66. [CrossRef]

30. National Research Council. *Sea-Level Rise for the Coasts of California, Oregon, and Washington: Past, Present, and Future*; The National Academies Press: Washington, DC, USA, 2012. [CrossRef]

31. Reguero, B.G.; Losada, I.J.; Méndez, F.J. A recent increase in global wave power as a consequence of oceanic warming. *Nat. Commun.* **2019**, *10*, 205. [CrossRef]

32. Wong, P.P.; Losada, J.; Gattuso, J.P.; Hinkel, J.; Khattabi, A.; McInnes, K.L.; Saito, Y.; Sallenger, A. Coastal systems and low-lying areas. In *Climate Change 2014: Impacts, Adaptation, and Vulnerability. Part A: Global and Sectoral Aspects. Contribution of Working Group II to the Fifth Assessment Report of the Intergovernmental Panel on Climate Change*; Field, C.B., Barros, D.J., Dokken, K.J., Mach, M.D., Mastrandrea, T.E., Bilir, M., Chatterjee, K.L., Ebi, Y.O., Estrada, R.C., Genova, B., et al., Eds.; Cambridge University Press: Cambridge, UK; New York, NY, USA, 2014; pp. 361–409.

33. UNISDR. *Words into Action Guidelines: National Disaster Risk Assessment*; The United Nations Office for Disaster Risk Reduction: Geneva, Switzerland, 2017; p. 303.

34. Neumann, B.; Vafeidis, A.T.; Zimmermann, J.; Nicholls, R.J. Future Coastal Population Growth and Exposure to Sea-Level Rise and Coastal Flooding—A Global Assessment. *PLoS ONE* **2015**, *10*, e0118571. [CrossRef]

35. Griggs, G.B.; Kinsman, N. *Beach Widths, Cliff Slopes, and Artificial Nourishment Along the California Coast. Shore & Beach*; ASBPA: Raleigh, NC, USA, 2016; Volume 84.

36. Muhs, D.R.; Simmons, K.R.; Kennedy, G.L.; Rockwell, T.K. The last interglacial period on the Pacific Coast of North America: Timing and paleoclimate. *GSA Bull.* **2002**, *114*, 569–592. [CrossRef]

37. Hapke, C.J.; Reid, D. *National Assessment of Shoreline Change, Part 4: Historical Coastal Cliff Retreat Along the CA Coast*; Report 2007; USGS: Reston, VA, USA, 2007.

38. Griggs, G.B.; Deepika, S.R. *California Coast from the Air: Images of Changing Landscape*; Mountain Press Publishing Company: Missoula, MO, USA, 2014; p. 167.

39. Griggs, G.B. *Between Paradise and Peril: The Natural Disaster History of the Monterey Bay Region*; Monterey Bay Press: Monterey, CA, USA, 2018; p. 198.

40. National Oceanic and Atmospheric Administration (NOAA). *Report on the National Significance of California's Ocean Economy*; NOAA Office for Coastal Management: Silver Spring, MD, USA, 2015; p. 39.

41. U.S. Census Bureau (Census). Census 2010 Summary File 1, Geographic Header Record G001. 2010. Available online: https://factfinder.census.gov/faces/tableservices/jsf/pages/productview.xhtml?src=bkmk (accessed on 15 December 2019).

42. National Ocean Economics Program (NOEP). National Ocean Economics Program Coastal Economy Data. 2015. Available online: http://www.oceaneconomics.org/Market/coastal/coastalEcon.asp?IC=N (accessed on 2 March 2015).

43. Loughney, M.M.; Melius, M.R. *Managing Coastal Armoring and Climate Change Adaptation in the 21st Century. Law & Policy Program, Environment and Natural Resources*; Stanford Law School: Stanford, CA, USA, 2015.

44. Barnard, P.L. Extreme Oceanographic Forcing and Coastal Response due to the 2015–2016 El Niño. *Nat. Commun.* **2016**. [CrossRef]

45. Reguero, B.G.; Menéndez, M.; Méndez, F.J.; Mínguez, R.; Losada, I.J. A Global Ocean Wave (GOW) calibrated reanalysis from 1948 onwards. *Coast. Eng.* **2012**, *65*, 38–55. [CrossRef]

46. Erikson, L.; Storlazzi, H.; Curt, D.; Golden, N.E. *Modeling Wave and Seabed Energetics on the California Continental Shelf*; USGS: Reston, VA, USA, 2014. [CrossRef]

47. Camus, P.; Fernando, J.; Mendez, R.M. A hybrid efficient method to downscale wave climate to coastal areas. *Coast. Eng.* **2011**, *58*, 851–862. [CrossRef]

48. Griggs, G.B.; Deepika, S.R. *Then and Now: Santa Cruz Coast*; Arcadia Publishing: Mount Pleasant, SC, USA, 2006; p. 97.

49. Moore, L.J.; Benumof, B.K.; Griggs, G.B. Coastal erosion hazards in Santa Cruz and San Diego counties, California. *J. Coast. Res.* **1998**, *28*, 121–139.

50. Griggs, G.B.; Johnson, R.F. Erosional processes and cliff retreat along the Northern Santa Cruz County Coastline, California. *Geology* **1979**, *32*, 67–76.

51. Young, A.P.; Guza, R.T.; O'Reilly, W.C.; Flick, R.E.; Gutierrez, R. Short-term retreat statistics of a slowly eroding coastal cliff. *Nat. Hazards Earth Syst. Sci.* **2011**, *11*, 205–217. [CrossRef]

52. Young, A.P. Decadal-scale coastal cliff retreat in southern and central California. *Geomorphology* **2018**, *300*, 164–175. [CrossRef]

53. Revell, D.L.; Battalio, R.; Spear, B.; Ruggiero, P.; Vandever, J. A methodology for predicting future coastal hazards due to SLR on the California Coast. *Clim. Chang.* **2011**, *109*, 251–276. [CrossRef]

54. Young, A.P.; Flick, R.E.; O'Reilly, W.C.; Chadwick, D.B.; Crampton, W.C.; Helly, J.J. Estimating cliff retreat in southern California considering sea level rise using a sand balance approach. *Mar. Geol.* **2014**, *348*, 15–26. [CrossRef]

Article

Influence of a Reef Flat on Beach Profiles Along the Atlantic Coast of Morocco

Mohammed Taaouati [1], Pietro Parisi [2], Giuseppe Passoni [3], Patricia Lopez-Garcia [2], Jeanette Romero-Cozar [2], Giorgio Anfuso [2], Juan Vidal [2] and Juan J. Muñoz-Perez [2,*]

[1] Department of Exact Sciences, National School of Architecture of Tetouan, 93000 Tetouan, Morocco; mtaaouati@gmail.com

[2] Departamento de Física Aplicada, Facultad de Ciencias Del Mar y Ambientales, Universidad de Cadiz, 11510 Puerto Real, Spain; pietroparisi177@gmail.com (P.P.); patricia.lopezgarcia@uca.es (P.L.-G.); jeanette.romero@uca.es (J.R.-C.); giorgio.anfuso@uca.es (G.A.); juan.vidal@uca.es (J.V.)

[3] Department of Electronics, Information Science, and Bioengineering, Politecnico di Milano, 20133 Milano, Italy; giuseppe.passoni@polimi.it

* Correspondence: juanjose.munoz@uca.es

Received: 7 February 2020; Accepted: 11 March 2020; Published: 12 March 2020

Abstract: The North Atlantic coast of Morocco is characterised by a flat rocky outcrop in the south (Asilah Beach) and a sandy beach free of rocky outcrops in the north (Charf el-Akab). These natural beaches were monitored for a period of two years (April 2005–January 2007) and two different profiles (one for each beach) were analysed based on differences in the substrate. Topographic data were analysed using statistics and empirical orthogonal functions (EOFs) to determine beach slope and volumetric changes over time. Several morphologic phenomena were identified (accretion/erosion and seasonal tilting of beach profiles around different hinge points), attesting to their importance in explaining variability in the data. Periods of accretion were similar in both profiles, but the volumetric rate of change was faster in the sand-rich (SR) profile than in the reef flat (RF) profile. Moreover, the erosion rate for the SR profile was greater than the RF profile (135.18 m^3/year vs. 55.39 m^3/year). Therefore, the RF acted as a geological control on the evolution of its profile because of wave energy attenuation. Thus, special attention should be given to the RF profile, which has larger slopes, less amounts of mobilised sand, and slower erosion/accretion rates than the SR profile.

Keywords: EOF; beach profiles; reef flat; coastal dynamics; sand rich; accretion; erosion rate

1. Introduction

The presence of rocky platforms on beaches is found worldwide. An RF beach is the name for beaches that are perched on hard landforms. The US Army Corps of Engineers [1] and Larson and Kraus [2] define this as a hard-bottom beach. Morphological changes on beaches due to the existence of an RF are not well studied. A few investigations have focused on shape changes, such as Black and Andrews [3] in New Zealand and New South Wales, and Sanderson and Eliot [4] in Australia. Other authors have studied temporal changes, reporting winter erosion and summer accretion rates over a limestone platform near Perth in south-west Australia [5]. Rock and coral landforms on beaches can dissipate wave energy, as confirmed by researchers in Galicia (north-west Spain [6]), St. Martin's Island (Bangladesh [7]), and the fringing reef along Kaanapali Beach in Maui [8].

Sea level rise and other anthropic phenomena induce coastal recession worldwide [9], and coastal researchers and engineers are interested in studying coastal evolution to properly design mitigation and/or remediation measures [10]. Short-term and long-term morphological variability must be considered in the design and evaluation of beach nourishments [11]. Evaluation over seasonal time scales (months or years) is important to determine the rate of erosion and therefore determine future land use in areas adjacent to beaches. Medium-term responses, such as seasonal oscillations in winter–summer profiles, provide information about the across-shore dimension of the berm and may play an important role in the location of beach services such as showers, litter bins, toilets, as well as ramps and bridges for wheelchair accessibility [12].

Levelling of beach profiles is a widely-used tool to monitor the evolution of the coast, and various formulae have been proposed to calculate a general expression (e.g., Dean's formula [13]), although some authors have questioned their validity when an underlying shoreface geology exists [14]. Thus, several researchers have presented results of the influence of coastal reefs on the spatial and temporal variability of beach morphology [15–19]. Other characteristics, such as wave attenuation over reef platforms [20], wave-setup and water-level fluctuations [21], interannual changes in beach morphology [22], modification of the A parameter of Dean's formula [23,24] or sediment flux [25] along reef-protected profiles, have also been discussed.

Nevertheless, few comparisons can be found between the behaviour of profiles on adjacent beaches subject to the same wave conditions but with different geological substrates or boundary conditions. It is worth noting that Muñoz-Perez and Medina [12] compared the behaviour of two beach profiles from Victoria Beach (Cadiz, Spain) over a five-year period where one profile was perched on a rock platform. The northernmost zone presented a rocky platform that emerged during low tide and acted as a geological boundary for profile development, whereas the southern zone had no such platform. Some differences in erosion and subsequent accretion rates were observed.

Thus, the aim of this paper is to compare how beach profiles change (volume and slope) over time on two adjacent beaches (under the same climatic conditions), one of which is a sand-rich beach (Charf el-Akab, SR) and the other beach is supported by a reef flat (Asilah, RF). Monitoring by beach profiling was performed to analyse their morphological differences over a period of two years (April 2005–January 2007) to observe seasonal changes between summer and winter in order to draw useful conclusions regarding the behaviour of beach morphology as it relates to differences in the seabed.

2. Study Area

Beach Location

The sites investigated in this paper are located along the North Atlantic coast of Morocco (Figure 1). Two adjacent beaches were chosen. The northern beach is Charf el-Akab (35°46′ N, 5°48′ W), close to Tanger. Immediately to the south is Asilah Beach, which takes its name from the homonymous city (35°28′ N, 6°2′ W). Both beaches have an NNE–SSW orientation and are composed of the same quartz sand. The main difference between the two sites is that Asilah presents an almost horizontal rocky platform situated around the low tide level, which influences the dynamics of the coast, while Charf el-Akab is a completely sandy beach.

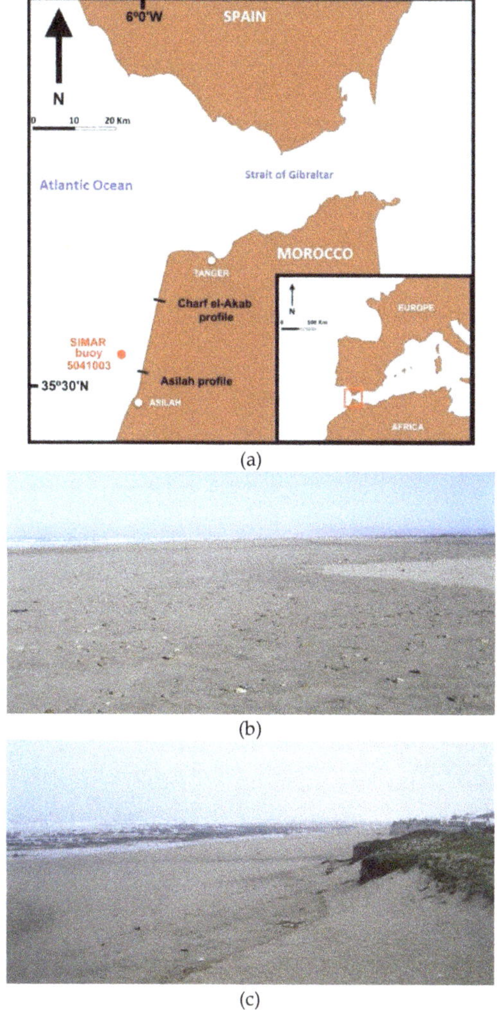

(a)

(b)

(c)

Figure 1. (**a**) Location of Charf el-Akab and Asilah beaches on the north-west coast of Morocco facing the Atlantic Ocean. Wave and climate data were collected from a virtual SIMAR buoy in front of Asilah (www.puertos.es); (**b**) view of Charf el-Akab sandy beach; (**c**) view of the Asilah beach supported by a reef flat. Photographs taken by the authors.

3. Methods

3.1. Meteorological Data

Meteorological data were collected from a wave prediction point (SIMAR point 5,041,003 from www.puertos.es) located in front of the monitored stretch of coast (Figure 1a). The area is mesotidal, with a tidal range of 2.7 m and a semidiurnal periodicity [26]. The hydrodynamic conditions are principally controlled by storms approaching from occidental quadrants [27]. The predominant winds, named "Chergui", blow from the east (i.e., from land) 27% of the time and are especially abundant in spring and summer, reaching maximum velocities of 130 km/h. Secondary winds ("Rharbi") blow from the west (i.e., from the Atlantic Ocean) 16% of the time. Rharbi winds are wet and prevail in winter and autumn [26].

The SIMAR database is obtained through numerical wave modelling from wind time series by solving the equation of energy balance. This virtual database does not come from direct measurements. However, it has been validated by numerous studies and used in practical applications along the Spanish coast [28].

Both beaches are subject to the same wave and wind climatic regimes because of their proximity to each other. The nearshore areas are uniform, and both beaches face the same direction. Figure 2 reports temporal series of wave height and period, wind speed, and wind direction from April 2005 to January 2007.

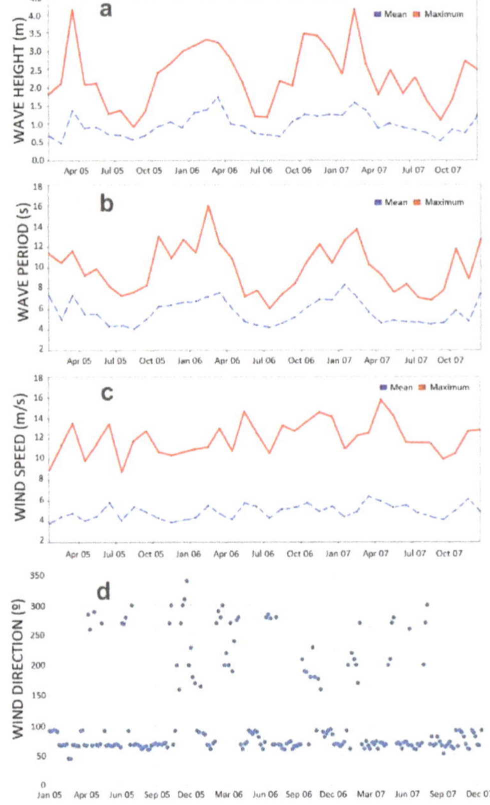

Figure 2. Temporal series of (**a**) wave height (m); (**b**) wave period (s); (**c**) wind speed (m/s); (**d**) most frequent wind direction: N = 0°, E = 90°, S = 180°, W = 270° (adapted from www.puertos.es). The dotted blue line refers to the average value, while the red one is the maximum value registered.

3.2. Field Data Surveying

The morphological changes of Charf el-Akab and Asilah beaches were studied through a topographic monitoring program carried out every two months during a two-year period, from April 2005 to January 2007. Data were collected on emerged beaches at low tide with a total station. The vertical datum (or zero elevation surface) matches the lowest low-water level (LLWL). Some fixed positions were selected and monitored at both beaches. Five profiles were taken in Charf el-Akab (Figure 1b) and seven in Asilah (Figure 1c). The beach profile spacing was 50 m, following recommendations found in the literature [29–32], whereas the distance between adjacent points along one profile was 5 m. However, because there were no appreciable differences between the profiles of the same beach (the variance ranges from 0.015 for Asilah to 0.061 for Charf el-Akab), only two mean profiles (one from each beach) were studied. Afterwards, analyses of the data collected were performed by statistical way (Statgraphics Centurion software) and EOF (multivariate statistical analysis package (MVSP)).

3.3. Statistics

Following Jimenez and Sanchez-Arcilla [33], a previous study by Anfuso et al. [34] chose to use least-squares linear regression to analyse the evolution of the profile. Similar methodologies to the one used by Anfuso et al. [34] were carried out to obtain the accretion/erosion volumes in this study. The analysis of the mean profiles was carried out using the statistical software to find differences between the behaviour of the RF and the SR profiles. The beach face slope and accretion/erosion volumes of sand per unit of beach length were calculated. The area between two profiles of adjacent dates is the accretion/erosion rate volume (m^3/m). The slope was obtained as the mean of slopes calculated at 5 m intervals along each profile. The use of EOFs enabled the identification of morphological changes [35] and allowed us to obtain additional results from the data that better explained the spatial and temporal variability of the beach profiles.

3.4. Empirical Orthogonal Functions (EOFs)

EOFs are a mathematical method and have been widely used in coastal geomorphology since Winant et al. [36] studied variability in beach profiles. Other researchers have applied this technique to different aspects of coastal morphology; for example, longitudinal variations in contour lines [37–39], sand transport in a transverse direction [40], or the distribution of sediment grain size along a transverse profile [41]. In addition, other phenomena have also been investigated using EOFs: responses to beach nourishment at different times and spatial scales [11], behavioural changes in profiles over a fortnightly tidal cycle [42], the capacity of this technique to identify modes of shoreline variability [38,39], and changes in coastal dune profiles [43].

EOFs, also known as principal components analysis (PCA), provide a technique for separating the spatial and temporal variability of beach-profile data; a detailed description of the method can be found in statistics textbooks [44]. In brief, if a function h = (x,t) represents the profile elevation at a particular position and time, such a function may then be defined as a linear combination of a few spatial, Xn (x) and temporal, Tn (t), eigenfunctions (and their associated eigenvalues) as follows:

$$h_{ij} = h\left(x_i, t_j\right) = \sum_{l=1}^{N} X_l(x_i) * T_l\left(t_j\right) * a_l = a_1 * X_1(x_i) * T_1\left(t_j\right) + a_2 * X_2(x_i) * T_2\left(t_j\right) + \dots \tag{1}$$

Eigenfunctions are ranked according to the percentage of variability they explain, defined as the mean squared value (MSV) of the data. In some cases [36], the mean value is of such importance in explaining the variability that it must be removed from the original data to allow for the better and clearer identification of other smaller, but important, changes. Then, the MSV becomes part of the variance. The first eigenfunction explains most of the MSV in the data, the second eigenfunction explains the greater part of the remaining MSV, and so on. The MVSP software was used to calculate the EOFs. Furthermore, according to Aubrey [45], assuming that a physical process provides most of the variability, the corresponding eigenfunction would be related to that physical process.

4. Results and Discussion

4.1. Topographic Profile Analysis

Topographic mean profiles carried out from April 2005 to January 2007 at the RF beach at Asilah and the SR beach at Charf el-Akab were investigated in order to assess how the profiles change over time and to compare the two types of beaches. A representation of the mean profiles over time is shown in Figure 3. Since it is not easy to see a rational behaviour or trend, as previously mentioned in Section 2, a statistical analysis was carried out with Statgraphics software to obtain the results presented in Table 1 (i.e., net sand volume variations and rates of accretion represented by positive values and erosion by negative values).

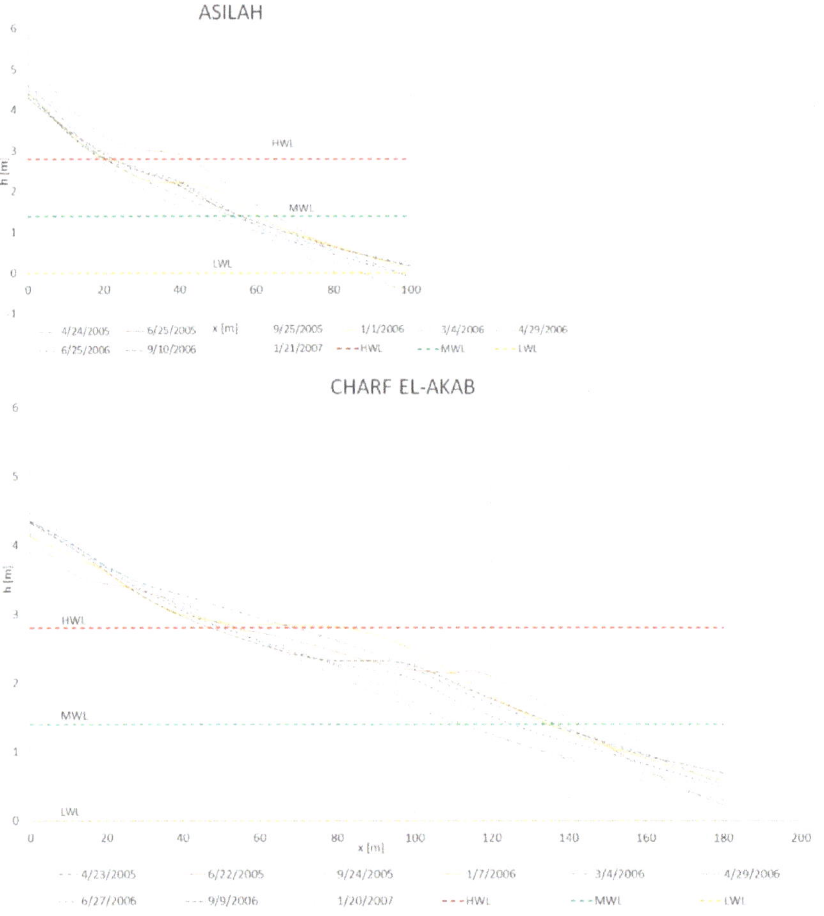

Figure 3. Topographic profiles. These are the mean profile of each beach over time from data collected in the field. The profiles are used to perform the statistical calculations and the complementary empirical orthogonal function (EOF) analyses. Profile dates are reported in the legend. High, mean, and low water levels are represented as HWL, MWL, and LWL, respectively. Levellings of the mean profile (h) are presented on the vertical axis.

Table 1. Volume rates of the Asilah (reef flat (RF)) and Charf el-Akab (sand-rich (SR)) beaches' mean profiles. The accretion volume between dates is represented as positive values (m³/m) and the volume eroded as negative values (m³/m). Cross-shore transport speed (m³/m per day) has been calculated as the net value of accretion less erosion (m³/m) divided by the time increment in days.

Date of Topographic Profile	Beach Slope (%)		Time (months)	Accretion (m³/m)		Erosion (m³/m)		Net (m³/m)		Cross-Shore Transport Speed (m³/m per day)	
	RF	SR		RF	SR	RF	SR	RF	SR	RF	SR
24 April 2005	3.6	2.1	2.00	46.04	27.31	−6.14	−11.29	39.91	16.02	0.67	0.27
23 June 2005	6.2	2.0	3.00	11.12	1.51	−32.56	−47.13	−21.44	−45.62	−0.24	−0.51
25 September 2005	3.2	1.8	3.20	4.55	5.70	−10.09	−36.18	−5.54	−30.48	−0.06	−0.32
3 January 2006	3.4	2.0	2.00	0.69	5.54	−23.11	−64.62	−22.42	−59.08	−0.37	−0.98
4 March 2006	3.1	1.8	2.00	15.33	12.04	−21.32	−4.34	−5.99	7.70	−0.10	0.13
29 April 2006	4.7	1.7	2.00	17.91	24.85	−1.46	−4.14	16.44	20.71	0.27	0.35
26 June 2006	3.6	1.8	2.50	9.74	18.33	−2.29	−4.38	7.45	13.95	0.10	0.19
10 September 2006	3.3	1.5	4.30	5.51	18.78	−18.20	−6.63	−12.69	12.15	−0.10	0.09
20 January 2007	4.8	2.2									

The results presented in Table 1 show the typical erosion/accretion cycle for both the SR and RF beaches. The accretion phase took place during the "summer" season (usually from April to September) due to the prevalence of relatively calm conditions. The erosion phase occurred during "winter" months (from October to March), when high-energy events caused erosion along the foreshore.

The erosion rate (m^3/m per year) was calculated as, first of all, the sum of the net volume eroded (negative values) in the 21 months of study. Then, the erosion rate was multiplied by the correction factor of 12/21 to compute the annual rate. Similarly, the accretion rate (m^3/m per year) was also estimated as the sum of the net accretion volume (positive values) across the entire time interval. The erosion rate for Asilah (RF) resulted in 38.90 m^3/m per year; this value was 77.25 m^3/m per year for Charf el-Akab (SR). The accretion rate was 36.46 m^3/m per year for Asilah (RF), and a similar value (40.30 m^3/m per year) was recorded for Charf el-Akab (SR). Therefore, cross-shore transport for Asilah (RF) ranged from −0.37 to 0.67 m^3/m per day and from −0.98 to 0.35 m^3/m per day for the SR beach. The slopes ranged from 3.1 to 6.2% for Asilah (RF) and from 1.5 to 2.2% for Charf el-Akab (SR).

Charf el-Akab (SR) lost twice the volume of sand per year than that for Asilah (RF). The RF dissipates the wave energy due to friction, causing less beach erosion. Nevertheless, Charf el-Akab recorded higher accretion rates than Asilah, but the erosion rate at Asilah was faster than its accretion speed; this favoured a negative sediment budget trend. Moreover, the slope of the RF beach was double (and sometimes triple) that of the SR beach (Figure 4). Once again, wave energy reduction (due to the friction on the RF) is the cause of the higher slope of the beach.

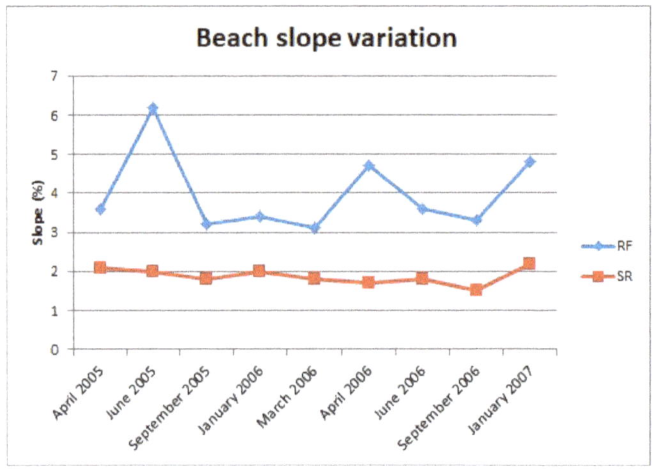

Figure 4. Slope (%) of the two kinds of beaches (reef flat, RF, is blue and sand-rich, SR, is red) during the April 2005–January 2007 period.

On the other hand, the slope of the regression line ("m") estimates the volumetric rate of change during the surveyed period. Therefore, the "m" values in Figure 5 express the erosion/accretion per month (m^3/month). The higher the slope of the fitted line, the clearer the profile trend. In this way, the SR beach presented high values of "m"; that is, clear tendencies. Both the SR and RF beaches have low R-squared values due to seasonal variability and episodes of erosion and accretion (Figure 5). The seasonal variability is clearly distinctive for the SR beach when summer periods have positive volumetric changes (accretion) and winter periods have negative trends (erosion). Even though the seasonal behaviour is similar for the RF beach, there were smaller changes in volume. The volume changes over time for the SR beach presented a more marked tendency. Anfuso et al. [34] stated that low correlation coefficients between volume changes and time indicate a high degree of beach variability.

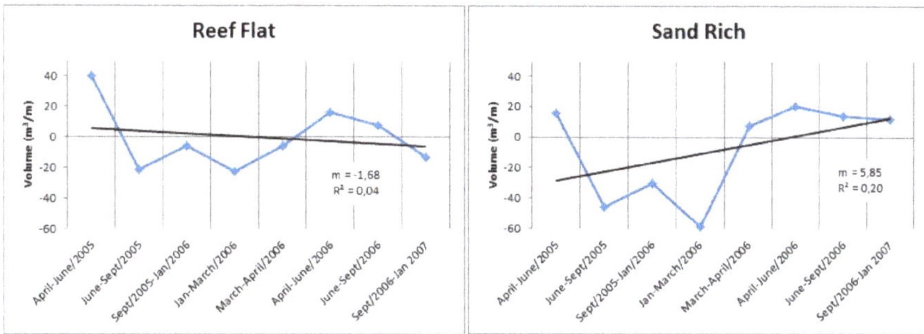

Figure 5. Evolution of sand volume of two types of beaches (reef flat, RF, and sand-rich, SR) during the April 2005–January 2007 period. Units of the linear regression slope (m) are in m^3/m per month.

Comparisons between SR and RF beaches from another country (Spain) were also carried out by Muñoz-Perez and Medina [12] using EOF methodology. The values are similar between RF and SR beaches from two different countries (Victoria Beach in southern Spain vs. Asilah and Charf el-Akab beaches in Morocco). The results are shown in Table 2 in order to verify that both countries present a similar behaviour depending on the kind of beach. The sand-rich profile of Victoria beach shows losses of less volume per year than Charf el-Akab. Nevertheless, the reef flat part of Victoria beach loses less volume per year than Asilah. Therefore, the rocky platform from Asilah experiences less change in volume because it is wider than the one at Victoria. Wide rocky platforms offer more friction, resulting in more wave energy dissipation. In the case of a sandy beach, the absence of a reef flat causes even more erosion because it is not protected from wave energy on the bottom. Significant differences between SR beaches and RF beaches indicate the importance of the presence and typology of the rock platform

Table 2. Comparison of erosion and accretion volume rates between reef flat and sand-rich profiles. The beaches are Victoria (Cadiz, Spain), Asilah and Charf el-Akab (Morocco).

		Slope Variation (%)	Erosion Rate (m^3/year)	Accretion Speed (m^3/day)
Reef flat profile	Victoria	3.8–7.8	29	0.33
	Asilah	3.1–6.2	55.4	−0.37 to 0.67
Sand-rich profile	Victoria	1.2–2.8	121	1.01
	Charf el-Akab	1.5–2.2	135.2	−0.98 to 0.35

4.2. EOF Analysis

To date, the analyses of the mean profiles have been statistically performed simply to identify differences between the RF and SR profiles on slope and accretion/erosion rates. As mentioned in Section 2, the EOF analysis allowed us to obtain more information from the data; in this way, it helped explain the variability in temporal and spatial profiles. The following results were obtained using EOFs applied to the profiles by subtracting the mean profile.

The first and second spatial components from the EOFs are plotted in Figure 6a (Asilah) and Figure 6c (Charf el-Akab), and the mean profile is presented with levelling on the right axis. Temporal components from EOFs are shown in Figure 6b,d. The variance described by the first component was bigger in the sand-rich profile than in the reef-protected profile, at 77.3% vs. 57.9%. Therefore, the second spatial component explained the greater weight in the variance for Asilah (33.6%) as compared to Charf el-Akab (9.6%). Each spatial component described how the data collected changed along the profile. Thus, the maximum and minimum points mark where either the accumulation or erosion of the beach was observed. Taking this into account, zero means there was no transport of material at that point, which is called the "rotation point." The temporal components describe the beach erosion/accumulation cycle.

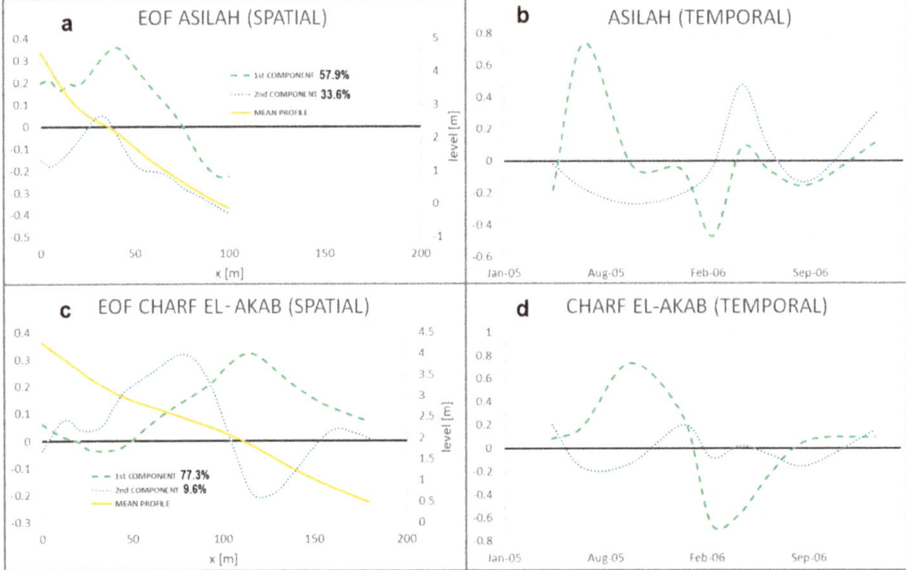

Figure 6. (**a**) Spatial EOF in Asilah (RF); (**b**) temporal EOF in Asilah (RF); (**c**) spatial EOF in Charf el-Akab (SR); (**d**) temporal EOF in Charf el-Akab (SR). The height of the mean profile is reported on the vertical axes. The percentages of variance described by the components are reported in the legends.

The first spatial component in Asilah beach (RF) crossed the rotation point at $x = 75$ m, which corresponded to $h = 0.67$ m in the mean profile. The second component was always negative except for a small part between $x = 25$ m and $x = 40$ m, which corresponded to $h = 2.68$ m and $h = 2.18$ m in the mean profile, respectively.

The first spatial component in the Charf el-Akab beach (SR) presented a small part where it changes sign between $x = 20$ m and $x = 50$ m and which corresponded to $h = 3.64$ m and $h = 2.88$ m in the mean profile, respectively. The second component presented two points that passed through the zero point: the first one at $x = 105$ m corresponded to $h = 2.05$ m in the mean profile, while the second one at $x = 155$ m corresponded to $h = 0.93$ m.

Oscillations around null axes can be observed in the temporal graphs and cannot be associated with seasonal variations. Indeed, two relative maximums and minimums were observed over the period of two years, indicating that the beach profiles changed only once every year. The peaks of the first component were observed in June 2005 and March 2006 for Asilah Beach, while the most important peak in the second component occurred in April 2006. The peaks of the first component in Charf el-Akab were recorded in September 2005 and March 2006, while the second component presented an oscillation without relevant peaks.

4.3. Physical Interpretation of the Changes

The first spatial component was associated with the general seasonal change in beach profiles as observed at Asilah; that is, typical of "storm" and "calm" conditions [13]. Similar characteristics were observed at beaches in Cadiz [12,46]. The accretion/erosion periods and the portion of the profile that reflected such changes were identified by the combined analysis of the spatial and temporal components. Therefore, if the rotation point for the first component in Asilah Beach (RF) was $x = 75$ m, which corresponded to h = 0.67 m in the mean profile, the accretion periods were above h = 0.67 m in summer and below this level in winter (Figure 7a). Moreover, Charf el-Akab Beach (SR) presented a small part in which the trend changed, between x = 20 m and x = 50 m, which corresponded to h = 3.64 m and h = 2.88 m, respectively. Therefore, great volumetric changes observed at Charf el-Akab took place from h = 2.88 m to the submerged zone, which corresponded with a period of accretion during 2005 and erosion during the first half of 2006. After that, the beach seemed to acquire equilibrium (Figure 7b).

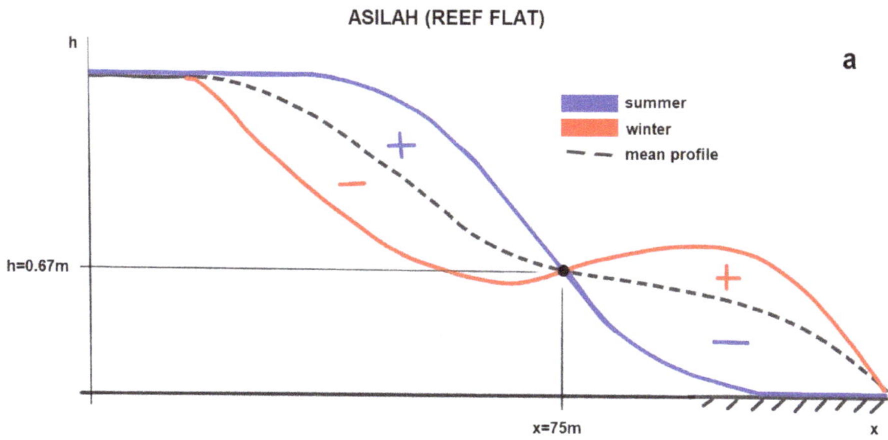

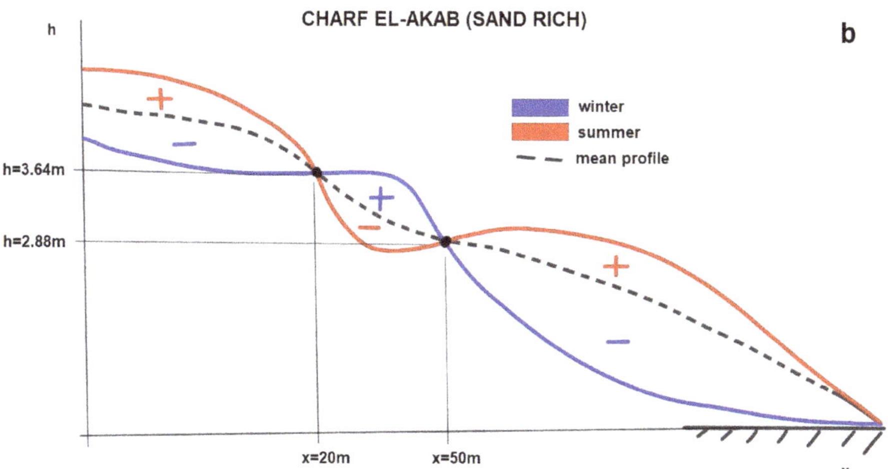

Figure 7. Sketch of seasonal changes in beach profiles at (**a**) Asilah and (**b**) Charf el-Akab beaches.

The behaviour of the second component was completely different. The increments of this component were associated, as for the first component, with a change in the significant wave height, but also with the prevalence of winds from the east. This condition produced aeolian transport on the beach, but not a large wave regime because the fetch was small and did not allow wave formation [47]. Thus, it is clear that the second component was affected by different variables and not solely controlled by sea and wave conditions. Hence, the second component was responsible for the shape of the profile and was related to different interactions among several variables, essentially sea regime and wind conditions. Thus, this component influenced the morphology of the profiles under the combined effects of wave and aeolian transport. Taking this into account, it is possible to state that the wind in Charf el-Akab did not produce notable changes in the beach profile because of the shield of topography and presence of some edification. Meanwhile, that was not the case for Asilah beach, which was affected by sand transport when east winds were strong. The presence of tall buildings in Victoria Beach is probably the reason why this behaviour was not observed in the Spanish coast. Beach morphological changes have been described in other papers [48], but more detailed knowledge about the interactions between wave transport and aeolian transport is less well described. Nickling and Davidson-Arnott [40] associated the shape of beach profiles with usable sand volume.

5. Conclusions

The aim of this paper is to assess differences in profile morphology (volume changes and slope variations of two beaches), taking into account the influence of the nature of the seabed (i.e., the presence or absence of a reef flat). Two kinds of beaches along the north-west coast of Morocco were studied; one is a sandy beach, and the other one is supported by a hard-bottom reef flat. The comparison of the profiles over time helps us to understand the importance and influence on the beach behaviour of the presence of a hard, rocky substrate. EOFs were used to determine the components that describe the behaviour of the beach profiles. The first component was associated with seasonal changes: erosion in winter months, when high-energy events cause significant erosion across the foreshore, and accretion in summer months because of calm conditions. This was particularly evident in the RF beach. On the other hand, the second component was associated with combined wave and wind sediment transport, as observed in many cases in which increments and movement of sand are associated with high winds and low wave heights. These results are similar to other studies based on the use of EOFs. In addition, the behaviour of the beaches is comparable with a previous study of Victoria Beach in the Gulf of Cadiz.

A typical yearly cycle of erosion and accretion can be observed at both beaches. Charf el-Akab (SR) beach lost twice as much sand per year as Asilah (RF) beach. When a beach has a reef flat, the beach profile suffers less erosion than a sand-rich beach where wave energy is not reduced by friction when in contact with a rocky bottom. For the same reason, the slope of the RF beach is twice that of the slope of the SR beach. Despite this, Charf el-Akab has higher yearly accretion rates than Asilah, but the erosion rate for Asilah was faster than the accretion rate, producing an erosional trend, whilst the Charf el-Akab beach seems to have reached a state of equilibrium.

Author Contributions: Conceptualization, G.P. and J.J.M.-P.; Data curation, M.T., P.P., P.L.-G. and J.R.-C.; Formal analysis, M.T., P.P., P.L.-G., J.R.-C. and J.V.; Funding acquisition, G.A.; Investigation, G.A.; Methodology, M.T., P.P., G.A. and J.J.M.-P.; Project administration, M.T. and G.A.; Resources, G.P.; Software, P.L.-G. and J.V.; Supervision, G.P. and G.A.; Writing—original draft, P.L.-G. and J.J.M.-P.; Writing—review & editing, G.P., J.R.-C., G.A, J.V. and J.J.M.-P. All authors have read and agreed to the published version of the manuscript.

Funding: This research received no external funding.

Conflicts of Interest: The authors declare no conflict of interest.

References

1. U.S. Army Corps of Engineers. Shore Protection Manual. In *Coastal Engineering Research Center*; Department of the Army, Waterways Experiment Station, Corps of Engineers: Vicksburg, MI, USA, 1984.
2. Larson, M.; Kraus, N.C. Representation of Non-Erodible (Hard) Bottoms in Beach Profile Change Modelling. *J. Coast. Res.* **2000**, *16*, 1–14.
3. Black, K.P.; Andrews, C.J. Sandy Shoreline Response to Offshore Obstacles Part 1: Salient and Tombolo Geometry and Shape. *J. Coast. Res.* **2001**, *29*, 82–93.

4. Sanderson, P.G.; Eliot, I. Shoreline Salients, Cuspate Forelands and Tombolos on the Coast of Western Australia. *J. Coast. Res.* **1996**, *12*, 761–773.

5. Doucette, J.S. Photographic monitoring of erosion and accretion events on a platform beach, Cottosloe, Western Australia. In *33rd International Association of Hydraulic Engineering and Research Biennial Congress*; IAHR: Vancouver, BC, Canada, 2009.

6. Rey, D.; Rubio, B.; Bernabeu, A.M.; Vilas, F. Formation, exposure, and evolution of a high-latitude beachrock in the intertidal zone of the Corrubedo complex (Ria de Arousa, Galicia, NW Spain). *Sediment. Geol.* **2004**, *169*, 93–105. [CrossRef]

7. Chowdhury, S.Q.; Fazlul Haq, A.T.M.; Hasan, K. Beachrock in st. martin's island, bangladesh: Implications of sea level changes on beachrock cementation. *Mar. Geod.* **1997**, *20*, 89–104. [CrossRef]

8. Eversole, D.; Fletcher, C.H. Longshore Sediment Transport Rates on a Reef-Fronted Beach: Field Data and Empirical Models Kaanapali Beach, Hawaii. *J. Coast. Res.* **2003**, *19*, 649–663.

9. Zhang, K.; Douglas, B.C.; Leatherman, S.P. Global warming and coastal erosion. *Clim. Chang.* **2004**, *64*, 41–58. [CrossRef]

10. Rangel-Buitrago, N.; Anfuso, G. Coastal storm characterization and morphological impacts on sandy coasts. *Earth Surf. Process. Landf.* **2011**, *36*, 1997–2010. [CrossRef]

11. Larson, M.; Hanson, H.; Kraus, N.C.; Newe, J. Short- and long-term responses of beach fills determined by EOF analysis. *J. Waterw. Port Coast. Ocean Eng.* **1999**, *125*, 285–293. [CrossRef]

12. Muñoz-Perez, J.J.; Medina, R. Comparison of long-, medium- and short-term variations of beach profiles with and without submerged geological control. *Coast. Eng.* **2010**, *57*, 41–51. [CrossRef]

13. Dean, R.G. Equilibrium Beach Profiles: Characteristics and Applications. *J. Coast. Res.* **1991**, *7*, 53–84.

14. Pilkey, O.H.; Young, R.S.; Riggs, S.R.; Smith, A.W.S.; Wu, H.; Pilkey, W.D. The Concept of Shoreface Profile of Equilibrium: A Critical Review. *J. Coast. Res.* **1993**, *9*, 255–278.

15. Muñoz-Pérez, J.J.; Medina, R. Short term variability of reef protected beach profiles: An Analysis Using EOF. In Proceedings of the Coastal Dynamics 2005—Proceedings of the Fifth Coastal Dynamics International Conference, Barcelona, Spain, 4–8 April 2005.

16. Gallop, S.L.; Bosserelle, C.; Eliot, I.; Pattiaratchi, C.B. The influence of coastal reefs on spatial variability in seasonal sand fluxes. *Mar. Geol.* **2013**, *344*, 132–143. [CrossRef]

17. Gallop, S.L.; Bosserelle, C.; Haigh, I.D.; Wadey, M.P.; Pattiaratchi, C.B.; Eliot, I. The impact of temperate reefs on 34years of shoreline and vegetation line stability at Yanchep, southwestern Australia and implications for coastal setback. *Mar. Geol.* **2015**, *369*, 224–232. [CrossRef]

18. Habel, S.; Fletcher, C.H.; Barbee, M.; Anderson, T.R. The influence of seasonal patterns on a beach nourishment project in a complex reef environment. *Coast. Eng.* **2016**, *116*, 67–76. [CrossRef]

19. Hoeke, R.; Storlazzi, C.; Ridd, P. Hydrodynamics of a bathymetrically complex fringing coral reef embayment: Wave climate, in situ observations, and wave prediction. *J. Geophys. Res. Ocean* **2011**, *116*, 10–1029. [CrossRef]

20. Johnson, H.K.; Karambas, T.V.; Avgeris, I.; Zanuttigh, B.; Gonzalez-Marco, D.; Caceres, I. Modelling of waves and currents around submerged breakwaters. *Coast. Eng.* **2005**, *52*, 949–969. [CrossRef]

21. Karunarathna, H.; Tanimoto, K. Numerical experiments on low-frequency fluctuations on a submerged coastal reef. *Coast. Eng.* **1995**, *26*, 271–289. [CrossRef]

22. Norcross, Z.M.; Fletcher, C.H.; Merrifield, M. Annual and interannual changes on a reef-fringed pocket beach: Kailua Bay, Hawaii. *Mar. Geol.* **2002**, *190*, 553–580. [CrossRef]

23. Dean, R. Equilibrium beach profiles: U.S. Atlantic and Gulf coasts. In *Department of Civil Engineering and College of Marine Studies*; University of Delaware: Newark, DE, USA, 1977.

24. Bernabeu-Tello, A.M.; Munoz-Pérez, J.J.; Medina-Santamaría, R. Influencia de un sustrato rocoso en la morfología del perfil de playa: Playa Victoria, Cádiz. *Cienc. Mar.* **2002**, *28*, 181–192. [CrossRef]

25. Roberts, H.H. Physical Processes and Sediment Flux Through Reef-Lagoon Systems. In Proceedings of the 17th International Conference Coastal Engineering, Sydney, Australia, 23–28 March 1980.

26. Taaouati, M.; Anfuso, G.; Nachite, D. Morphological Characterization and Evolution of Tahadart Littoral Spit, Atlantic Coast of Morocco. In *Sand and Gravel Spits*; Springer: Berlin/Heidelberg, Germany, 2015; pp. 289–306.

27. Jaaidi, E.B.; Cirac, P. La Couverture Sédimentaire Meuble du Plateau Continental Atlantique Marocain Entre Larache et Agadir [The soft Sedimentary Cover of the Moroccan Atlantic Continental Shelf between Larache and Agadir]. *Bull. Inst. Geol. Bassin. Aquitaine Bordx.* **1987**, *42*, 33–51.

28. Tomás, A.; Méndez, F.J.; Medina, R.; Losada, I.J.; Menéndez, M.; Liste, M. *Bases de datos de oleaje y nivel del mar, calibración y análisis: el cambio climático en la dinámica marina en españa. El Clima entre el Mar y la Montaña*; Asociación Española de Climatología y Universidad de Cantabria: Cantabria, Spain, 2004.

29. Browder, A.E.; Dean, R.G. Monitoring and comparison to predictive models of the Perdido Key beach nourishment project, Florida, USA. *Coast. Eng.* **2000**, *39*, 173–191. [CrossRef]

30. Lippmann, T.C.; Holman, R.A. The spatial and temporal variability of sand bar morphology. *J. Geophys. Res.* **1990**, *95*, 10–1029. [CrossRef]

31. U.S. Army Corps of Engineers. Engineering and Design, Hydrographic Surveying. In *Engineering Manual*; DEPARTMENT OF THE ARMY U.S. Army Corps of Engineers: Washington, DC, USA, 2002.

32. Muñoz-Perez, J.J.; Payo, A.; Roman-Sierra, J.; Navarro, M.; Moreno, L. Optimización del espaciado del perfil de playa: Una herramienta aplicada para el seguimiento costero. *Sci. Mar.* **2012**, *76*, 791–798.

33. Jiménez, J.A.; Sánchez-Arcilla, A. Medium-term coastal response at the Ebro delta, Spain. *Mar. Geol.* **1993**, *114*, 105–118. [CrossRef]

34. Anfuso, G.; Benavente, J.; Del Río, L.; Gracia, F.J. An approximation to short-term evolution and sediment transport pathways along the littoral of Cadiz Bay (SW Spain). *Environ. Geol.* **2008**, *56*, 69–79. [CrossRef]

35. Muñoz-Perez, J.J.; Gomez-Pina, G.; Enriquez, J. Comments on "An approximation to short-term evolution and sediment transport pathways along the littoral of Cadiz Bay (SW Spain)" by Anfuso and et al. (Environ Geol 56:69-79). *Environ. Earth Sci.* **2009**, *59*, 477–479. [CrossRef]

36. Winant, C.D.; Inman, D.L.; Nordstrom, C.E. Description of seasonal beach changes using empirical eigenfunctions. *J. Geophys. Res.* **1975**, *80*, 1979–1986. [CrossRef]

37. Losada, M.; Medina, R.; Vidal, C.; Roldan, A. Historical evolution and morphological analysis of "El Puntal" Spit, Santander (Spain). *J. Coast. Res.* **1991**, *7*, 711–722.

38. Muñoz-Pérez, J.J.; Medina, R.; Tejedor, B. Evolution of longshore beach contour lines determined by the EOF method. *Sci. Mar.* **2001**, *65*, 393–402. [CrossRef]

39. Miller, J.K.; Dean, R.G. Shoreline variability via empirical orthogonal function analysis: Part I temporal and spatial characteristics. *Coast. Eng.* **2007**, *54*, 111–131. [CrossRef]

40. Medina, R.; Losada, M.A.; Dalrymple, R.A.; Roldan, A. Cross-shore sediment transport determined by EOF method. *Proc. Coast. Sediments* **1991**, *91*, 2160–2174.

41. Medina, R.; Losada, M.A.; Losada, I.J.; Vidal, C. Temporal and spatial relationship between sediment grain size and beach profile. *Mar. Geol.* **1994**, *118*, 195–206. [CrossRef]

42. Muñoz-Pérez, J.J.; Medina, R. Profile changes due to a fortnightly tidal cycle. In Proceedings of the Coastal Engineering 2000—Proceedings of the 27th International Conference on Coastal Engineering, Sydney, Australia, 16–21 July 2000; pp. 3063–3075.

43. Muñoz-Perez, J.J.; Navarro, M.; Roman-Sierra, J.; Tejedor, B.; Rodriguez, I.; Gomez-Pina, G. Long-term evolution of a transgressive migrating dune using reconstruction of the EOF method. *Geomorphology* **2009**, *112*, 167–177. [CrossRef]

44. Jackson, J.E. *A User's Guide to Principal Components*; John Wiley & Sons: New York, NY, USA, 1991.

45. Aubrey, D.G. Seasonal patterns of onshore/offshore sediment movement. *J. Geophys. Res.* **1979**, *84*, 6347–6355. [CrossRef]

46. Anfuso, G.; Martinez del Pozo, J.A.; Garcia, F.J.; Lopez-Aguayo, F. Long-shore distribution of morphodynamic beach states along an apparently homogeneous coast in SW Spain. *J. Coast. Conserv.* **2003**, *9*, 49–56. [CrossRef]

47. Nickling, W.; Davidson-Arnott, R. Aeolian Sediment Transport on Beaches and Coastal Sand dunes. In *Canadian Symposium on Coastal Sand Dunes, Canadian Coastal Science and Engineering Association*; Department of Geography, University of Guelph: Guelph, ON, Canada, 1990.

48. Arens, S.M.; Wiersma, J. The Dutch foredunes: Inventory and classification. *J. Coast. Res.* **1994**, *10*, 189–202.

Article

Mangrove Forests Evolution and Threats in the Caribbean Sea of Colombia

Diego Andrés Villate Daza [1], Hernando Sánchez Moreno [2], Luana Portz [3,*],
Rogério Portantiolo Manzolli [3], Hernando José Bolívar-Anillo [2,*] and Giorgio Anfuso [4]

[1] Grupo de Investigaciones Marino Costeras GIMAC, Escuela Naval de Suboficiales ARC, Barranquilla 080002, Colombia; godievi@gmail.com

[2] Laboratorio de Investigación en Microbiología, Universidad Simón Bolívar, Barranquilla 080002, Colombia; hsanchez13@unisimonbolivar.edu.co

[3] Department of Civil and Environmental, Universidad De La Costa, Barranquilla 080002, Colombia; rportant1@cuc.edu.co

[4] Departamento de Ciencias de la Tierra, Facultad de Ciencias del Mar y Ambientales, Universidad de Cádiz, 11510 Puerto Real (Cádiz), Spain; giorgio.anfuso@uca.es

* Correspondence: lportz1@cuc.edu.co (L.P.); hbolivar1@unisimonbolivar.edu.co (H.J.B.-A.)

Received: 11 March 2020; Accepted: 11 April 2020; Published: 15 April 2020

Abstract: Colombia has approximately 379,954 hectares of mangrove forests distributed along the Pacific Ocean and the Caribbean Sea coasts. Such forests are experiencing the highest annual rate of loss recorded in South America and, in the last three decades, approximately 40,000 hectares have been greatly affected by natural and, especially, human impacts. This study determined, by the use of Landsat multispectral satellite images, the evolution of three mangrove forests located in the Colombian Caribbean Sea: Malloquín, Totumo, and La Virgen swamps. Mangrove forest at Mallorquín Swamp recorded a loss of 15 ha in the period of 1985–2018, associated with alterations in forest hydrology, illegal logging, urban growth, and coastal erosion. Totumo Swamp lost 301 ha in the period 1985–2018 associated with changes in hydrological conditions, illegal logging, and increased agricultural and livestock uses. La Virgen Swamp presented a loss of 31 ha in the period of 2013–2018 that was linked to the construction of a roadway, alterations of hydrological conditions, illegal logging, and soil urbanization, mainly for tourist purposes. Although Colombian legislation has made efforts to protect mangrove ecosystems, human activities are the main cause of mangrove degradation, and thus it is mandatory for the local population to understand the value of the ecosystem services provided by mangroves.

Keywords: mangrove; coastal dynamic; salinization; *Rhizophora mangle*; *Avicennia germinans*; *Laguncularia racemosa*

---

1. Introduction

Mangrove forests are composed of unique plant species, that is, halophilic trees and shrubs that have specific morphological, physiological, and reproductive characteristics that enable them to survive in a critical interface among terrestrial, estuarine, and near-shore marine ecosystems in tropical and subtropical regions around the world. They are considered one of the most productive natural ecosystems on earth because of their relevant ecosystem services and ecological functions, such as being a nesting habitat for fishes, birds, marine mammals, crustaceans, amphibians, and reptiles. They also act as effective nutrient filters, support numerous rural economies, and protect coastal communities from storms and floods by acting as windbreaks and wave barriers, reducing coastal erosion [1–7]. Last but not least, mangrove forests, due to their great biomass (above- and below-ground) and capacity of accumulation of sediments, are able to store more carbon (on average 22 ± 6 Tg year^{-1}) than terrestrial forests, making them one of the most carbon-rich ecosystems in the tropics, with an estimated value of USD 194,000 per hectare per year [8].

Mangrove forests currently occupy less than 14 million hectares, representing <1% of the world's coastal areas, of which more than two-thirds are located in 18 countries: Indonesia, Brazil, Australia, Mexico, Nigeria, Malaysia, Myanmar, Bangladesh, Cuba, India, Papua New Guinea, Colombia, Guinea Bissau, Mozambique, Madagascar, the Philippines, Thailand, and Vietnam [1,3]. It is estimated that approximately 35% of mangrove forests disappeared during the last two decades of the 20th century, mainly due to their direct conversion to different land uses [1,6,9,10] such as aquaculture, agriculture, urbanization, and impacts due to alterations in the hydrology of river basins and changes in fluvial sediment inflow, among others [3,5,7,11,12]. Although the rate of mangrove forest loss has decreased significantly in the last two decades, it is still worrying, with rates of up to 3.1% per year in some countries—this could lead to a loss of their functionality in less than 100 years. In addition, only 6.9% of the world mangroves are protected, and hence it is mandatory to establish new areas of protection in an effort to reduce the rate of loss [1,4,10,13]. It is estimated that between 0.02 and 0.12 Pg per year of carbon have been released into the atmosphere as a consequence of mangrove degradation, which represents 10% of the total emissions resulting from deforestation [7]. Therefore, global net loss of mangroves would require the successful rehabilitation of about 100,000 ha per year unless the necessary measures were taken to halt current mangrove losses [14]. Further, mangroves work as a transitional intertidal ecosystem that is particularly vulnerable to the effects of climate change, mainly those linked to rising sea level, surface water warming, warming and changes in the composition of the atmosphere, and changes in rainfall, among others [5,12]. In different regions of Latin America and the Caribbean, climate change is considered to be the main driver of environmental impacts on mangroves [12]. Natural and anthropogenic stressors may interact in an additive or synergistic manner, which could lead to accelerated and massive alterations of these ecosystems [11]. However, mangrove forests are considered as a highly resilient ecosystem that has the capacity to adapt and adjust to changing conditions [6], and hence it can play a fundamental role in the design of climate change adaptation strategies.

Mangrove forests observed in Latin America and the Caribbean represent around 26% of the total amount recorded at world scale. They cover an area between 3.58 and 4.54 × 10^6 ha, of which 80% is found in six countries: Brazil, Mexico, Cuba, Colombia, Venezuela, and Honduras [12]. On the western side of South America, the largest mangrove forest cover is found in the tropical zone of the Colombian Pacific coast and in northern Ecuador [15]. Colombia is characterized as being the only country of South America with coasts in the Pacific Ocean and the Caribbean Sea with an extension of 1200 and 1800 km, respectively. Differences in precipitation and tidal range between both coasts favor the existence of almost continuous strips of mangroves along the Pacific coast, whereas in the Caribbean this ecosystem is closely linked to freshwater sources [16].

Overall, Colombia has a mangrove forest cover of around 379,954 ha, with 292,724 and 87,230 ha respectively located on the Pacific and Caribbean coasts [15,17,18]. They show a total amount of eight species: *Rhizophora mangle*, *Rhizophora harrisonii*, *Rhizophora racemosa*, *Laguncularia racemosa*, *Conocarpus erectus*, *Avicennia germinans*, *Avicennia harrisoni*, *Pelliciera rhizophorae*, and *Mora oleífera*.

On the Caribbean coast of Colombia, in the Department of Atlántico (Figure 1), there are currently around 613.3 hectares of mangrove forests located in different municipalities [19] and, along the coast of the Department of Bolívar, they currently occupy a surface of around 7000 hectares [20]. This paper determines the evolution, during the last decades, of the most extended and representative mangrove forests on the Colombian Caribbean coast between Barranquilla and Cartagena de Indias (departments of Atlántico and Bolívar)—Mallorquín, La Virgen, and Totumo mangrove swamps (Figure 1). This study took into consideration both natural changes (due to coastal erosion/accretion, etc.) and those produced by anthropic activities [21], which have influenced the evolution of the aforementioned mangrove forests, in order to design adequate plans for their environmental improvement and sound conservation strategies. The present paper investigates their evolution and their human (deforestation, road construction, etc.) and natural impacts (salinity variations, coastal erosion, etc.) on three mangrove forests located at the departments of Atlántico and Bolívar, located in the Caribbean coast of Colombia (Figure 1). Such areas are characterized by five mangrove species that are common along the Colombian Caribbean Sea [21]: *Rhizophora mangle*, *Avicennia germinans*, *Laguncularia racemosa*, *Conocarpus erectus*, and *Pelliciera rhizophorae*.

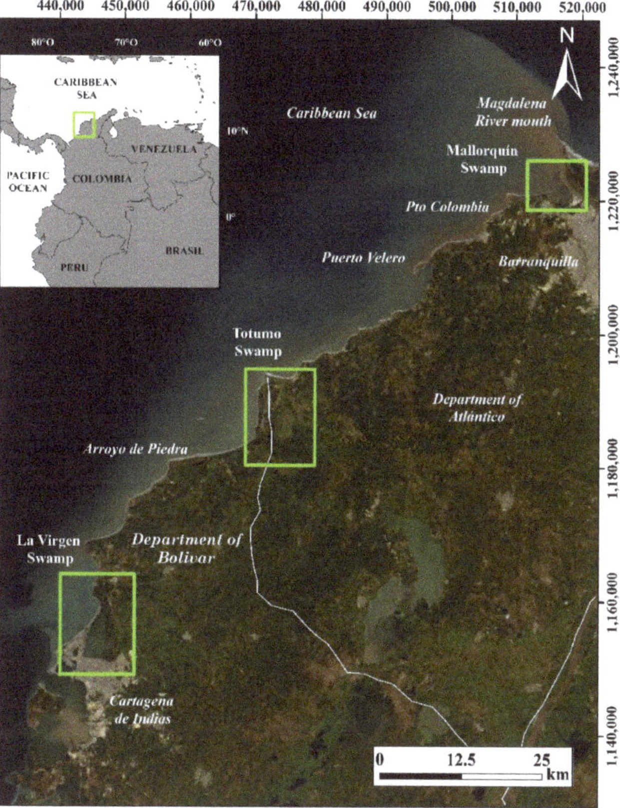

Figure 1. Location map with studied mangrove forests.

2. Study Area

The Caribbean coast of Colombia is a tropical environment with seasonal variations in rainfall (Figure 2) from the dry season (December–March) and the transitional seasonal (April–July) to the rainy season (August–November) [22,23].

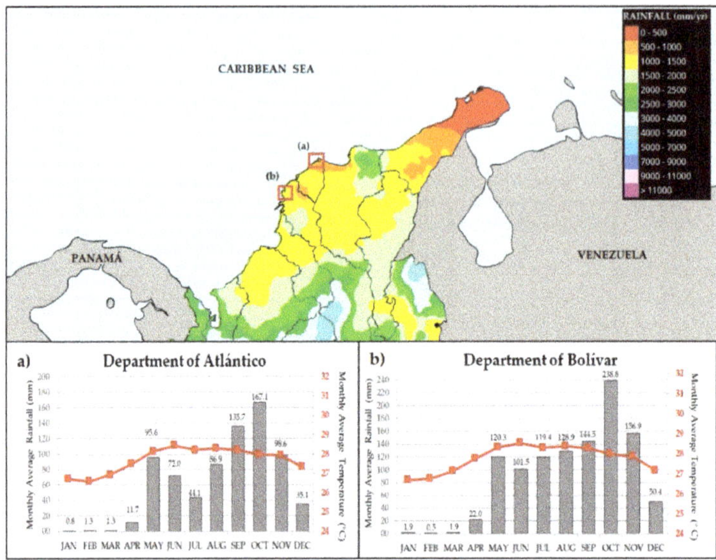

Figure 2. Rainfall and temperature variations at Colombian Caribbean coast (data recorded in the 1980–2010 period- IDEAM [24]): (**a**) Department of Atlántico; (**b**) Department of Bolivar.

The main regulator of rain cycles throughout the Colombian territory is the Intertropical Convergence Zone [25,26]. This low pressure equatorial system follows the synchronization of the sun [27] and a southern movement (between 27.5° S and 27.5° N), having the largest amount of solar energy received by the planet [28]. Other oscillations, such as Maiden and Julian, as well as the variability associated with the El Niño and La Niña phenomena (Southern Oscillation—ENSO) in the Pacific, contribute especially to the intensity of the precipitation anomalies on the Colombian Caribbean coast [26]. Tide has mixed semi-diurnal periodicity and microtidal variability, with maximum values of 60 cm [22]. Wind average velocity is <12 m/s, with the strongest winds blowing from the northeast in December–March and weakest values associated with easterly approaching winds, usually blowing in September–November. The dominant sea surface current is the *Caribbean current* that flows during almost all the year from east to west; an opposite current, the *Darien* or *Colombia* current, flows from Panama northeastward [29]. Significant wave height is between 1 and 2 m and wave climate is dominated by swell waves approaching from the northeast from November to May, and sea (smaller) waves, approaching from the northwest, west-southwest, and southwest, during the remainder of the year. Predominant longshore sediment transport is southwestwardly directed and a minor reversal takes place during rainy periods when southerly winds achieve more importance, giving rise to short erosive waves [30]. Coastal erosion is essentially linked to the impact of hurricanes and cold fronts—the former events impacting the coast from June to November, and latter events from January to March [31].

Mallorquín Swamp is located in the northwest part of the Department of Atlántico (Figure 1), on the western bank of the Magdalena River, close to Barranquilla. It is a shallow estuarine coastal lagoon with an area of around 650 ha, surrounded by floodplains and sand dunes. Three species of mangroves have been reported in Mallorquín Swamp: *A. germinans*, *R. mangle*, and *L. racemosa* with a maximum average height of 15.7 m. The most abundant species is *A. germinans* (71%), followed by *L. racemosa* (21%), and finally *R. mangle* (8%) [32]. Salinity range was 9‰–22‰ in surface water and 14‰–50‰ in interstitial water, that is, at a depth between 50 and 100 cm [32]. The mangrove leaves had cuts and perforations associated with herbivory. The presence of climbing plants was also observed—they generate overweight and strangulation of stems and branches [32]. The system has been affected by coastal erosion, increased sedimentation processes, and anthropogenic activities linked to the urbanization of nearby areas and contamination processes due to solid waste and sewage discharges [33–37].

Totumo Swamp (Figure 1), located between the departments of Bolívar and Atlántico, is composed by a main body of water with an approximate extension of 1361.06 ha and presents several mangrove patches on the borders of the swamp. Totumo Swamp presents average salinity values from 0.1‰ to 2‰, and thus it is considered a fresh water body [38]. The vegetation adjacent to Totumo Swamp belongs to the dry tropical forest and very dry thorny scrub. Lastly, along the swamp flood plains, an abundant vegetation of *Typha dominguensis* was observed [38]. The swamp is partially protected by a 5 km long spit, namely, Galerazamba spit, made up of a wide beach with a gentle slope from 2° to 5°, which has undergone major morphological changes since its initial description by Francisco J. Fidalgo in 1805 [39–41]. The spit encloses inland lagoons (e.g., La Redonda) that are fed by rainfalls, as well as directly by the sea during the stormy season. La Redonda lagoon, inside the Galerazamba spit, has an average salinity of 23‰, suitable for mangrove development [38]. On and nearby the spit, there are several mangrove areas composed of *L. racemosa* (77%), which is located on the edge of the lagoon, and *C. erectus* (23%), which is found at the mouth of the swamp. Mangroves have a shrubby growth, with maximum average heights of 6.1 m for *L. racemosa* and 5.7 m for *C. erectus*. The vegetation is currently under anthropogenic pressure, mainly due to the expansion of the agricultural frontier that has been increasing in recent years, negatively impacting the mangrove forest by illegal logging and successive land reclamation. Presently, the original brackish vegetation is changing towards a typical freshwater vegetation ecosystem because of the reduced inputs of marine waters due to the construction of gates (that are usually closed) at the lagoon inlet entrance [19,38].

North of Cartagena de Indias (Department of Bolívar, Figure 1), there is one of the most important coastal wetland of Colombia, namely, **La Virgen Swamp,** which has an approximate surface of 20 km^2 and a mangrove forest cover of around 824 ha, with a main drainage network consisting of 8 streams in the rural area and 20 channels in the urban perimeter of Cartagena de Indias. Four species of mangrove have been reported in La Virgen Swamp: *A. germinans* (67%), *R. mangle* (30%), *L. racemose*, and *C. erectus* that, together, make up 3% of the swamp. *R. mangle* is located on the internal border of lagoons and channels; *A. germinans* is found in the less intervened sites, reaching the limit with the tropical dry forest; *L. racemosa* is mainly found on the edges of abandoned ponds; and *C. erectus* is only found at the border between the mangrove forest and the mainland vegetation [42]. Salinity varies considerably, and its most frequent peaks are in the range of 0‰ to 35‰ [42]. The average height of the mangrove plants varies from 1 to 10 m, according to the forest sector [21,43].

The area is subject to various threats such as illegal logging; artificial filling; and terracing for the implementation of fish farming, waste disposal, urbanization, and pollution; for example, La Virgen Swamp receives about 60% of the Cartagena de Indias wastewaters, around 114,000 m^3/day. Further, because of the reduced capacity of water exchange between the lagoon and the Caribbean Sea, several problems have arisen in the last decades, such as eutrophication, increased salinity, and fish mortality [20,44,45].

3. Methodology

The methodology applied to assess the mangrove area changes was supported by multi-date remote-sensed data. Using satellite images, it is possible to identify, calculate, and monitor mangrove areas, as well as the surfaces affected by erosion processes [46–49].

3.1. Data Used

To determine the extent of mangrove ecosystems, we used Landsat 5 and 8 multispectral satellite images (TM—thematic mapper, and OLI—operational land) with a spatial resolution of 30 m in their optical channels available at the United States Geological Survey [50]. The dates of the images varied between 1985 and 2018 (Table 1). The dataset was composed of the OLI sensor bands, already subjected to a complex algorithm of adjustment of the atmospheric effects that was based on parameters estimated from the same sensor bands and the application of the model of the second simulation of the satellite signal with the code known as solar vector spectrum (6SV) [51].

The images presented a good resolution due to the lack of cloudiness and atmospheric disturbances. Similarly, Google Earth Pro images and thematic cartography available in DIMAR-CIOH [52], INVEMAR [53], and a layer from the Humboldt Foundation cartographic base were used.

3.2. Digital Images Analysis

The manipulation of Landsat images was carried out using ARGIS 10.4. Initially, the coordinate system was defined by processing the georeferencing according to the parameters defined for land mapping in Colombia with the Magna Sirgas Datum Bogotá coordinate system.

The analysis, interpretation, and quantification procedures of the images used were performed using band composition, as well as the normalized difference vegetation index (NDVI), which was calculated using the following equation:

$$NDVI = (IVP - V)/(IVP + V) \tag{1}$$

where IVP and V represent the reflectance values in the near-infrared and red infrastructure bands, respectively. The NDVI varies from -1 to $+1$, with negative values and zero representing areas without vegetation [54]. The procedure for delimiting the areas identified as mangrove forests was carried out by manual vectoring on the classified images. This method requires a great deal of interpretation and is time consuming, as each polygon must be determined individually.

Ortho-rectified satellite images were also used to map shoreline position. Shoreline migration was analyzed in ARGIS 10.4 by the Digital Shoreline Analysis System—DSAS 5.0 [55].

Table 1. Details of multispectral remote sensing data and other documents used in this study.

Location Area	Sensor and/or Document	Spatial Resolution (m)/or Scale	Year/Month/Day of Acquisition	Source	Cloud Cover (%)
Mallorquín Swamp	Landsat 5-OLI	30	1985/01/24	USGS	<5
	Landsat 5-OLI	30	1998/05/20	USGS	<5
	Landsat 8-OLI	30	2013/04/01	USGS	<5
	Landsat 8-OLI	30	2018/12/05	USGS	<10
Totumo Swamp	Landsat 5-OLI	30	1985/01/24	USGS	<5
	Landsat 5-OLI	30	1998/09/20	USGS	<5
	Landsat 8-OLI	30	2013/05/13	USGS	<5
	Landsat 8-OLI	30	2018/12/05	USGS	<5
La Virgen Swamp	Landsat 5-OLI	30	1985/01/24	USGS	<5
	Landsat 5-OLI	30	1998/09/20	USGS	<5
	Landsat 8-OLI	30	2013/05/13	USGS	<5
	Landsat 8-OLI	30	2018/12/05	USGS	<5
All areas	Unidentified	4	All years	Google Earth Pro	<5
	Lidar-Spot [1]	1:50,000	All years	DIMAR-CIOH	Unidentified
	Layer Colombia [2] Mangrove	1:5,000,000	2005	INVEMAR	Unidentified
Mallorquín Swamp/Totumo Swamp	Spot-Aster [3]	30	1986	INVEMAR	Unidentified
			2004	INVEMAR	Unidentified

[1] Atlas Geomorfológico del Caribe Colombiano [52]. [2] Sistema de información para la gestión de los manglares de Colombia SIGMA [56]. [3] Ordenamiento Ambiental de la Zona Costera del Departamento del Atlántico [53].

3.3. Field Visits

Detailed field visits were made to the three mangrove forests investigated in this study in order to determine actual predominant plant species, their distribution, their conditions (leaf characteristics and plant average height), and the typology of anthropic activities (road emplacement, expansion of urbanized areas, disposal of waste materials, illegal logging, etc.) and their effects on the environment. Characteristics of mangroves in past investigated time spans were essentially reconstructed from unpublished reports carried out by environmental departmental authorities and research institutes.

4. Results and Discussion

4.1. Mangrove Swamps Evolution

The evolution of three most relevant mangrove forests along the Department of Atlántico and Bolívar, that is, the Mallorquín, Totumo, and La Virgen swamps, is presented in the following sub-sections.

4.1.1. Mallorquín Mangrove Swamp

The mangrove forest of Mallorquín Swamp presented small changes in the period of 1985–1998, a reduction of around 51 ha from 1998 to 2013, and an increase of around 34 ha during the 2013–2018 period (Figure 3, Table 2).

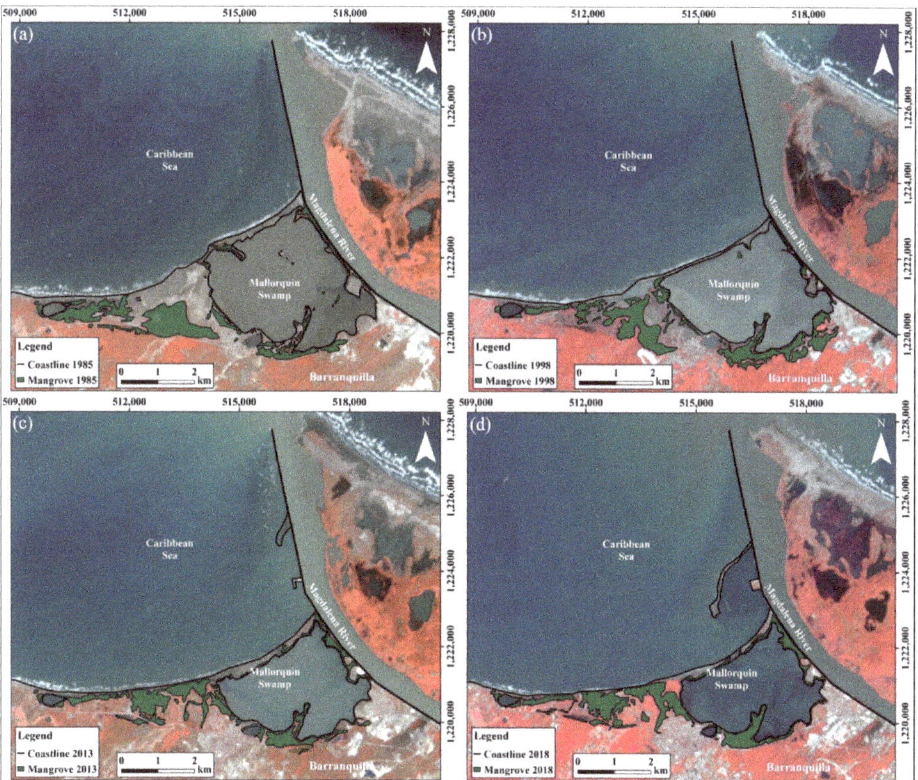

Figure 3. Evolution of the mangrove forest (green) in the Mallorquín swamp (**a**) 1985, (**b**) 1998, (**c**) 2013, and (**d**) 2018.

Table 2. Variation of the mangrove forest cover (ha) in the Mallorquín, Totumo, and La Virgen swamps.

Year	Mangrove Cover (ha)		
	Mallorquín	Totumo	La Virgen
1985	302	496	725
1998	304	215	685
2013	253	229	941
2018	287	195	910

Until the beginning of the 1940s, the swamp had an estuarine regime constituted of different coexisting and connected systems, that is, the Cantagallo, Mallorquín, La Playa, and Manatíes swamps, which presented a wide variety of ecosystems and fishing areas [57,58]. After the construction of the western jetty at the Magdalena River mouth, during the 1925–1935 time span, the sand bar that encloses Mallorquín Swamp suffered—between 1939 and 1987—an erosion rate of 65 m/year, finally acquiring its current configuration. The swamp recorded a strong hydrologic imbalance due to a deficit in fresh water supplies that were essentially limited to the León stream and direct rainfall [53,59]. The deficiency in water supplies significantly affected the different ecosystems, impacting the fishing activities and the mangrove forest health [57,58].

Concerning the period from **1985 to 1998**, fairly unrepresentative changes were reported, that is, a very small increase (Table 2) as a result of the mangrove loss observed at the sand bar that encloses the swamp and the mangrove cover increase recorded in the southern part of the swamp (Figure 3a,b). Despite such small changes, the swamp reflected a stressful environmental condition, mainly due to the diminution of hydrodynamic processes. The effects of such an unfavorable situation continued in following years, and were more evident at the beginning of the 2000s.

In this same period, the area of Mallorquin Swamp was reduced considerably. The bar migrated landward on average 29.5 m/year, resulting in a period of relevant morphological changes and loss in mangrove cover.

Changes in the hydrodynamic conditions, as observed since the end of the 1980s by Galvis et al. [19,60], were linked to the few continental and marine water supplies, because of the reduced fresh water supplies from the León stream and the lack of permeability of the western jetty at the Magdalena River mouth. This brought to the modification of the mangrove ecosystem, and *A. germinans* and eurihaline vegetation became the predominant species compared with *R. mangle*, reflecting a process of salinization that continued in following years, reaching interstitial salinity values from 20‰ to 30‰, as reported by Galvis et al. and Ulloa-Delgado et al. [60,61]. However, the loss of *R. mangle* was not only associated with the salinity increase but also with erosion processes that led to the loss of trees and beach surfaces, which made the implantation of new *R. mangle* propagules impossible.

In the period from **1998** to **2013**, around 51 ha of forest were lost (Table 2) as a result of urban area expansion in the southern part of the swamp [19,61], where illegal forest cutting by local inhabitants, accumulation of solid waste materials, and artificial infilling works were observed (Figure 3b,c). On the other hand, mangrove growth was recorded in the sand bar because of the reduction of coastal erosion (Figure 4) that allowed bar stabilization and propagules implantation and growth.

The most exploited species was *L. racemosa*, used for the construction of huts to provide shade for tourists visiting local beaches. As for solid waste materials, which prevent the proper development of seedlings and pneumatophores and hence favor a decrease in the self-healing capacity of the forest [62], they essentially came from the Magdalena River that drains a basin of 257,430 km^2, wherein 724 municipalities are located, representing 80% of the Colombian population [63]. Further mangrove losses were related to the diminution of fresh water supplies, especially during the 1997–1998 period in which Colombia was greatly impacted by the El Niño phenomenon that led to a generalized decrease of rainfall and river discharges. During the most critical dry period (October 1997 to January 1998), the flow of the Magdalena River fluctuated between 45% and 70% in its lower basin [64], considerably decreasing its water contribution to Mallorquín Swamp and the consequent increase of salinity. The very narrow strip of *R. mangle* observed in 1998 in the southern, northeastern, and northwestern parts of the swamp was later replaced by a monospecific forest of *A. germinans*, with the presence of few individuals of *L. racemosa* and *C. erectus* and xerophytic plants (*Prosopis juliflora*, *Capparis odorantissima*, *Sarcostema* sp., and *Stigmaphyllon* sp.) [61]. In general, the forest presented mature individuals, but of little height (average: 3.86 m) and with leaves characteristic of xerophytic plants [61]. Such a trend continued in 2005, as reported by INVEMAR [53] and GTA [57], but slightly decreased in 2007 when salinity achieved values between 20‰ and 35‰ that allowed *A. germinans* to survive [65] and led to a slow recovery in the general mangrove forest cover [53]. In addition, in 2007, reforestation programs were initiated by the fishing communities with the support of local environmental authorities [53]. The stabilization of the hydrological conditions and the reforestation programs had positive effects that were observed in the following period investigated.

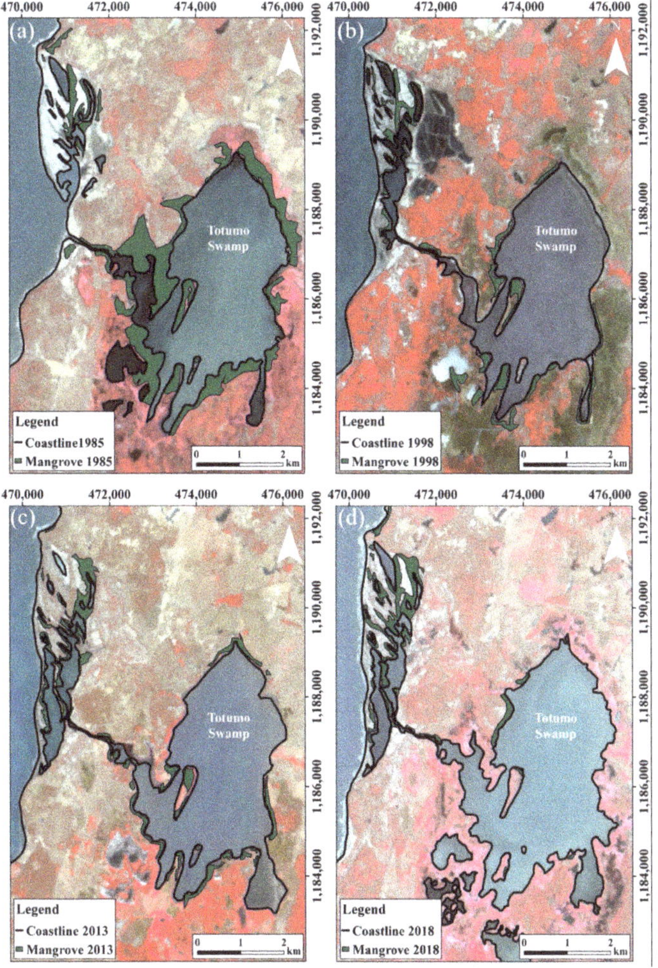

Figure 4. Evolution of the mangrove forest (green) in the Totumo swamp: (**a**) 1985, (**b**) 1998, (**c**) 2013, and (**d**) 2018.

In the period from **2013** to **2018**, an increase in mangrove coverage of around 34 ha was located on the west part of the sand bar and in the southern part of the swamp (Figure 3c,d). In this period, the stability of the sandy bar was observed, as a consequence of the growth of a spit, as being connected to the Magdalena western jetty, located to the north of the swamp. This new formed feature protected the bar from incoming waves (Figure 3c,d).

Additionally, the recovery of the mangrove could be related to the reforestation campaigns carried out by different entities [53] and the maintenance of the different channels that communicate the swamp with the Magdalena River. In 2016, the Corporación Autónoma Regional del Atlántico (Local Environmental Authority) carried out activities aimed at the environmental recovery of the swamp by directly cleaning the channels and by installing meshes in the boxcoulverts to reduce waste materials entering in the channels communicating the swamp with the Magdalena River [58]. Presently, an ongoing project is being developed to declare Mallorquín Swamp a protected area [66].

4.1.2. Totumo Mangrove Swamp

The mangrove forest of Totumo Swamp presented a reduction of ca. 280 ha in the period from **1985 to 1998**, an increase of ca. 13 ha in the period from **1998 to 2013** and a loss of ca. 33 ha during the **period from 2013 to 2018** (Figure 4 and Table 2).

The large loss of mangrove cover recorded in the first period (from **1985 to 1998**) was related to the transformation of the forest into a freshwater vegetation ecosystem [38]. Most relevant changes were recorded in the northern part of the swamp (Figure 4a,b). Totumo Swamp communicated with the Caribbean Sea through a natural inlet that allowed the entry of salt water, generating ideal conditions for the growth of mangrove [38,53]. However, in the 1970s, gates emplaced at the inlet entrance, 20 m in width, greatly limited sea water entrance, turning much of Totumo Swamp into a freshwater system (Figure 4) [38,67], which receives supplies from numerous streams, such as Punta Antigua, Lata, Calabria, and Bombo [61]. As a result, during the rainy season, salinity was close to zero and, during dry seasons (i.e., January–May and July–September), salinity increased up to 15‰ [68]. Mangrove plants presented shrubby and stunted forms and did not exceed 2 m in height. There were observed specimens of *R. mangle*, *L. racemosa*, *C. erectus*, and *A. germinans*, the latter being the most abundant. A considerable decrease of *R. mangle* was observed at the mouth of the swamp, and *Typha dominguensis* was recorded on the west bank, alternating with *C. erectus* with rhizomes on the roots, stems, and lower branches [61].

An increase of around 13 ha in mangrove surface was recorded in the period from **1998 to 2013** as a result of mangrove growth on the western side of the swamp, on the Galerazamba spit, as well as a decrease on the eastern side (Figure 4b,c).

Galerazamba spit has undergone relevant morphological changes since its original description by Francisco J. Fidalgo in 1805 [39–41]. Anfuso et al. [69] reported spit evolution from 1947 to 2013, recording an accretion of around 0.7 km^2 and a down drift (i.e., southward) migration of about 80 m. Hence, the increase in the mangrove area was directly related to the growth of the spit that enclosed numerous small lagoons, among which the larger one was La Redonda Swamp, fed by rainfalls and by the sea during stormy season [39,40].

On the eastern side of the swamp, INVEMAR [19] observed in 2004 that the mangrove cover was reduced to a strip of approximately 5 to 10 m in width, mainly composed of *C. erectus* and *L. racemosa*. At this area, the natural mangrove cover was progressively displaced by pastures and macrophytes and human activities as livestock and agriculture [53]. On the southwestern side of the swamp, only a very narrow line of *C. erectus*, with plants about 7.5 m high, was recorded; no natural regeneration or flowering of mangroves was observed [19].

The mangrove forest of Totumo Swamp degraded in the period from **2013** to **2018**, with a loss of around 33 ha (Figure 4c,d). The decrease was mainly observed on the eastern side, whereas the western side showed only a small increase associated with mangrove growth in the lagoons at Galerazamba split. The overall reduction in mangrove cover was undoubtedly related to the loss of water exchange with the Caribbean Sea. This turned Totumo Swamp into a freshwater ecosystem, wherein water is nowadays mainly used for agricultural and livestock activities. The economic interests linked to the maintenance of such activities retain the gates at the inlet entrance that was closed to prevent seawater entrance, thus prohibiting the restoration of the original swamp ecosystem.

4.1.3. La Virgen Mangrove Swamp

La Virgen mangrove forest presented a reduction of around 40 ha in the period from **1985 to 1998**, an increase of around 255 ha in the period from **1998 to 2013**, and again a reduction of around **34 ha** in the period from **2013 to 2018** (Figure 5, Table 2).

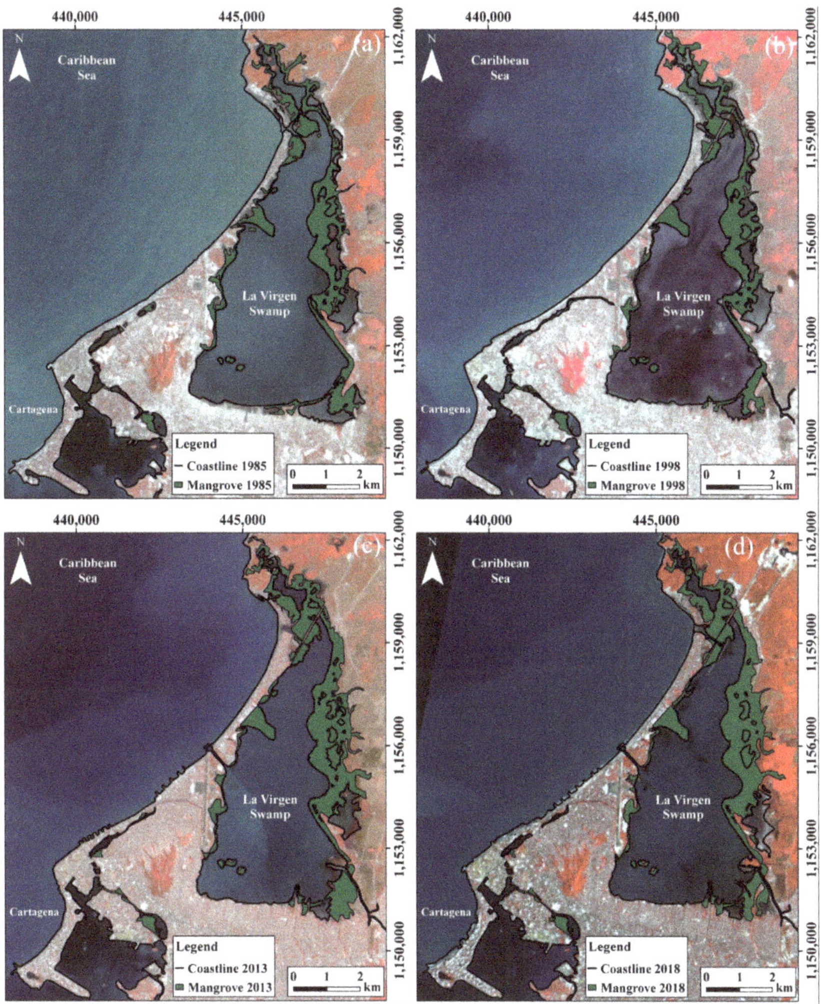

Figure 5. Evolution of the mangrove forest (green) in La Virgen swamp: (**a**) 1985, (**b**) 1998, (**c**) 2013, and (**d**): 2018.

The mangrove forest of La Virgen presented a reduction of approximately 40 ha in the period from **1985 to 1998** (Table 2). This reduction was mainly observed in the northern part of the swamp (Figure 5a,b) and was associated with different factors that were reported since 1984 by CARDIQUE [70], among them are (i) the accumulation of waste materials, mainly plastics, observed especially at the roots of *R. mangle* [21]; (ii) the illegal logging—for the construction of houses by the communities living on the banks of the swamp—of *L. racemose*, which consequently suffered a relevant pressure, with mortality values of 92.9%; and (iii) the land filling for urban infrastructure emplacement, such as the construction of a connection runway to the local airport and the roadway Cartagena-Barranquilla, among others [45,70]. In the following years, the mentioned roadway greatly affected water exchanges between the swamp and the Caribbean Sea, reducing the existing inlets to a single point located at La Boquilla [43,71]. Therefore, water exchange capacity was significantly reduced, limiting the oxygenation process and altering the constant flow of sediments and organic material, mainly due to the periodic infilling of the inlet, which generated problems such as eutrophication, increased salinity, and fish mortality [44]. In addition, the road construction at the border of the mangrove forest will probably lead to the replacement of native vegetation by other vegetation types, as observed in Punta Mala Bay, Panama [72]. During this investigated period, along and nearby the mentioned roadway, the construction of many resorts and hotels also took place, which was done to satisfy the tourism market demand, as well as the implementation of aquaculture activities (e.g., cultivation of tarpon fish) [71].

In the southern sector, the loss of mangrove areas occurred mainly through landfill processes and, consequently, substitution by urbanization (Figure 5).

Concerning the period from **1998 to 2013**, an increase of 255 ha was observed essentially along the eastern border of the swamp (Figure 5b,c). This mangrove surface increase was linked to the execution of a project, at the end of 2000, which stabilized the La Bocana area to guarantee the permanent water entrance from the Caribbean Sea. The project consisted in the emplacement of different structures at the inlet entrance and along the feeding channels (sand traps, gates, etc.) that favored the recirculation of water within the swamp, improving the level of oxygenation and salinity; in this way, the lagoon recovered the natural ecosystem services [73]. INVEMAR [19] recorded an increase of mangrove vegetation in 2004, which was dominated by *R. mangle* and *A. germinans* (that together accounted for 97% of the total amount), followed by *L. racemosa* and *C. erectus*. *L. racemosa* was rare and was usually found on the edges of abandoned ponds, whereas *C. erectus* was the least frequent and was only located at lagoon borders, along with the mainland vegetation. In addition, in 2004, a reforestation program was carried out, wherein 401,564 plants (of four species of mangroves) were planted in an area of 44.3 ha [19]. Summing up, the reestablishment of the hydrodynamic of the swamp and the implementation of reforestation programs seemed to be the main reasons for the increase in mangrove coverage between 1998 and 2013.

Concerning the last investigated period, from 2013 to 2018 (Figure 5c,d), the swamp presented a reduction of mangrove forests of approximately 30 ha, essentially due to the conclusion of the construction, in 2015, of the Cartagena-Barranquilla roadway, as well as a new roadway that included the emplacement of a viaduct 5.39 km in length, 4.73 km of which passes over La Virgen swamp [74]. The environmental assessment prior to the construction of the viaduct stated that the structure would impact around 52 ha of swamp surface and 41 ha of mangroves [75]. The National Environmental Licensing Authority (ANLA) authorized in 2015 the logging of 1673 trees, among which 1158 were mangrove trees [75]. The construction company has to compensate such actuations by creating a new green area of 177 ha, which could in the next few years increase the total forest coverage [75]. In 2015, the Institute Humboldt and the Foundation Omacha observed, as is the case in the northern zone of the swamp, there is only *R. mangle* (64%) on the banks and *A. germinans* (33%) on the mainland, whereas four species of mangrove were recorded in the eastern sector: *A. germinans* (65%), *R. mangle* (33%), *C. erectus* (0,6%), and *L. racemosa* (0.3%). Inside the forest, there were no permanent flooding events, and hence *A. germinans* dominated (with 97%), and both *C. erectus* and *L. racemosa* were rare and interspersed with *A. germinans* [76]. Further losses of mangrove coverage in the period between 2013 and 2018 were observed in the southern and southeastern areas of the swamp that have been progressively filled in and occupied by human illegal settlements and aquaculture ponds that are generally linked to the illegal occupancy by people displaced by violent events from different territories in Colombia and in Venezuela [43,45]. These different factors could together be the cause of the mangrove loss recorded in the analyzed time frame. Nowadays, the swamp shows around 775 ha of mangroves, including *R. mangle*, *A. germinans*, and *L. racemosa* [45].

4.2. Environmental and Natural Concerns

Oceanographic, climatic, and geomorphological aspects; soil conditions; level and duration of flooding; salinity; and sediment load determine the structure and floristic composition of mangrove forests [77].

Salinity is one of the most important abiotic factors influencing the structure, seedling establishment, and function of mangrove ecosystems, and small changes in salinity can lead to abrupt ecological changes [6,78]. Most mangrove species can grow in fresh water, but growth is stimulated under saline conditions, with optimal seawater concentrations ranging from 5‰ to 50‰, depending on the mangrove species [78,79]. High salinity soils limit seed germination and reduce plant growth [80]. The low rates of water loss from the leaves, as a strategy for surviving under hypersaline growth conditions, limit rates of carbon gain affecting the plant growth; as a result of this, individuals of low stature and stunted growth (scrub mangroves) are observed [79]. Therefore, the degree of tolerance to salinity determines the dominance of species within the mangrove forest. Mangrove forests dominated by *R. mangle* can cope with wave strength and salinities close to those of sea water (about 35‰), those dominated by *A. germinans* can tolerate and develop under salinity conditions of 60–65‰ and sandier substrates. Although *L. racemosa* tolerates salinities similar to *R. mangle*, it requires a more open canopy, which allows a greater availability of light for regeneration, and therefore is generally associated with disturbed areas [77]. The effect of the increased salinity on the mangrove species was clearly observed in Mallorquín Swamp, which suffered a loss of diversity—*A. germinans* and *R. mangrove* were replaced by *A. germinans* of low stature and a few individuals of *L. racemosa* [19]. The same condition was also described by Sánchez et al. [65] in a mangrove forest located in the municipality of Puerto Colombia (Colombian Caribbean Sea). The authors detected that in soils with high salinity (76‰ at 0.3 m), the dominant species was *A. germinans*; meanwhile, *R. mangle* was absent or very scarce. A similar trend, that is, the predominance of *A. germinans* associated with a few individuals of *L. racemosa*, was also observed in the forest along the Indian River Lagoon (Florida, USA), which is developed on high salinity soils due to low sea water inputs and high evaporation rates [79]. Mangrove mortality due to hypersalinity conditions was reported in the Ciénaga Grande de Santa Marta (Colombia), where sites with dead or dwarfed vegetation had an average soil salinity of 74‰, with values between 52‰ and 100‰. Basal area and forest biomass volume were inversely correlated with soil salinity [81]. Currently, the Mallorquin swamp presents salinity values between 14‰–50‰ in interstitial water, which allow the growth of *A. germinans*, *R. mangle*, and *L. racemosa* [32]. In addition, new sandy bodies are being formed in front of Mallorquín Swamp, where mangrove seedlings are observed (Figures 3d and 6a). This could generate ideal conditions prone to the increase of this mangrove forest. However, the presence of solid waste is one of the main problems that this forest presents; therefore, cleaning and educational campaigns should be carried out in the surrounding communities and also should be devoted to the tourists that visit the forest (Figure 6b).

Figure 6. (**a**) Mangrove seedlings growing in the new sand bar, and (**b**) plastic waste observed in the mangrove forest of Mallorquín Swamp.

The mangrove, as a facultative halophyte plant, can tolerate fresh water conditions for a limited time but not during its entire life cycle [82,83]. The flowing of seawater into the mangrove forest system reduces the possibility of survival for plant species that are not salt tolerant [82]. The mangrove forest of Totumo Swamp was transformed on its eastern side into an ecosystem dominated by freshwater plants (*Typha dominguensis* and *Eichhornia crassipes*) due to the loss of seawater input that brought salinity to zero (Figure 7a); meanwhile, mangrove remnants were recorded on its western side that is characterized by brackish water conditions (Figure 7b). La Virgen Swamp was affected mainly by the construction of a road that impedes the normal exchange of water with the Caribbean Sea. The swamp receives about 60% of the city wastewater, which has an approximate volume of 114,000 m^3/day [44,45]. The lack of communication with the Caribbean Sea generated an increase in wastewater concentration, generating eutrophication processes that affected the mangrove forest.

Figure 7. (**a**) Eastern side of Totumo Swamp characterized by fresh water conditions; (**b**) mangrove remnants (*Rhizophora mangle*) in the western side of Totumo Swamp characterized by brackish water conditions.

Deforestation is considered one of the main anthropogenic causes of mangrove degradation [84]. Colombia has the highest annual rate of deforestation in South America, showing values between 1.1% and 0.6% for 1980–1990 and 2000–2005 periods versus average values of 0.69% and 0.18% observed in South America [85]. In Colombia, mangrove areas have been converted mainly into agricultural land, ponds for aquaculture, and urban development (mainly for tourism). Illegal logging has been reported for the three mangrove forests investigated in this paper (Mallorquín, La Virgen, and Totumo), but each one was associated with different activities.

In Mallorquín Swamp, the logging was particularly linked to an increase in urban growth, mainly in the southern part of the swamp (Figure 3d). In La Virgen Swamp (Figure 8a) urban growth for tourism, road construction (Figure 8b), aquaculture, and the use of wood for house building were the main factors associated with the felling of the mangrove forest. In Totumo Swamp, the expansion of the agricultural frontier and the increase in livestock activity on the margins of the swamp were the main factors associated with the felling of the mangrove forest. With respect to land uses in Totumo Swamp, they include extensive livestock farming (5209.43 ha), annual or seasonal crops (53.75 ha), pasture cultivation (3049.85 ha), and aquaculture (155.82 ha) [38].

Figure 8. (a) Mangroves (*R. mangle*) in good condition in the west sector of La Virgen Swamp; (b) mangrove cutting due to the construction of the viaduct over La Virgen Swamp.

Deforestation modifies the mangrove phytogeographic landscape and reduces its biological diversity [84]. Deforestation not only decreases the number of mangrove specimens, but also produces ecological effects on the forest, such as stunted and shrubby growth, canopy opening, stem mortality, decreased regeneration of harvested species, and changes in species composition, among others [86]. The extraction of woody and non-woody products has degraded many areas of mangroves, resulting in the development of plants of low height and thin diameter [85], as reported mainly in the Mallorquín and Totumo swamps. Mangrove deforestation is linked to the lack of adequate methods to assess their economic value; this has led people to consider them as worthless wastelands and as unhealthy risk areas that should be adapted to alternative and more lucrative uses [84].

Mangrove forests are also subject to natural disturbance related to changes in the amount and seasonality of rainfall. This is one of the most relevant factors in the Colombian Caribbean coast, an example being the intensification of the El Niño phenomenon, which generates decreases in the productivity of mangrove forests dominated by *R. mangle* and, in the long term, their replacement by *A. germinans* [77]. This was observed in the mangrove forest of Mallorquín Swamp that, due to alterations in its hydrology caused by anthropic activities (e.g., construction of the western jetty at the Magdalena River mouth) and natural phenomena such as a relevant reduce of precipitations due to El Niño (1997–1998), recorded not only alterations/degradation in its cover, but also in its composition becoming a monospecific forest consisting only of *A. germinans*.

Mortality events associated with climate extremes (e.g., tropical cyclones and El Niño and La Niña phenomena) could increase in the coming years because mangroves are important sentinels of global climate change processes [87].

5. Conclusions

The results obtained in the framework of this paper show how, during different periods of time, mangrove forests at localities investigated in the Colombian Caribbean Sea have been impacted by diverse and complex anthropogenic activities and natural disturbances. Although natural disturbances such as the El Niño phenomenon have greatly affected, at times, the cover and the structure of mangrove forests, human activities were the main cause of degradation and loss. Alterations in the hydrology of swamps, which lead to increases or decreases in salinity, urban growth, illegal logging, expansion of agricultural frontiers, and road construction, were the main causes associated with the loss of mangrove cover.

Although losses of mangrove covers were observed in all investigated sites, it is worth noting that different activities carried out by local environmental authorities, together with local inhabitants, favored mangrove forest cover stabilization and, at places and times, led to an increase, such as in the Mallorquín and La Virgen swamps. In Totumo Swamp, different economic interests, essentially linked to agricultural activities, prevented mangrove forest recover. At many places, illegal activities such as logging continue to affect the mangrove forests investigated in this paper, and thus it is mandatory for the Colombian environmental authorities to develop strategies aimed not only at protecting and recovering these ecosystems, but also at raising awareness among the local inhabitants concerning the ecological value of these ecosystems, as well as their importance in coastal adaptation and mitigation function of climate change-related processes.

Author Contributions: Conceptualization, H.S.M., D.A.V.D. and H.J.B.-A.; Data curation, G.A., R.P.M.; Formal analysis, L.P., D.A.V.D., H.S.M. and H.J.B.-A.; Methodology, D.A.V.D., L.P., R.P.M. and G.A.; Writing–original draft, L.P., R.P.M., H.J.B.-A. and G.A. All authors have read and agreed to the published version of the manuscript.

Funding: This research received no external funding.

Acknowledgments: This research is a contribution to the Andalusia PAI Research Group RNM-328, the RED PROPLAYAS network, Universidad Simón Bolívar (Barranquilla, Colombia), Escuela Naval de Suboficiales ARC Barranquilla, Universidad de la Costa (CUC), and the Center for Marine and Limnological Research of the Caribbean CICMAR (Barranquilla, Colombia).

Conflicts of Interest: The authors declare no conflict of interest.

References

1. Sanderman, J.; Hengl, T.; Fiske, G.; Solvik, K.; Adame, M.F.; Benson, L.; Bukoski, J.J.; Carnell, P.; Cifuentes-Jara, M.; Donato, D.; et al. A global map of mangrove forest soil carbon at 30 m spatial resolution. *Environ. Res. Lett.* **2018**, *13*, 055002. [CrossRef]

2. Carugati, L.; Gatto, B.; Rastelli, E.; Martire, M.L.; Coral, C.; Greco, S.; Danovaro, R. Impact of mangrove forests degradation on biodiversity and ecosystem functioning. *Sci. Rep.* **2018**, *8*, 13298. [CrossRef] [PubMed]

3. Barbier, E.B. The protective service of mangrove ecosystems: A review of valuation methods. *Mar. Pollut. Bull.* **2016**, *109*, 676–681. [CrossRef] [PubMed]

4. Chauhan, H.B. Mangrove Inventory, Monitoring, and Health Assessment. In *Coastal Wetlands: Alteration and Remediation*; Frinkl, C., Makouski, C., Eds.; Springer: Cham, Switzerland, 2017; pp. 573–630.

5. Jennerjahn, T.C.; Gilman, E.; Krauss, K.W.; Lacerda, L.D.; Nordhaus, I.; Wolanski, E. Mangrove Ecosystems under Climate Change. In *Mangrove Ecosystems: A Global Biogeographic Perspective*; Rivera-Monroy, V., Lee, S., Kristensen, E., Twilley, R., Eds.; Springer: Cham, Switzerland, 2017; pp. 211–244.

6. Osland, M.; Feher, L.; Lopéz-Portillo, J.; Day, R.; Suman, D.; Guzmán, J.; Rivera-Monroy, V. Mangrove forests in a rapidly changing world: Global change impacts and conservation opportunities along the Gulf of Mexico coast. *Estuar. Coast. Shelf Sci.* **2018**, *214*, 120–140. [CrossRef]

7. Thomas, N.; Lucas, R.; Bunting, P.; Hardy, A.; Rosenqvist, A.; Simard, M. Distribution and drivers of global mangrove forest change, 1996–2010. *PLoS ONE* **2017**, *12*, e0179302. [CrossRef] [PubMed]

8. Costanza, R.; de Groot, R.; Sutton, P.; van der Ploeg, S.; Anderson, S.J.; Kubiszewski, I.; Farber, S.; Turner, R.K. Changes in the global value of ecosystem services. *Glob. Environ. Chang.* **2014**, *26*, 152–158. [CrossRef]

9. Jia, M.; Wang, Z.; Zhang, Y.; Mao, D.; Wang, C. Monitoring loss and recovery of mangrove forests during 42 years: The achievements of mangrove conservation in China. *Int. J. Appl. Earth Obs. Geoinf.* **2018**, *73*, 535–545. [CrossRef]

10. Polidoro, B.A.; Carpenter, K.E.; Collins, L.; Duke, N.C.; Ellison, A.M.; Ellison, J.C.; Farnsworth, E.J.; Fernando, E.S.; Kathiresan, K.; Koedam, N.E.; et al. The loss of species: Mangrove extinction risk and geographic areas of global concern. *PLoS ONE* **2010**, *5*, e10095. [CrossRef]

11. Feller, I.C.; Friess, D.A.; Krauss, K.W.; Lewis, R.R. The state of the world's mangroves in the 21st century under climate change. *Hydrobiologia* **2017**, *803*, 1–12. [CrossRef]

12. De Lacerda, L.D.; Borges, R.; Ferreira, A.C. Neotropical mangroves: Conservation and sustainable use in a scenario of global climate change. *Aquat. Conserv. Mar. Freshw. Ecosyst.* **2019**, *29*, 1347–1364. [CrossRef]

13. De Almeida, L.T.; Olímpio, J.L.S.; Pantalena, A.F.; de Almeida, B.S.; de Oliveira Soares, M. Evaluating ten years of management effectiveness in a mangrove protected area. *Ocean Coast. Manag.* **2016**, *125*, 29–37. [CrossRef]

14. Lewis, R.; Brown, B.; Flynn, L. Methods and Criteria for Successful Mangrove Forest Rehabilitation. In *Coastal Wetlands: An Integrated and Ecosystem Approach*; Perillo, G., Wolanski, E., Cahoon, D., Hopkinson, C., Eds.; Elsevier: Amsterdam, The Netherlands, 2019; pp. 863–887.

15. Palacios, M.L.; Cantera, J.R. Mangrove timber use as an ecosystem service in the Colombian Pacific. *Hydrobiologia* **2017**, *803*, 345–358. [CrossRef]

16. Polanía, J.; Urrego, L.E.; Agudelo, C.M. Recent advances in understanding Colombian mangroves. *Acta Oecologica* **2015**, *63*, 82–90. [CrossRef]

17. Instituto de investigaciones marinas y costeras (INVEMAR). *Diagnóstico y Evaluación de la Calidad de Aguas Marinas y Costeras en el Caribe y Pacífico Colombianos*; INVEMAR: Santa Marta, Colombia, 2015; pp. 1–315.

18. Bolívar-Anillo, H.J.; Sánchez, H.; Fernandez, R.; Villate, D.; Anfuso, G. An Overview on Mangrove Forests Distribution in Colombia: An Ecosystem at Risk. *J. Aquat. Sci. Mar. Biol.* **2019**, *2*, 16–18.

19. Instituto de investigaciones marinas y costeras (INVEMAR). *Actualizacion y Ajuste del Diagnóstico y Zonificación de los Manglares de la Zona Costera del Departamento del Atlantico, Caribe Colombiano*; INVEMAR: Santa Marta, Colombia, 2005; pp. 1–191.

20. Corporacion Autonoma Regional del Canal del Dique. *Zonificación Definitiva de las Áreas de Manglar del Departamento de Bolívar para la Conservación y el Manejo Sostenible*; CARDIQUE: Cartagena de Indias, Colombia, 2001; pp. 1–70.

21. Corporación Autónoma Regional del Canal del Dique. *Diagnostico, Zonificación y Planificación Estrategica de las Areas de Manglar de Bolívar*; CARDIQUE: Cartagena de Indias, Colombia, 1998; pp. 1–176.

22. Andrade, C. Cambios Recientes del Nivel del Mar en Colombia. In *Deltas de Colombia: Morfodinámica y Vulnerabilidad Ante el Cambio Global*; Restrepo, J., Ed.; EAFIT Univ. Press: Medellin, Colombia, 2008; pp. 103–122.

23. Solano, J.M.; Barros Henríquez, J.A.; Roncallo Fandiño, B.; Arrieta Pico, G. Requerimientos hídricos de cuatro gramíneas de corte para uso eficiente del agua en el Caribe seco colombiano. *Cienc. y Tecnol. Agropecu.* **2015**, *15*, 83. [CrossRef]

24. Instituto de Hidrología, Meteorología y Estudios Ambientales (IDEAM). Atlas Climatológico de Colombia. Available online: http://atlas.ideam.gov.co/visorAtlasClimatologico.html (accessed on 19 July 2019).

25. Hurtado Montoya, A.F.; Mesa Sánchez, Ó.J. Cambio climatico y variabilidad espacio-temporal de la precipitación en Colombia. *Rev. EIA* **2015**, *11*, 131–150.

26. Andrade Amaya, C.A.; Barton, E.D. Sobre la existencia de una celda de circulación atmosférica sobre el Caribe y su efecto en las corrientes de Ekman del Caribe suroccidental. *Boletín Científico CIOH* **2013**, *31*, 73–94. [CrossRef]

27. Philander, S. Atlantic Ocean Equatorial Currents. In *Ocean Currents*; Steele, J., Thorpe, S., Turekian, K., Eds.; Elsevier: London, UK, 2001; pp. 188–191.

28. Instituto de Hidrología, Meteorología y Estudios Ambientales (IDEAM)-Universidad Nacional de Colombia. *La Variabilidad Climática y el Cambio Climático en Colombia*; IDEAM: Bogotá, Colombia, 2018; pp. 1–53.

29. Thomas, Y. *Climatología Marina, Presión Atmosférica, Viento y Olas para las Aguas Territoriales Bajo Jurisdicción Colombiana. 8–19 N y 69–84 W.*; Datos TOPEXPOSEIDON; Laboratorie de Géographie Physique (CNRS): París, France, 2006; p. 69.

30. Correa, I.; Morton, R. Coasts of Colombia U.S. Department of the Interior USGS, St. Petersburg, Florida, USA, 2011. Available online: http://coastal.er.usgs.gov/coasts-colombia/ (accessed on 19 July 2019).

31. Ortiz-Royero, J.C.; Otero, L.J.; Restrepo, J.C.; Ruiz, J.; Cadena, M. Cold fronts in the Colombian Caribbean Sea and their relationship to extreme wave events. *Nat. Hazards Earth Syst. Sci.* **2013**, *13*, 2797–2804. [CrossRef]

32. Instituto de investigaciones marinas y costeras (INVEMAR). *Evaluación de la Calidad Ambiental de los Manglares de La Ciénaga Mallorquín, Departamento del Atlántico*; INVEMAR: Santa Marta, Colombia, 2015; pp. 1–32.

33. Arrieta, L.; de la Rosa, J. Estructura de la comunidad íctica de la ciénaga de Mallorquín, Caribe Colombiano. *Bol. Investig. Mar. y Costeras* **2003**, *32*, 231–242. [CrossRef]

34. Páez, C. Análisis de las dimensiones del desarrollo sostenible en la ciénaga de Mallorquín. *Modul. Arquit.* **2015**, *14*, 63–84.

35. Castro-Rodríguez, E.; León-Luna, I.; Pinedo-Hernández, J. Biogeochemistry of mangrove sediments in the Swamp of Mallorquin, Colombia. *Reg. Stud. Mar. Sci.* **2018**, *17*, 38–46. [CrossRef]

36. Fuentes-Gandara, F.; Pinedo-Hernández, J.; Marrugo-Negrete, J.; Díez, S. Human health impacts of exposure to metals through extreme consumption of fish from the Colombian Caribbean Sea. *Environ. Geochem. Health* **2018**, *40*, 229–242. [CrossRef] [PubMed]

37. Portz, L.; Manzolli, R.; Andrade, C.F.; Villate Daza, D.; Bolivar, D.B.; Alcantara-Carrio, J. Assessment of heavy metals pollution (Hg, Cr, Cd, Ni) in the sediments of Mallorquin lagoon-Barranquilla, Colombia. *J. Coast. Res.* **2020**, in press.

38. Instituto Colombiano de Desarrollo Rural. *Plan de Manejo y Ordenación Pesquera del Humedal Ciénaga del Totumo*; Universidad Jorge Tadeo Lozano: Cartagena de Indias, Colombia, 2011; pp. 1–242.

39. Carvajal, A. Caracterización físico-biótica del litoral del departamento del Atlántico. In *Caracterización Físico-Biótica del Litoral Caribe Colombiano*; DIMAR-CIOH, Ed.; Dirección General Marítima: Cartagena de Indias, Colombia, 2009; pp. 97–110.

40. Orejarena Rondón, A.F.; Afanador Franco, F.; Ramos de la Hoz, I.; Conde Frías, M.; Restrepo López, J.C. Evolución morfológica de la espiga de Galerazamba, Caribe colombiano. *Boletín Científico CIOH* **2015**, 123–144. [CrossRef]

41. Anfuso, G.; Rangel-Buitrago, N.; Correa, I. Evolution of sandspits along the Caribbean coast of Colombia: Natural and human influences. In *Sand and Gravel Spits*; Randazzo, G., Jackson, D.W.T., Cooper, J.A.G., Eds.; Springer: New York, NY, USA, 2015; pp. 1–19.

42. Instituto Humboldt-Fundación Omacha. *Aplicación de Criterios Biológicos y Ecológicos para la Identificación, Caracterización y Establecimiento de Límites de Humedales en la Ventana de Estudio: Ciénaga de La Virgen*; Instituto Humboldt-Fundación Omacha: Bogotá, Colombia, 2015; pp. 1–50.

43. Agudelo, C. Estructura de los Bosques de Manglar del Departamento de Bolívar y su Relación con Algunos Parametros Abioticos. Bachelor's Degree Thesis in Marine Biology, Univeridad Jorge Tadeo Lozano, Bogotá, Colombia, 2000.

44. Maldonado, W.; Baldiris, I.; Díaz, J. Evaluación de la calidad del agua en la Ciénaga de la Virgen (Cartagena, Colombia) durante el período 2006–2010*. *Cienc. Exactas y Apl.* **2011**, *9*, 79–87.

45. Carbal Herrera, A.; Muñoz Carbal, J.; Solar Cumplido, L. Valoración económica integral de los bienes y servicios ambientales ofertados por el ecosistema de manglar ubicado en la Ciénaga de la Virgen. Cartagena-Colombia. *Sabercienc. y Lib.* **2015**, *10*, 125–146.

46. Pavithra, B.; Kalaivani, K.; Ulagapriya, K. Remote sensing techniques for mangrove mapping. *Int. J. Eng. Adv. Technol.* **2019**, *8*, 27–30.

47. Heumann, B.W. Satellite remote sensing of mangrove forests: Recent advances and future opportunities. *Prog. Phys. Geogr.* **2011**, *35*, 87–108. [CrossRef]

48. Khairuddin, B.; Yulianda, F.; Kusmana, C. Degradation Mangrove by Using Landsat 5 TM and Landsat 8 OLI Image in Mempawah Regency, West Kalimantan Province year 1989–2014. *Procedia Environ. Sci.* **2016**, *33*, 460–464. [CrossRef]

49. Lymburner, L.; Bunting, P.; Lucas, R.; Scarth, P.; Alam, I.; Phillips, C.; Ticehurst, C.; Held, A. Mapping the multi-decadal mangrove dynamics of the Australian coastline. *Remote Sens. Environ.* **2020**, *238*, 111185. [CrossRef]

50. U.S. Geological Survey. Available online: https://earthexplorer.usgs.gov (accessed on 15 July 2019).

51. Vermote, E.; Justice, C.; Claverie, M.; Franch, B. Preliminary analysis of the performance of the Landsat 8/OLI land surface reflectance product. *Remote Sens. Environ.* **2016**, *185*, 46–56. [CrossRef] [PubMed]

52. Afanador Franco, F.; Carvajal Díaz, A.F.; Franco Arias, D.A.; Orozco Quintero, F.J.; Pacheco Gómez, J.D.; Santos Barrera, Y. *Atlas Geomorfológico del Litoral Caribe Colombiano*; Dirección General Marítima: Cartagena de Indias, Colombia, 2013; pp. 1–216.

53. Instituto de investigaciones marinas y costeras (INVEMAR). *Ordenamiento Ambiental de la Zona Costera del Departamento del Atlántico*; INVEMAR: Santa Marta, Colombia, 2007; pp. 1–583.

54. Rouse, J.W.; Hass, R.H.; Schell, J.A.; Deering, D.W.; Harlan, J.C. *Monitoring The Vernal Advancement and Retrogradation (Greenwave Effect) of Natural Vegetation*; Final Report; NASA/GSFC: Greenbelt, MA, USA, 1974; pp. 1–137.

55. Thieler, E.R.; Himmelstoss, E.A.; Zichichi, J.L.; Ergul, A. *The Digital Shoreline Analysis System (DSAS) Version 4.0-an ArcGIS Extension for Calculating Shoreline Change (ver. 4.4 July 2017)*; U.S. Geological Survey Open-File Report 2008–1278; USGS: Reston, VA, USA, 2017.

56. Instituto de investigaciones marinas y costeras (INVEMAR). El Sistema de Información para la Gestión de los Manglares en Colombia SIGMA. Available online: http://sigma.invemar.org.co/ (accessed on 15 July 2019).

57. Grupo de investigación en tecnologías del agua GTA. *Análisis Sobre el Manejo Integrado del Recurso Hídrico en la Ciénaga de Mallorquín*; GTA: Barranquilla, Colombia, 2005; pp. 1–323.

58. Corporación Autónoma Regional del Atlántico. *Documentación del Estado de las Cuencas Hidrográficas en el Departamento del Atlántico*; CRA: Barranquilla, Colombia, 2007; pp. 1–114.

59. Martinez, J.O.; Pilkey, O.H.; Neal, W.J. Rapid formation of large coastal sand bodies after emplacement of Magdalena river jetties, northern Colombia. *Environ. Geol. Water Sci.* **1990**, *16*, 187–194. [CrossRef]

60. Galvis, O.; Ramiréz, F.; Vacca, V.; Herrera, O.; Bolaño, F.; Arteta, V.; Rodríguez, L.; Jiménez, M.; Vergara, A. Primera fase del programa para la rehabilitación integral de la ciénaga de Mallorquín. *Dugandia* **1989**, *1*, 10–12.

61. Ulloa-Delgado, G.; Sanchez-Paez, H.; Gil-Torres, W.; Pino-Rengifo, J.; Rodriguez-Cruz, H.; Alvarez-Leon, R. *Conservación y Uso Sostenible de los Manglares del Caribe Colombiano*; Sanchez-Paez, H., Ulloa-Delgado, G., Alvarez-León, R., Eds.; Ministerio del Medio Ambiente: Bogotá, Colombia, 1998; pp. 1–224.

62. Ivar do Sul, J.A.; Costa, M.F.; Silva-Cavalcanti, J.S.; Araújo, M.C.B. Plastic debris retention and exportation by a mangrove forest patch. *Mar. Pollut. Bull.* **2014**, *78*, 252–257. [CrossRef] [PubMed]

63. Restrepo, J.D.; Kjerfve, B. Magdalena river: Interannual variability (1975–1995) and revised water discharge and sediment load estimates. *J. Hydrol.* **2000**, *235*, 137–149. [CrossRef]

64. Corporación Andina de Fomento. *El fenomeno El Niño 1997–1998: Memoria, Retos y Soluciones: Colombia*; CAF: Caracas, Venezuela, 2000; pp. 1–242.

65. Sánchez, H.; Bolívar-Anillo, H.J.; Villate-Daza, D.; Escobar-Olaya, G.; Anfuso, G. Influencia de los impactos antrópicos sobre la evolución del bosque de manglar en Puerto Colombia (Mar Caribe colombiano). *Rev. Latinoam. Recur. Nat.* **2019**, *15*, 1–16.

66. Contraloria distrital de Barranquilla. *Informe del Estado de los Recursos Naturales y del Ambiente de Barranquilla*; Contraloria Distrital: Barranquilla, Colombia, 2018; pp. 1–49.

67. Niño, L. Los acuerdos de pesca responsable en el humedal ciénaga del Totumo (Atlántico-Bolívar). In *Proceedings of the Biodiversidad y Turismo para un Desarrollo Sostenible*; Olivero, J., de León, J., Eds.; AECID: Cartagena de Indias, Colombia, 2011; pp. 111–125.

68. Narváez, J.; Acero, A.; Blanco, J. Variación morfométrica en poblaciones naturalizadas y domesticadas de la Tilapia del Nilo Oreochromis niloticus (Teleostei: Cichlidae) en el norte de Colombia. *Rev. la Acad. Colomb. Cienc.* **2005**, *29*, 383–394.

69. Anfuso, G.; Rangel-Buitrago, N.; Correa, I. Evolution of four different sandy features along the Caribbean littoral of Colombia. In *Sand and Gravel Spits Coastal Research Library*; Randazzo, G., Jackson, D., Cooper, A., Eds.; Springer: New York, NY, USA, 2015; pp. 1–21.

70. Corporación Autónoma Regional del Canal del Dique. *Actualización de la Zonificación de Manglares en la Jurisdicción de CARDIQUE.*; CARDIQUE: Cartagena de Indias, Colombia, 2007; pp. 1–148.

71. Torregroza, E.; Gómez, A.; Borja, F. Aplicación del sistema de información geográfico quantum GIS en la regionalización ecológica de la cuenca ciénaga de la Virgen (Cartagena de Indias-Colombia). *RITI J.* **2014**, *2*, 1–13.

72. Benfield, S.L.; Guzman, H.M.; Mair, J.M. Temporal mangrove dynamics in relation to coastal development in Pacific Panama. *J. Environ. Manage.* **2005**, *76*, 263–276. [CrossRef]

73. Moor, R.; van Maren, M.; van Laarhoven, C. A controlled stable tidal inlet at Cartagena de Indias, Colombia. *Terra Aqua* **2002**, *88*, 3–14.

74. Vélez, S. Estudio de caso: Concesión vial Cartagena-Barranquilla y Circunvalar de la Prosperidad. Análisis de política pública. *Derecho y Econ.* **2018**, *49*, 173–207.

75. Autoridad Nacional de Licencias Ambientales. *Resolución 1290 de 2015*; ANLA: Bogotá, Colombia, 2015; pp. 1–171.

76. Instituto Humboldt-Fundación Omacha. *Caracterización Biológica y Ecológica de las Comunidades de Plantas Acuáticas, Plantas Terrrestres y Macroinvertebrados, y Caracterización Físico-Química de las Aguas de la Ventana de Estudio de la Ciénaga de la Virgen*; Instituto Humboldt-Fundación Omacha: Bogotá, Colombia, 2015; pp. 1–109.

77. Mira, J.D.; Urrego, L.E.; Monsalve, K. Determinantes naturales y antrópicos de la distribución, estructura y composición florística de los manglares de la Reserva Natural Sanguaré, Colombia. *Rev. Biol. Trop.* **2019**, *67*, 810–824. [CrossRef]

78. Ball, M.C. Interactive effects of salinity and irradiance on growth: Implications for mangrove forest structure along salinity gradients. *Trees-Struct. Funct.* **2002**, *16*, 126–139. [CrossRef]

79. Sobrado, M.A.; Ewe, S.M.L. Ecophysiological characteristics of *Avicennia germinans* and *Laguncularia racemosa* coexisting in a scrub mangrove forest at the Indian River Lagoon, Florida. *Trees-Struct. Funct.* **2006**, *20*, 679–687. [CrossRef]

80. Mckee, K.; Kerrylee, R.; Saintilan, N. Response of salt marsh and mangrove wetlands to changes in atmospheric response of salt marsh and mangrove wetlands to changes in atmospheric CO_2, climate, and sea level. In *Global Change and the Function and Distribution of Wetlands*; Middleton, B., Ed.; Springer: Dordrecht, The Netherlands, 2012; pp. 63–96.

81. Cardona, P.; Botero, L. Soil characteristics and vegetation structure in a heavily deteriorated mangrove forest in the Caribbean Coast of Colombia. *Biotropica* **1998**, *30*, 24–34. [CrossRef]

82. Krauss, K.W.; Ball, M.C. On the halophytic nature of mangroves. *Trees-Struct. Funct.* **2013**, *27*, 7–11. [CrossRef]

83. Wang, W.; Yan, Z.; You, S.; Zhang, Y.; Chen, L.; Lin, G. Mangroves: Obligate or facultative halophytes? A review. *Trees-Struct. Funct.* **2011**, *25*, 953–963. [CrossRef]

84. Tovilla-Hernández, C.; Espino, G.; Orihuela-Belmonte, D. Impact of logging on a mangrove swamp in South Mexico: Cost/benefit. *Rev. Biol. Trop.* **2014**, *49*, 571–580.

85. Blanco, J.F.; Estrada, E.A.; Ortiz, L.F.; Urrego, L.E. Ecosystem-Wide Impacts of Deforestation in Mangroves: The Urabá Gulf (Colombian Caribbean) Case Study. *ISRN Ecol.* **2012**, *2012*, 958709. [CrossRef]

86. Walters, B.B. Ecological effects of small-scale cutting of Philippine mangrove forests. *For. Ecol. Manage.* **2005**, *206*, 331–348. [CrossRef]

87. Sippo, J.Z.; Lovelock, C.E.; Santos, I.R.; Sanders, C.J.; Maher, D.T. Mangrove mortality in a changing climate: An overview. *Estuar. Coast. Shelf Sci.* **2018**, *215*, 241–249. [CrossRef]

Article

Dune Systems' Characterization and Evolution in the Andalusia Mediterranean Coast (Spain)

Rosa Molina [1], Giorgio Manno [2,*], Carlo Lo Re [2] and Giorgio Anfuso [1]

[1] Department of Earth Sciences, Faculty of Marine and Environmental Sciences, University of Cádiz, Polígono Río San Pedro s/n, 11510 Puerto Real, Spain; r.molina.gil@gmail.com (R.M.); giorgio.anfuso@uca.es (G.A.)

[2] Department of Engineering, University of Palermo, Viale delle Scienze, Bd. 8, 90128 Palermo, Italy; carlo.lore@unipa.it

*   Correspondence: giorgio.manno@unipa.it; Tel.: +39-091-23896524

Received: 26 May 2020; Accepted: 21 July 2020; Published: 23 July 2020

Abstract: This paper deals with the characterization and evolution of dune systems along the Mediterranean coast of Andalusia, in the South of Spain, a first step to assess their relevant value in coastal flood protection and in the determination of sound management strategies to protect such valuable ecological systems. Different dune types were mapped as well as dune toe position and fragmentation, which favors dune sensitivity to storms' impacts, and human occupation and evolution from 1977 to 2001 and from 2001 to 2016. Within a GIS (Geographic Information System) project, 53 dune systems were mapped that summed a total length of ca. 106 km in 1977, differentiating three dune environments: (i) Embryo and mobile dunes (Type I), (ii) grass-fixed dunes (Type II) and (iii) stabilized dunes (Type III). A general decrease in dunes' surfaces was recorded in the 1977–2001 period (-7.5×10^6 m^2), especially in Málaga and Almería provinces, and linked to dunes' fragmentation and the increase of anthropic occupation ($+2.3 \times 10^6$ m^2). During the 2001–2016 period, smaller changes in the level of fragmentation and in dunes' surfaces were observed. An increase of dunes' surfaces was only observed on stable or accreting beaches, both in natural and anthropic areas (usually updrift of ports).

Keywords: dune characterization; anthropic occupation; fragmentation index; dune surface

---

1. Introduction

Human interest in coastal processes and evolution has greatly increased in recent decades due to the increment of human developments recorded in coastal areas [1] and the impacts of extreme events, such as hurricanes and storms [2,3], the effects of which are enhanced by sea level rise and other climatic change-related processes, such as the increasing height of extreme waves, or changes in the tracks, frequency and intensity of storms [4–7]. Coastal development, which is essentially linked to tourism—one of the world's largest industries [8]—continues to increase, and some 50% of the world's coastline is currently under pressure from excessive development [1]. In Europe, the rapid expansion of urban artificial surfaces in coastal zones during the 1990–2000 period [9] has occurred in the Mediterranean and South Atlantic areas, namely Portugal (34% increase) and Spain (18%), followed by France, Italy and Greece.

Activities and infrastructures related to tourism and other human developments too (e.g., fishing and industrial activities) are significantly affected by the impacts of storms and hurricanes that, over the past century, have caused huge economic losses along with high mortality rates along the world's coastlines [10–13]. Coastal erosion and flooding processes have reduced beach and dune ridges' width and produced the loss of associated touristic, aesthetic and natural values [14–17]. Beach erosion/accretion cycles are often recorded at an inter-annual time scale and are related to seasonal wave climate variations due to temporal and spatial distributions of high-latitude storms and hurricanes/typhoons [3,18–22]. Erosion is observed after storm events, at high latitudes recorded during winter months, but beach recovery takes place during fair weather conditions, which is known as "seasonal" beach behavior [10,23], and in general happens at longer time intervals, from weeks to months [24,25]. Hence, natural beach recovery guarantees the reformation of a wide beach and its associated protection function and touristic use, but dune response to erosive events is very different. Meanwhile, dune erosion is always very rapid and time located accretion is a process that usually takes place with low rates over a long time, from several months to years in the Andalusia Atlantic littoral [26]. Hence, the determination of coastal dune characteristics, behavior and evolution need special attention in the attempt to reduce erosive/flooding processes' impacts on natural and urbanized coasts. Several recent investigations [27,28] have identified dunes as one of the most relevant coastal ecosystems as a natural defense able to reduce flood sensitivity/vulnerability and hence, dunes' maintenance/emplacement has been considered as an effective coastal protection measure included among possible "Disaster Risk Reduction" (DRR) strategies in several European directives [28–32]. In fact, dune ridges protect large sections of low-lying coasts against flooding during extreme storms [9,33,34], and hence, lateral dune continuity and level of fragmentation are extremely relevant [35–38].

This paper is the first one that deals with the characterization and evolution of all dune systems along the Mediterranean coast of Andalusia (South of Spain), and this is a first step to assess their great ecological value, sensitivity and relevance in coastal flood protection. Different dune types have been mapped as well as their level of fragmentation (by means of a new index proposed in this paper) and present human occupation and evolution from 1977 to 2016. Results obtained are of relevance to enhance the general database on dune characteristics along the Mediterranean coast of Andalusia and the possibility of use ecosystem-based solutions in coastal protection along with, or instead of, traditional engineering approaches [27,28].

2. Study Area

Located in South Spain, the Mediterranean coast of Andalusia administratively belongs to the provinces of Cádiz, Málaga, Granada and Almería (Figure 1). It has a prevailing rectilinear E-W outline, with two NE-SW easterly facing sectors, i.e., the Gibraltar Strait and the Almería easternmost coastal sector (Figure 1). It is a micro-tidal environment (tidal range < 20 cm) with a total length of ca. 546 km, including rocky sectors (ca. 195 km) and intermediate to reflective beaches (ca. 350 km) [39], usually composed of medium to coarse dark sands and/or pebbles. Dune systems, which have a total length of ca. 76 km, are especially observed along the provinces of Cádiz and Almería (Figures 1 and 2) [25,40,41].

The Betic Range, a tectonically active mountain chain that, at places, reaches relevant elevations higher than 2200 m above sea level (m a.s.l.) close to the coast, determines coastal orography and morphology, forming cliffs, embayments and promontories. Several small coastal plains are especially extended at the mouth of short rivers and *ramblas* (seasonal streams) draining the chain, the most important being the Guadiaro, Guadalhorce, Guadalfeo, Adra and Andarax rivers (Figures 1 and 2). In the past decades, river basin regulation plans devoted to water management for tourist and agricultural purposes have enhanced the construction of dams and reservoirs that have reduced sediment supplies to the coast and have promoted coastal retreat in most deltas [39,42–44].

Large coastal towns are Málaga (>500,000 inhabitants), Almería (ca. 200,000 inhabitants) and the tourist towns along the western part of Costa del Sol area, namely Marbella (150,000 inhabitants), Fuengirola (80,000) and Torremolinos (70,000). Málaga is the province that has experienced the most important coastal occupation, in particular due to the construction of structures related to national and international tourism [45]. Main commercial ports are located at Almería, Algeciras, Cádiz and Málaga, and several marinas at Costa del Sol [46–48].

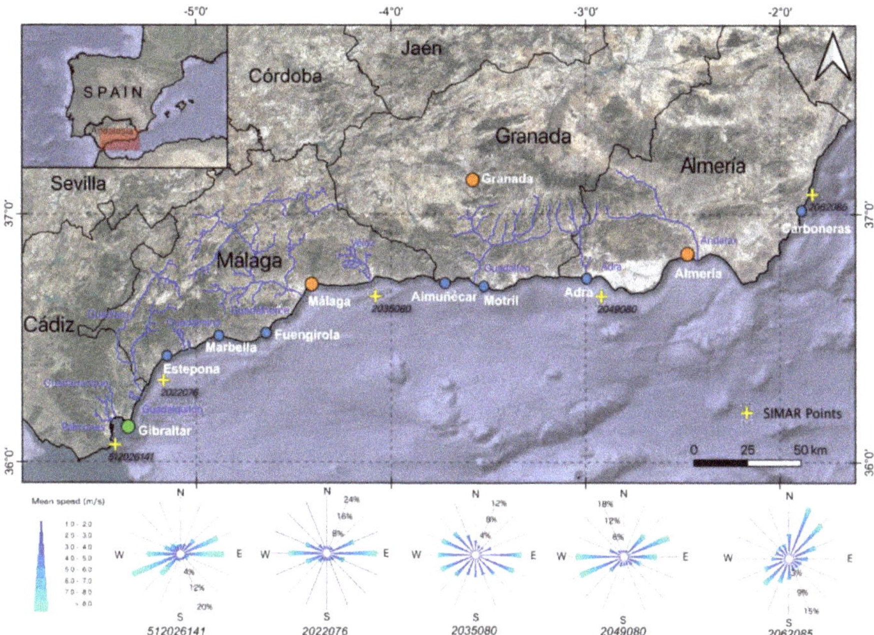

Figure 1. Location of the study area and average wind speed roses at 5 points of SIMAR (SImulación MARina), from the wave reanalysis model by Puertos del Estado (PdE).

Concerning weather characteristics, the provinces of Cádiz, Málaga and Granada have a Mediterranean climate with sub-tropical characteristics, with coastal orientation and the Betic Chain favoring average annual temperature of ca. 13 °C and, in July and August, the average is 19 °C. Annual rainfall ranges from 400 to 900 mm, with the most abundant values observed at Gibraltar Strait. The province of Almería presents a Mediterranean climate with sub-desert characteristics, i.e., rainfall is extremely limited (ca. 200 mm/year), average annual temperature is 21 °C and in July–August, temperature is 26 °C [49].

The coast is generally exposed to winds blowing from E to W and from NNE to SW in the easternmost part of Andalusia (i.e., at Carboneras, Figure 1), with minimum and maximum velocities ranging from 0.4 to 9.0 m/s [50]. The wave climate and storm energy are very variable [13,50]: the coast of Cádiz province is mainly affected by eastern storms, Málaga, Granada and (partially) Almería provinces are exposed both to western and eastern storms, whereas the easternmost portion of the coast of Almería province is primarily exposed to eastern storms [13,50].

The mean duration of storm events ranges from 0.9 to 7.0 days, despite their intensity. Waves show a clear seasonal behavior with storm conditions being recorded during November–March, i.e., the winter season [42,50,51], with mean values of significant wave height that reaches 5.18 m in extreme storm conditions [50]. A storm characterization for the studied area was developed by Molina et al. [50], using the Energy Flux parameter to classify storm events into five classes, from weak (Class I) to extreme (Class V). They observed that the most energetic area was the central part of the Mediterranean Andalusian coast, i.e., the coast between Málaga and Almería provinces, highly exposed to storms belonging to all classes, and specially to most energetic ones that can have a great impact on both natural and urbanized sectors [50].

Due to shoreline orientation, predominant easterly winds (Figure 1) and associated storm waves give rise to sea wave conditions generating a prevailing westward littoral drift [51]; meanwhile, an opposing drift is particularly important in certain coastal sectors and/or periods [42,50].

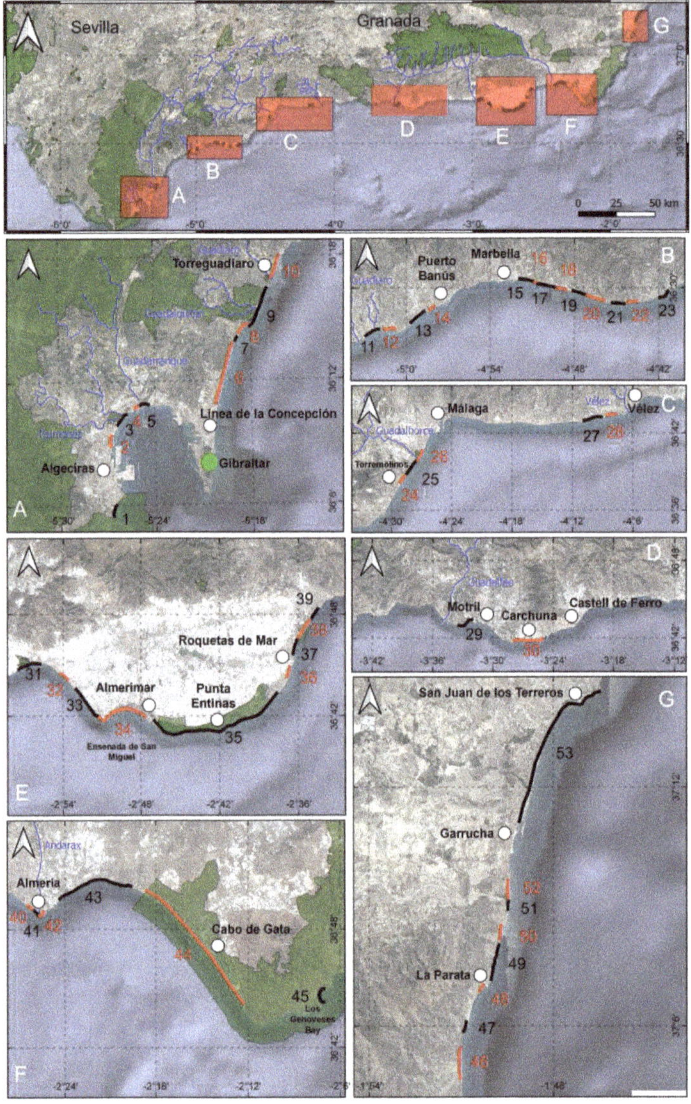

Figure 2. Location of the studied dune systems. Natural protected areas are in green. The capital letters A–G in the main subplot refer to the zoomed areas in the other subplots.

3. Materials and Methods

Aerial orthophotographs dated 1977, 2001 and 2016 (Table 1) were used to map different dune systems, and to quantify their surface evolution, level of fragmentation and the progression of human occupation. Aerial orthophotographs were obtained from the Web Map Services (WMS) of the Open Geospatial Consortium (OGC) of the Andalusia Regional Administration [52].

Table 1. Aerial orthophotos used [52], 2001 and 2016 orthophotos are from Plan Nacional de Ortofotografía Aérea (PNOA).

Year	Scale	Spatial Resolution(m)	Flight
1977	1:5000	0.5	Iryda flight 1977
2001	1:10,000	0.5	PNOA 2001
2016	1:5000	0.25	PNOA 2016

Orthophotographs were elaborated within a GIS project (reference system WGS84–UTM 30N) by means of the ArcMap application from ArcGIS Desktop, Release 10 Redlands, CA: Environmental Systems Research Institute. All dune systems with a minimum longshore seaward front of 100 m in length were mapped, summing a total of ca. 106 km in 1977, including 53 systems (Figure 2). Within each system, three dune environments were mapped according to Sanjaume Saumel and Gracia Prieto [25], who defined, on the base of the most important coastal dune habitats in Spain [53], i.e., (i) embryo and mobile dunes (Type I), (ii) grass-fixed dunes (Type II) and (iii) stabilized dunes (Type III). Coastal dune habitats described at the study area corresponded to the Sites of Community Importance (SCI's) of the European Commission Habitat Directive listed in Table 2.

Table 2. Sites of Community Importance (SCI's) described in the study area and their correspondence with the dune typologies mapped in this work [54].

Sites of Community Importance (SCI's)	Classification
2110—Embryonic shifting dunes 2120—Shifting dunes along the shoreline with *Ammophila arenaria* ("white dunes")	I—Embryo and mobile dunes
2130—Fixed coastal dunes with herbaceous vegetation ("grey dunes") 2150—Atlantic decalcified fixed dunes 2210—*Crucianellion maritimae* fixed beach dunes 2230—*Malcolmietalia* dune grasslands 2240—*Brachyopodietalia* dune grasslands with annuals	II—Grass-fixed dunes
2250—Coastal dunes with *Juniperus* spp. 2260—*Cisto-Lavanduletalia* dune sclerophyllous scrubs 2270—Wooded dunes with *Pinus pinea* and/or *Pinus pinaster*	III—Stabilized dunes

The first group (Type I) comprises embryo and mobile dunes, which are the first band of colonizing vegetation and the first important continuous sandy relief. The second group (Type II) comprises grass-fixed dunes, which develop in a more stable soil and form a more continuous plant cover based on lawns or even some woody plants and bushes. The third group (Type III) comprises the stabilized dunes, is the innermost band of the dune system, and is made up of fully fixed vegetated dunes, with structured and stabilized soils. Its vegetation evolves into forests and a dense and diverse vegetation cover is developed.

The main characteristics used to distinguish between each dune type was the color and vegetation density, so that as systems evolve, color darkens and plant density increases, i.e., embryo and mobile dunes are often called "white" or "yellow" dunes and fixed dunes are called "grey" dunes because of their characteristic color. Díez-Garretas et al. [55], in their study on spatio-temporal changes of coastal ecosystems in Southern Iberian Peninsula (Spain), used a similar classification, taking into account the phytosociological plant communities present at the location studied and the habitat code. They related the habitat code with the ecological units present in their study, including mobile dunes, semi-fixed dunes and stable dunes. Pintó et al. [56] recognized the distinct habitats present in their study area and related them to the sea-to-land ecological gradient and the Habitats of Community Interest. Their classification is more detailed, attending more to morphological than ecological criteria.

The position, evolution and fragmentation of the dune toe position was also reconstructed, and the latter aspect favors dune sensitivity to storms' impacts [40,57–59]. Further, the total surface of each one of the 53 system and dunes' surfaces occupied by human structures/interventions was calculated.

The proxy used to map the dune toe was the seaward dune vegetation line, manually detected by a GIS operator [13,39,48,60–62]. To calculate dune fragmentation, a database was obtained for each dune system containing three shape files: the first file included a polyline of the total length of the dune toe line, the second file included the length of all breaks observed along the dune toe line of each system, and a third file, which was the result of the differences between the two previous shape files. The level of fragmentation was calculated by determining the ratio between the length of all breaks and the whole dune toe length at each dune system and year. These values were normalized according to a constant length of 100 m by dividing the total length of all breaks in the shorefront dune toe ("l") by the entire length of the dune toe ("L"):

$$F = \frac{l}{L} \tag{1}$$

The F Index is a new index proposed for the first time in this paper vaguely based on the coefficient of infrastructural impact "K" [63]. It was applied along unitary coastal sectors of 100 m in length in order to reduce the importance of dune seaward length. The F Index was calculated for the systems present in all investigated periods (37 out of 53 systems), that is, the dune systems that disappeared in the second period were not taken into account to avoid interpreting a decrease in fragmentation when, in the reality, the entire system was lost. Values of the F Index used to express the fragmentation level were classified into three classes using the Natural Breaks Function [64], from Class 1 ("Null or very low fragmentation", 0.00 < F < 0.06), Class 2 ("Medium fragmentation", 0.06 < F < 0.16) to Class 3 ("High fragmentation", 0.16 < F < 0.41).

4. Results

4.1. Dune Systems' Distribution and Evolution

Of the 53 dune systems investigated in the Mediterranean coast of Andalusia, 10 belonged to natural protected areas and, from an administrative point of view, 10 were located in Cádiz province, 18 in Málaga, 2 in Granada and 23 in Almería province (Figure 2, Appendix A Table A1). In Cádiz province, dune systems were equally divided between the Algeciras Bay and an exposed, rectilinear shoreline, including both natural and urbanized areas (Figure 2A). In Málaga province, they were mainly located at the westernmost part of the littoral (Figure 2B), and south of the Guadalhorce river mouth and west of the Vélez river delta (Figure 2C). In Granada province, only 2 dune systems were observed, located in an area updrift of the port of Motril and at Carchuna (Figure 2C). In Almería province, 5 systems were located close to delta areas, namely at Adra and, especially, at Andarax river delta (Figure 2E,F), and 8 were located at rectilinear coastal sectors limited by ports, promontories or river deltas (Figure 2E–G). Very developed dune systems were located in the relevant protected area of Punta Entinas-El Sabinar (Figure 2E); meanwhile, several systems were located at the easternmost area of Almería province and the most relevant system was observed in a large pocket beach (Los Genoveses) (Figure 2F,G).

A total of 15 dune systems disappeared from 1977 to 2016, 7 of them located in Málaga, 7 in Almería and 1 in Cádiz provinces. Dune systems' extension was changing during the periods studied without a clear trend; meanwhile, a clear decrease in size was evident for the three largest dune systems (Appendix A Table A2). Surfaces of "Embryo and mobile dunes" (Type I), "Grass-fixed dunes" (Type II) and "Stabilized dunes" (Type III) were calculated within each one of the 53 dune systems and per each time span considered. The progressive decrease of typologies I and II was observed, meanwhile, "Stabilized dunes" (Type III) recorded a decrease from 1977 to 2001 and a slight increase from 2001 to 2016 (Figure 3).

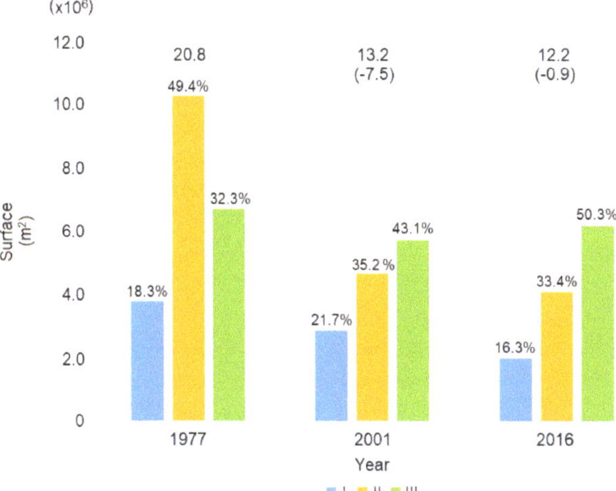

Figure 3. Yearly surface values of each dune typology, i.e., "Embryo and mobile dunes" (Type I), "Grass-fixed dunes" (Type II) and "Stabilized dunes" (Type III). Value on top represents the surface of all dune typologies and in brackets the surface that was lost with respect to the previous year is reported.

The distribution of the different dune typologies within each dune system varied during the studied period (Figure 4). At all provinces (but Cádiz), a reduction of all dune typologies was recorded during the 1977–2001 period; meanwhile, a decrease in all provinces (but Almería) of types I and III and an increase of Type II was recorded in the 2001–2016 period (Figure 4).

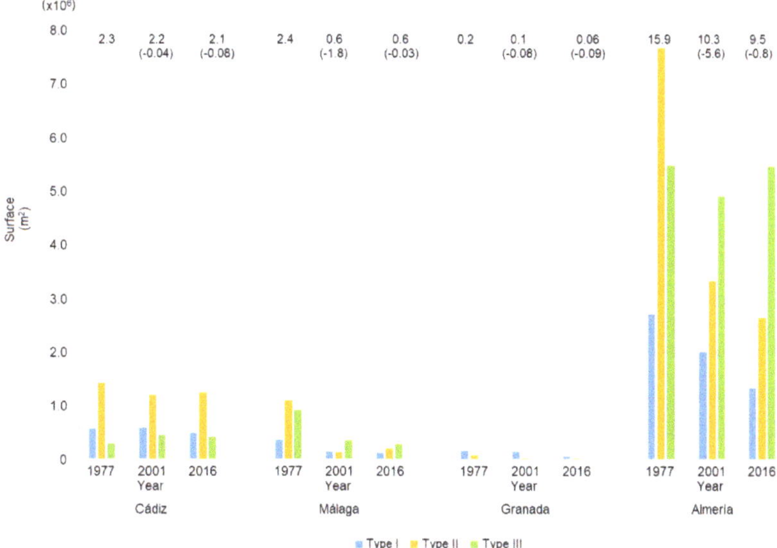

Figure 4. Distribution of dune surface typologies, i.e., "Embryo and mobile dunes" (Type I), "Grass-fixed dunes" (Type II) and "Stabilized dunes" (Type III), by province during the studied periods. The value on top represents the sum of dune surfaces (in m^2) and in brackets the value of lost dune surface with respect to the previous year is presented.

The comparison of the total amount of eroded/accreted surfaces recorded in each system for the periods 1977–2001, 2001–2016 and 1977–2016 showed a clear negative balance for 49 systems and a positive one for 4 systems (Figure 3). The dune systems that showed a positive balance were located at Playa del Rinconcillo (System no. 2, 38,884.4 m^2), at the Guadarranque (no. 5, 19,262.8 m^2) and at the Guadalquitón (no. 9, 260,531.7 m^2) rivers' mouths in Cádiz province (Figure 2A), and in the Albufera de Adra (no. 31, 6708.5 m^2), a natural protected area at the Adra river delta (Almería province, Figures 1 and 2E). The most eroded dune systems were located at Ensenada de San Miguel (System no. 34, −557,765.0 m^2), Punta Entinas–El Sabinar (no. 35, −4,166,157.9 m^2) and Vera (no. 53, −567,841.2 m^2) areas, in Almería province.

Comparing the evolution of each system in the 1977–2001 and 2001–2016 periods, it was observed that 3 systems recorded accretion and 23 erosion in both periods, and the others showed different behaviors. Regarding the distribution of these records, the 3 accreting dune systems were located in Cádiz province and most of the dune systems that recorded erosion or disappeared were located in Málaga province, and dunes in Granada province presented erosion for both periods. In Almería province, the systems located in Almería Bay and at the easternmost part of the province presented a negative trend for both periods, and the group located from the Adra river delta to Ensenada de San Miguel presented erosion and then accretion. The two dune systems located at Roquetas de Mar disappeared in 2001–2016 (Figures 2E and 5).

Figure 5. Dune systems urbanized in Málaga and Almería provinces. System no. 14 Playa Nueva Andalucía in 1977 (**A**) and 2016 (**B**), and System no. 36 Playa de Roquetas in 1977 (**C**) and 2016 (**D**).

4.2. Anthropic Occupation Evolution

Surfaces occupied by human structures/interventions were calculated within each one and per each year of the 53 dune systems (Figure 6). The greatest increase (ca. 2.3×10^6 m^2) was observed in the 1977–2001 period.

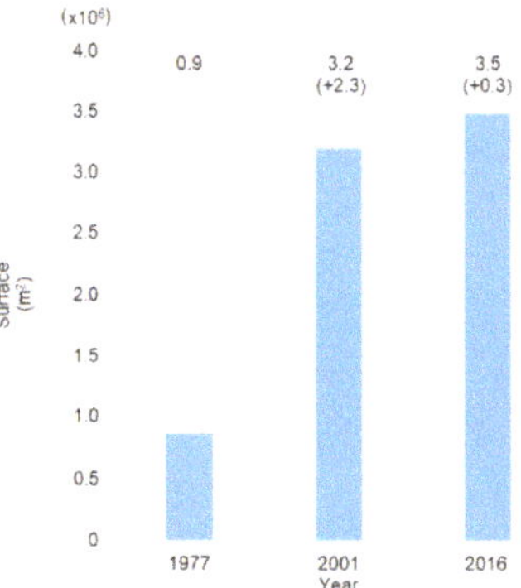

Figure 6. Evolution of the surfaces occupied by anthropic interventions during the studied periods. The number on top of histograms represents the total value of surface occupied, and in brackets shows the increment recorded among successive periods.

Comparing the evolution of human occupation of each system in the 1977–2001 and 2001–2016 periods, it was observed that 21 out of 53 systems presented an increase of human occupation in both periods and 2 systems a decrease due to the removal of small installations (Figure 7). Further, 4 systems recorded an increase in the first period and a decrease in the second due to the urbanization of a part of the dune system, and the removal of small installations, and the opposite was true for 1 system due to coastal erosion problems since the shoreline retreatment forced the removal of human structures (Figure 7).

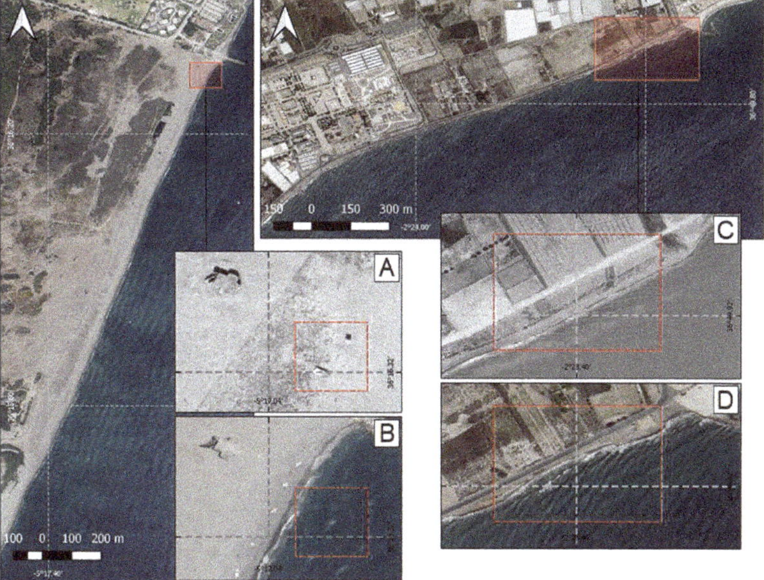

Figure 7. Removal of small constructions in Cádiz and croplands in Almería provinces. System no. 9 Guadalquitón 1977 (**A**), 2016 (**B**), with a decrease of the occupation area of 112.90 m² in 1977–2001 and 263.31 m² in 2001–2016. System no. 43 Las Algaidas-Las Marinas in 1977 (**C**) and 2016 (**D**), with a decrease of 117.14 m² in 1977–2001 and 184.07 m² in 2001–2016.

At places in System no. 2 Playa del Rinconcillo (in Cádiz province), the increase of dune surface and anthropic occupation was linked to the formation of a new beach at the northern side of the port of Algeciras (Figure 8). Summing up, the decrease of occupation was essentially due to the removal of buildings and was always very small.

Figure 8. System no. 2 Playa del Rinconcillo registered an increase of dune surface of 732.26 m^2 in 2001–2016. (**A**) 1977, (**B**) 2001 and (**C**) 2016.

4.3. Dune Fragmentation

Analysis of the dune toe fragmentation was carried out for such systems (37 out of 53) that were observed in all investigated periods and a general increase of fragmentation was evident (Figure 9), confirming the trend observed for the evolution of human occupation.

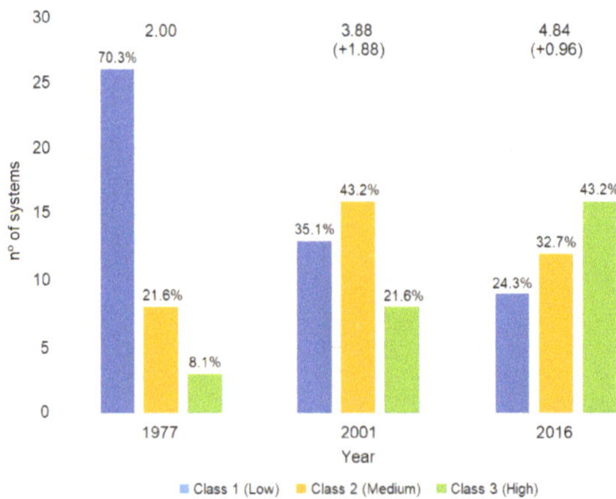

Figure 9. Fragmentation Index. The number on top of histograms represents the total value of fragmentation per year and in brackets is the increment recorded among successive periods.

Considering the 1977–2016 timespan, 23 dune systems presented an increase of fragmentation, 3 systems recorded a decrease (they were located in erosive coastal sectors within natural protected areas) and 11 presented no variations. The two systems that recorded a major increment of fragmentation were Punta del Río (no. 41), in Almería province, and Playa de las Chapas (no. 20), in Málaga province, with an increase of +0.25 and +0.22, respectively. Conversely, the two systems that recorded the major decrease of fragmentation were Playa de Río Real (no. 16) and Playa de la Misericordia (no. 26) in Málaga province, with a decrease of −0.20 and −0.09, respectively. Comparing the evolution of fragmentation at each dune system in the 1977–2001 and 2001–2016 periods, it was observed that only 7 out of 23 presented an increase of the fragmentation at both periods, in general due to coastal zone urbanization (Figure 10A–C). Only 1 dune system showed a decrease of fragmentation (no. 16) (Figure 10D–F), and 8 presented no variation in both periods.

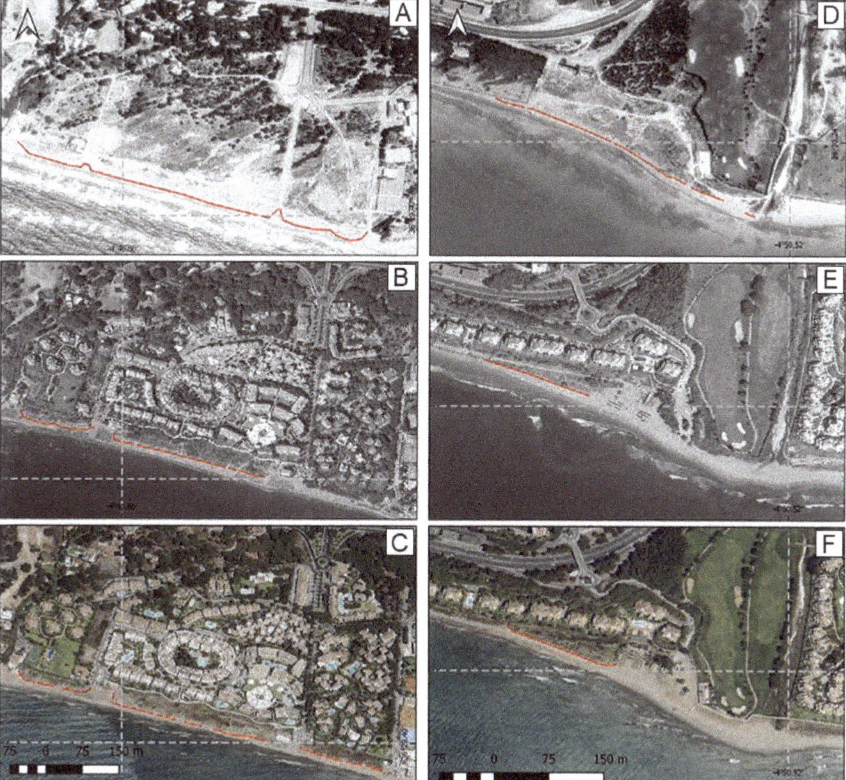

Figure 10. System no. 20 Playa de las Chapas, in Málaga province, recorded an increase of fragmentation due to the increment of anthropic pressure, changing from (**A**) Class 1 (F = 0.003) in 1977 to (**B**) Class 2 (F = 0.12) in 2001 and to (**C**) Class 3 (F = 0.22) in 2016. System no. 16 Playa de Río Real, in Málaga province, where the modification of the coastal zone and the emplacement of a touristic urbanization produced the destruction of the already fragmented dune system. Remnant dune systems presented a lower Fragmentation Index, changing from (**D**) Class 3 (F = 0.20) in 1977 to (**E**) Class 2 (F = 0.15) in 2001 and to (**F**) Class 1 (F = 0.0) in 2016. Red line represents dune toe position.

Other dune systems presented a different behavior at both periods: 4 systems recorded an increase in the first period and a decrease in the second and the opposite was true for 1 system (Figure 11). In general, the increase in fragmentation occurred along with the increase of urbanization and anthropic pressure, while the opposite was observed in natural protected areas. At places where systems were already fragmented, their erosion implied a reduction in their fragmentation since: (i) very fragmented sectors often disappeared and the remaining ones presented low fragmentation (Figure 10D–F and Figure 11A,B) and (ii) coastal erosion produced the loss of the most fragmented part of dune toe (Figure 11B,C). An increase of fragmentation in 7 dune systems was due to the increment of erosion processes and/or the formation of pedestrian pathways.

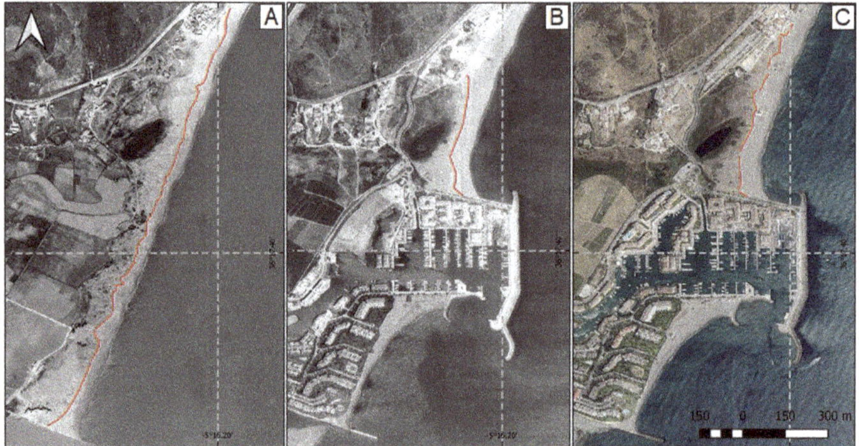

Figure 11. System no. 10 Torreguadiaro (in Cádiz province) recorded a slight decrease of fragmentation in the 1977–2001 period and an increase in the 2001–2016 period due to the reduction of a high fragmented dune sector that was replaced by the port structure. The dune system located in front of the natural protected Torreguadiaro Lagoon was not modified. In the 2001–2016 period, the growth of new dunes at the northern part of the system and the increase of pathways at both sides of it was observed. This resulted in a decrease of fragmentation that changed from (**A**) Class 1 (F = 0.08) in 1977 to (**B**) Class 2 (F = 0.03) in 2001 and an increase of fragmentation to (**C**) Class 3 (F = 0.20) in 2016. Red line represents dune toe position.

5. Discussion

5.1. Erosional Dune Systems

Erosion or complete disappearance of dune systems can be produced by human activities or natural processes [65–67]. Anthropic impacts were related to: (i) urban developments, mainly due to the coastal tourist demand, and the associated opening of pathways on dune ridges, which was especially evident in Málaga province (Figure 5) [41,46,48,55,68], (ii) dunes' occupation due to the demand for agricultural uses, as observed at different locations in Andalusia, and reported by References [67,69–73] in other Mediterranean Spanish areas (in Catalonia, [56]) or on the Mediterranean coast of Morocco [74] and (iii) the decrease of sediments' inputs to coastal environments due to the construction of ports and harbors, as observed along the study area by Malvárez et al. [46] and Manno et al. [48], and the reduction of the sedimentary load of rivers due to the construction of dams in river basins, especially in Málaga and Almería provinces [43,46,75], also observed in other Mediterranean rivers, e.g., for the Ebro [76] and the Arno [77] rivers.

Among natural processes, there are the impacts of chronic erosion processes and of extreme storms, the impacts of which are often enhanced by climatic change-related processes, e.g., an increase of storm intensity and frequency and Sea Level Rise [11,35,37,67,78–83]. Specifically, for the studied area, storm characterization was described by Guisado et al. [42] and Molina et al. [50]; meanwhile, it seems that Sea Level Rise is not relevant at the studied area [84–86].

Of the 53 dune systems studied, all but 4 recorded a reduction of their surface, or even disappeared, and this was especially evident where the systems were affected by hard human interventions [41,55,68–73] and, secondarily, by shoreline erosion [71,87,88]. The greatest loss of dune surface was recorded in the 1977–2001 period due to the massive urban occupation of coastal areas, although in the 2001–2016 period, a decrease in the loss of dunes' surfaces was observed because the main causes of their destruction recorded in the previous period partially ceased. Cases of disappearance due to urban occupation were still observed, especially in Málaga province [41], but the anthropic pressure derived from the tourist use of beaches and the decrease in river contributions were not so evident as in the 1977–2001 period [46,68–73].

The loss of dune surface was at places and times linked to the progressive fragmentation of the dune toe (i.e., the increase of dune discontinuity), which is a factor that has to be considered in order to estimate coastal and dune vulnerability [40,57–59] since a fragmented dune system is more vulnerable to temporary flooding and hence, it is less effective against storm surges [35,40,58,59,89–91]. In this study, the most fragmented (and hence susceptible sectors) were observed at the west side of the Andarax river delta in Almería province (no. 41, Figures 1 and 2), which was the most fragmented dune system located in a natural area (Appendix A Table A2) and the system at Las Chapas beach in Málaga province (no. 20, Figures 1, 2 and 10), located in a strongly developed urban area.

At almost all sectors, dunes' fragmentation was mainly due to the opening of pathways and to their progressive expansion due to marine- and wind-induced erosion processes, as also observed by Gracia et al. [40], Pintó et al. [56], Rangel-Buitrago and Anfuso [58] and Rizzo et al. [59]. Due to the accuracy of the orthophotos used in this study, dune discontinuities caused by overwash processes were only detected at few places (Figure 12). Such processes were distinguished from other types of fragmentation due to the absence of vegetation at the areas presenting the characteristic shape of a washover fan; meanwhile, pathways showed narrow rectilinear shapes.

Summing up, the majority of the dune systems that showed an aerial decrease were affected by anthropic factors, highlighting the importance of urban and agricultural occupations that were very relevant in Málaga and Almería administrative districts.

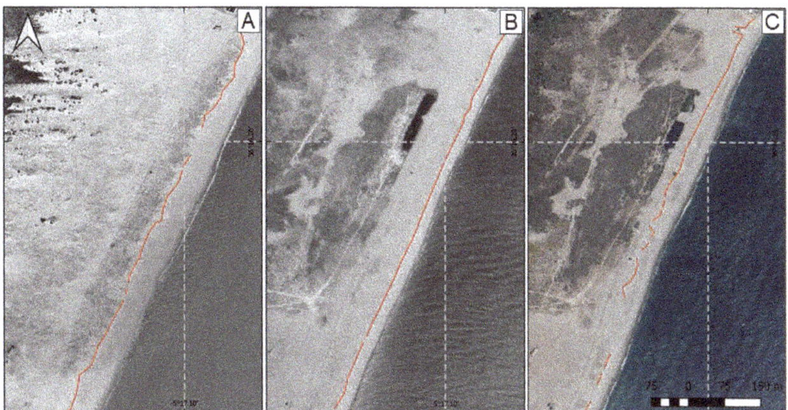

Figure 12. Details of the central area of System no. 9 Guadalquitón (Cádiz province), where washover fans and pathways were observed. Red line represents dune toe position. (**A**) 1977, (**B**) 2001 and (**C**) 2016.

5.2. Accretional Dune Systems

Conditions for dune formation and development were discussed by a large number of authors who agree that the temporal variation of the sedimentary contribution and the wind regimes are the most important factors controlling the beach–dune system relationship [65,92–95].

The increase of systems located in the Bay of Algeciras (Figure 2A) was associated with the sedimentation processes recorded in such beaches [96] that receive the sediment supplies of the Palmones and Guadarranque rivers [73]. Such beaches are located next to two large coastal protection structures that promote sedimentation processes. In the case of the system observed at El Rinconcillo in 2001, it began to form after the expansion of the port of Algeciras (Figure 8). Instead, systems at Guadalquitón and Albufera de Adra (Figure 2) were located in areas that registered an important erosion [39,96] and a significant human occupation linked to urban development in the case of Guadalquitón and intense agricultural occupation in Albufera de Adra. The Guadalquitón dune system recorded the highest increase during the 1977–2001 period, and it was due to the degradation of the vegetation that facilitated the inland dune migration. The formation of large mobile dunes in this area was also due to strong east winds (Figure 1), especially on the east-facing beaches [73]. In the case of the Albufera de Adra system, an important loss of dune surface in the 1977–2001 period was caused by shoreline erosion and the significant anthropic pressure (intense agricultural activities) [39,43,96]; however, the sedimentation produced at the north side of the system [39,96] supported the development of mobile dunes.

5.3. Dune Types' Evolution

Several dune systems studied in this paper were described by different authors [41,55,68–73,87,88], but none of them provided a description of all dune systems along the whole Mediterranean coast of Andalusia.

Unlike tidal-influenced coasts, in which the sedimentary contribution can be obtained through periodic exposure of the intertidal plain, on micro-tidal coasts such as the Mediterranean one, the beach itself is the main source of sedimentary contribution to the dune systems. In addition, when the beaches are composed of gravels, as is the case of many beaches of Málaga and Almería, it is more difficult to ascertain the source of sandy sediment necessary for the dune systems, so the rivers become the main sediment suppliers of the system [72]. As stated before, river systems at the Mediterranean coast of Andalusia are mostly short or seasonal streams and, in general, provide a coarse grain size on the beaches. Further, the accentuated relief observed nearby the coast and the presence of reflective beaches represent great limitations for the development of coastal dunes [72].

Further points to be taken into consideration are the intensity and direction of predominant winds that, to be effective in dune formation, should be shore normal. Due to their coastal orientation, which is normal to predominant winds, the provinces of Cadiz (especially) and Almeria constitute areas favorable for dune formation. According to Bardají et al. [72] and Gracia et al. [73], the central part of the Andalusia coastline is parallel to predominant winds that give rise to a relevant longshore transport that supplies different dune systems, e.g., at Artola-Cabopino [72,88].

Analyzing the evolution of each dune typology is of relevance since each typology represents a clear evolution state from Embryo and mobile dunes (Type I) to Stabilized dunes (Type III) [65,92]. The increment over the 1977–2016 period of the Stabilized dunes (Type III) (Figure 3) was due to the progressive evolution of Grass-fixed dunes (Type II), a natural process described by Hesp [65,92].

Surface variations of the different types of dunes' systems were relatively homogeneous (Figure 4). With the exception of the province of Cádiz, the rest of the provinces showed a decrease of the three types of dunes in the first period and, in the second period, a decrease of types I and III and an increase of Type II in all provinces except Almería. The general decrease recorded in the period 1977–2001 was mainly due to urban occupation, intensive agricultural exploitation and the extraction of sand—such activities were not regulated until the approval of the Coastal Law in 1988 [41,46,47,55,68–73]. Dune destruction was especially evident in Málaga and Almería provinces, where entire dune systems disappeared: in Málaga province, a total surface of 1,766,711 m^2 was lost, of which ca. 1×10^6 m^2 were Type II dunes and ca. 600,000 m^2 were Type III dunes, and in Almería, ca. 56,300,000 m^2 of dune surface was lost, of which ca. 4,360,000 m^2 were Type II dunes. Some examples of papers that quantified the loss of dune surfaces in specific areas were by Viciana Martínez-Lage [71], who quantified a loss of 262 ha of dunes in Punta Entinas–El Sabinar, in Almería province, due to sand extractions, or Gómez Zotano [41], who quantified a reduction of 44.5% of the dune surface in Saladillo area, in Málaga province, during the 1956–2007 period.

The increase, in the 2001–2016 period, of the Type II in Málaga province was linked to the degradation of Type III dunes, especially evident in an area west of Marbella (Figure 2B) that was greatly impacted by urban developments, a quite common trend in Málaga province [41,46,47]. The increase of Type III in Almería was due to the stabilization of Type II dunes, especially in the area from Albufera de Adra to Almerimar and at Cabo de Gata (Figure 2E,F), which are areas where the shoreline is stable [39]. Overall, in Cádiz province, a slight increase of Type III was observed, and the other dunes' types recorded small variations (Figure 4). Such behavior was due to the low human pressure, the stable or even accreting conditions of the area [39,96] and the action of strong east winds (Figure 1) that favored dunes' growth and mobility [73].

6. Conclusions

This study analyzed the evolution of the dune systems along the Mediterranean coast of Andalusia, focusing on their characterization, level of fragmentation and anthropic occupation, for the 1977–2001 and 2001–2016 periods. Within a GIS project, there were 53 dune systems mapped that summed a total length of ca. 106 km in 1977 and ca. 76 km in 2001 and 2016.

Of the 53 dune systems, all but 4 recorded a reduction of their surface, or even disappeared, and this was especially evident in 1977–2001 when dune systems were affected by hard human interventions, such as the emplacement of buildings and touristic constructions, especially at Málaga province, and agricultural expansion at Almería province, and secondarily, at places by shoreline erosion processes.

Dunes' loss was at places and times linked to the progressive fragmentation of the dune toe, mainly due to the opening of pathways and to their progressive expansion due to marine- and wind-induced erosion processes. An increase of dunes' surface was observed in both natural and anthropic areas in Cádiz and Almería provinces, in accreting and stable beaches, usually on the updrift side of ports or due to strong east winds on the east-facing beaches.

Concerning the evolution of the Embryo and mobile dunes (Type I), Grass-fixed dunes (Type II) and Stabilized dunes (Type III), most of the provinces showed a decrease of the three types of dunes in the 1977–2001 period and, in the 2001–2016 period, a decrease of types I and III and an increase of Type II in all provinces. The increase of Type II dunes was linked to the degradation of Type III, observed in the 2001–2016 period at very anthropized areas; meanwhile, an increase of Type III was observed in stable and accreting areas.

Results obtained could be used to enhance the general database on dune characteristics along the Mediterranean coast of Andalusia and the possibility of utilizing ecosystem-based solutions in coastal protection, along with, or instead of, measures based on traditional engineering approaches. The methodology used in this study could be applied in other locations with a similar database.

Author Contributions: Data curation, R.M., G.M. and C.L.R.; Formal analysis, R.M., G.M., C.L.R. and G.A.; Methodology, R.M., G.M., C.L.R. and G.A.; Software, R.M., C.L.R. and G.M.; Geological and geomorphological supervision, G.M.; Hydraulic analysis, C.L.R.; Supervision, G.A.; Writing—original draft preparation, R.M., G.M., C.L.R. and G.A.; Writing—review and editing, G.A. All authors have read and agreed to the published version of the manuscript.

Funding: This research received no external funding.

Acknowledgments: This work is a contribution to the Andalusia Research Group PAI RNM-328 and to the PROPLAYAS network and has been partially developed at the Centro Andaluz de Ciencia y Tecnología Marinas (CACYTMAR), Puerto Real (Cádiz, Spain).

Conflicts of Interest: The authors declare no conflict of interest.

Appendix A

Table A1. Name and protection typology of each dune system and balance for the entire studied period.

System No.	Name	Protection	1977–2016		
			Dune Surface (m^2)	Occupation (m^2)	Fragmentation
1	Playa de Getares	Estrecho [1]	−70,477.70	41,768.39	0.16
2	Playa del Rinconcillo		38,884.38	732.26	
3	Marismas del Río Palmones	Marismas del Río Palmones [2]	−47,670.51	3352.14	0.12
4	Guadarranque W		−59,167.52	57,534.25	
5	Guadarranque E		19,262.80	2136.58	0.22
6	La Línea de la Concepción		−102,603.90	30,682.41	0.04
7	La Alcaidesa S		−3796.38	3967.45	0.03
8	La Alcaidesa N		−14,072.90	20.88	0.17
9	Guadalquitón		260,531.70	−376.21	0.09
10	Torreguadiaro	Laguna de Torreguadiaro [3]	−145,337.38	89,006.50	0.11
11	Playa del Saladillo W		−65,592.18	25,389.60	0.15
12	Playa del Saladillo E		−52,295.06	37,351.43	0.01
13	Playa de San Pedro de Alcántara		−97,452.20	28,411.01	
14	Playa Nueva Andalucía		−29,702.08	26,131.94	
15	Playa del Pinillo		−27,738.26	9490.44	0.06
16	Playa de Río Real		−16,322.02	3643.28	−0.20
17	Playa de los Monteros		−373,156.19	355,081.76	0.14
18	Playa del Alcate		−49,725.72	32,996.16	0.16
19	Playa Real de Zaragoza		−200,549.55	41,358.39	0.10
20	Playa de las Chapas		−171,835.74	152,819.73	0.22
21	Playa de Artola	Dunas de Artola [4]	−21,575.89	5068.88	0.16
22	Cabopino-Calahonda		0.00	0.00	
23	Torrenueva - Mijas		−64,545.36	64,344.35	
24	Playa de Canuela - Torremolinos		−44,014.67	970.12	
25	Playa de San Julián		−412,157.55	2085.54	0.10
26	Playa de la Misericordia	Desembocadura del Guadalhorce [2]	−98,481.69	84,116.14	−0.09
27	Arroyo de los Íberos		−54,160.94	90.78	
28	El Hornillo		−19,501.79	2698.63	
29	Playa del Poniente - Motril		−59,813.96	15,990.88	0.05
30	Carchuna		−115,558.06	38,433.86	0.00
31	Albufera de Adra	Albufera de Adra [5]	6708.50	0.00	−0.01
32	Playa de Balerma N		−17,100.05	994.82	0.01
33	Playa de Balerma S		−294,158.23	260,649.68	0.10
34	Ensenada de San Miguel		−557,765.03	259,966.94	0.06
35	Punta Entinas-El Sabinar	Punta Entinas – El Sabinar [2,5]	−4,166,157.86	239,842.20	0.03
36	Playa de Roquetas S		−65,618.01	32,255.48	
37	Playa de Roquetas N		−36,963.45	66.72	
38	Playa de los Bajos	Arrecife Barrera de Posidonia [4]	−18,995.79	408.82	−0.07
39	Playa Urbanización de Aguadulce		−14,309.26	5344.12	
40	Playa Ciudad Luminosa		−11,118.95	3916.61	

Table A1. *Cont.*

System No.	Name	Protection	1977–2016		
			Dune Surface (m²)	Occupation (m²)	Fragmentation
41	Punta del Río W		−2429.22	0.00	0.25
42	Punta del Río E		−8830.07	0.00	0.16
43	Las Algaidas-Las Marinas		−109,308.76	−1102.88	0.01
44	Cabo de Gata	Cabo de Gata-Níjar [1]	−395,189.20	404,678.11	0.14
45	Los Genoveses	Cabo de Gata-Níjar [1]	−31,645.57	−40.32	0.12
46	Playa de Bolmayor		−27,747.41	−301.21	0.04
47	Playa Venta del Bancal		−11,300.77	3577.15	0.08
48	Playa Cueva del Lobo		−11,748.40	4166.26	
49	El Cantal		−35,578.88	15,662.30	0.02
50	Playa del Descargador		−15,373.75	3011.05	
51	Playa de Rumina		−12,937.51	41.41	
52	Playa Marina de la Torre		−29,181.18	365.40	0.00
53	Vera		−567,841.25	233,429.88	0.10

Typologies of protection: [1] Natural Park, [2] Natural Site, [3] Special Plan for the Protection of the Physical Environment, [4] Natural Monument, [5] Natural Reserve. Fragmentation index was not calculated for periods where the dune system disappeared.

Table A2. Results obtained at each dune system.

System No.	1977			2001			2016		
	Dune Surface (m²)	Occupation (m²)	Fragmentation	Dune Surface (m²)	Occupation (m²)	Fragmentation	Dune Surface (m²)	Occupation (m²)	Fragmentation
1	144,188.37	9597.20	0.03	77,746.64	51,237.13	0.18	73,710.67	51,365.59	0.19
2	0.00	0.00		12,339.80	0.00		38,884.38	732.26	
3	205,724.64	3178.45	0.23	157,943.70	6516.26	0.31	158,054.13	6530.60	0.36
4	59,167.52	1633.27		0.00	59,167.52		0.00	59,167.52	
5	6316.76	1510.25	0.00	22,658.14	3717.80	0.13	25,579.56	3646.84	0.22
6	684,586.98	2295.32	0.07	719,675.84	3095.54	0.07	581,983.08	32,977.74	0.11
7	5370.58	0.00	0.00	2877.63	0.00	0.00	1574.20	3967.45	0.03
8	31,114.12	70.97	0.03	21,839.17	58.06	0.09	17,041.22	91.85	0.20
9	947,015.74	807.87	0.06	1,167,066.53	694.96	0.02	1,207,547.44	431.65	0.15
10	194,657.12	504.84	0.08	51,787.89	89,372.03	0.04	49,319.73	89,511.34	0.19
11	78,121.80	273.30	0.04	18,609.35	25,662.89	0.08	12,529.62	25,662.89	0.18
12	82,690.38	0.00	0.04	40,109.17	37,351.43	0.05	30,395.32	37,351.43	0.05
13	97,452.20	6377.47		36,487.89	22,027.47		0.00	34,788.48	
14	29,702.08	0.00		0.00	26,131.94		0.00	26,131.94	
15	39,444.19	0.00	0.04	21,835.12	5905.71	0.17	11,705.93	9490.44	0.10
16	19,307.50	0.00	0.20	2956.27	3643.28	0.15	2985.49	3643.28	0.00
17	412,867.48	0.00	0.03	38,602.94	348,097.72	0.12	39,711.29	355,081.76	0.17
18	67,942.45	0.00	0.03	6744.90	32,996.16	0.05	18,216.73	32,996.16	0.19
19	286,323.39	5333.48	0.11	82,951.01	46,657.53	0.15	85,773.84	46,691.87	0.21
20	199,167.71	1830.11	0.00	21,955.35	154,649.83	0.13	27,331.97	154,649.83	0.22
21	306,301.24	6506.44	0.00	246,613.13	11,448.04	0.22	284,725.35	11,575.32	0.16
22	0.00	0.00		4837.91	0.00		0.00	0.00	
23	64,545.36	201.01		0.00	64,545.36		0.00	64,545.36	
24	44,014.67	2722.12		0.00	3692.24		0.00	3692.24	
25	430,219.96	799.40	0.10	33,563.96	797.38	0.16	18,062.41	2884.93	0.20
26	153,324.54	0.00	0.11	63,109.77	84,116.14	0.02	54,842.86	84,116.14	0.02
27	54,160.94	635.79		0.00	726.58		0.00	726.58	
28	19,501.79	0.00		0.00	2698.63		0.00	2698.63	
29	68,178.56	1102.48	0.10	8705.37	17,108.96	0.29	8364.60	17,093.36	0.15

Table A2. *Cont.*

System No.	1977			2001			2016		
	Dune Surface (m²)	Occupation (m²)	Fragmentation	Dune Surface (m²)	Occupation (m²)	Fragmentation	Dune Surface (m²)	Occupation (m²)	Fragmentation
30	165,721.54	115.73	0.04	142,747.51	38,473.11	0.01	50,163.48	38,549.59	0.04
31	21,286.91	0.00	0.04	8193.76	0.00	0.05	27,995.40	0.00	0.03
32	27,651.90	39.11	0.05	9352.84	826.33	0.07	10,551.85	1033.93	0.05
33	440,474.93	701.17	0.01	125,373.21	257,051.80	0.09	146,316.70	261,350.84	0.11
34	806,346.40	5696.20	0.03	199,426.12	264,867.83	0.09	248,581.37	265,663.15	0.10
35	9,389,288.69	653,813.65	0.03	6,030,071.83	895,307.46	0.03	5,223,130.83	893,655.85	0.06
36	65,618.01	0.00		0.00	32,255.48		0.00	32,255.48	
37	36,963.45	0.00		0.00	66.72		0.00	66.72	
38	36,092.39	218.00	0.08	30,662.57	311.90	0.05	17,096.59	626.82	0.01
39	14,309.26	0.00		4399.52	3429.40		0.00	5344.12	
40	11,118.95	1182.32		0.00	5098.93		0.00	5098.93	
41	17,274.33	0.00	0.08	9693.33	0.00	0.10	14,845.11	0.00	0.08
42	16,891.37	0.00	0.03	13,529.19	0.00	0.08	8061.30	0.00	0.16
43	582,481.36	15,199.95	0.05	477,646.62	10,837.79	0.12	473,172.60	14,097.07	0.17
44	3,305,029.98	118,675.34	0.03	2,950,549.49	321,804.73	0.11	2,909,840.78	523,353.44	0.07
45	231,633.59	95.61	0.05	198,067.08	55.29	0.12	199,988.03	55.29	0.17
46	42,559.55	561.69	0.03	17,828.63	444.56	0.11	14,812.14	260.48	0.07
47	13,831.67	0.00	0.00	1729.15	3496.24	0.00	2530.90	3577.15	0.08
48	11,748.40	0.00		6694.07	697.64		0.00	4166.26	
49	45,547.32	4386.12	0.05	9285.69	15,269.76	0.01	9968.44	20,048.42	0.07
50	15,373.75	165.32		2855.83	1601.84		0.00	3176.38	
51	12,937.51	1214.63		4958.36	2923.88		0.00	1256.04	
52	29,896.87	0.00	0.00	11,900.67	365.40	0.06	715.69	365.40	0.00
53	722,709.90	21,902.89	0.03	152,125.47	240,022.68	0.06	154,868.65	255,332.76	0.13

Fragmentation index was not calculated for periods where the dune system disappeared.

References

1. Finkl, C.W.; Kruempel, C. Threats, obstacles and barriers to coastal environmental conservation: Societal perceptions and managerial positionalities that defeat sustainable development. In Proceedings of the First International Conference on Coastal Conservation and Management in the Atlantic and Mediterranean, Algarve, Portugal, 17–20 April 2005; pp. 3–28.
2. Meyer-Arendt, K.J. Grand Isle, Louisiana: A historic US Gulf Coast resort adapts to hurricanes, subsidence and sea level rise. In *Disappearing Destinations: Climate Change and Future Challenges for Coastal Tourism*; Jones, A., Phillips, M., Eds.; CAB International: Wallingford, UK, 2011; pp. 203–217, ISBN 978-1-84593-548-1.
3. Anfuso, G.; Loureiro, C.; Taaouati, M.; Smyth, T.A.G.; Jackson, D.W.T. Spatial variability of beach impact from post-tropical cyclone Katia (2011) on Northern Ireland's North coast. *Water* **2020**, *12*, 1380. [CrossRef]
4. Jones, A.; Phillips, M. *Disappearing Destinations: Climate Change and Future Challenges for Coastal Tourism*; CAB International: Wallingford, UK, 2011; ISBN 978-1-84593-548-1.
5. Cid, A.; Menéndez, M.; Castanedo, S.; Abascal, A.J.; Méndez, F.J.; Medina, R. Long-term changes in the frequency, intensity and duration of extreme storm surge events in southern Europe. *Clim. Dyn.* **2016**, *46*, 1503–1516. [CrossRef]
6. Nguyen, T.T.X.; Bonetti, J.; Rogers, K.; Woodroffe, D. Indicator-based assessment of climate-change impacts on coasts: A review of concepts, methodical approaches and vulnerability indices. *Ocean Coast. Manag.* **2006**, *123*, 18–43. [CrossRef]
7. Wolf, J.; Woolf, D.; Bricheno, L. Impacts of climate change on storms and waves relevant to the coastal and marine environment around the UK. *MCCIP Sci. Rev.* **2020**, 132–157. [CrossRef]
8. Klein, Y.L.; Osleeb, J.P.; Viola, M.R. Tourism-generated earnings in the coastal zone: A regional analysis. *J. Coast. Res.* **2004**, *20*, 1080–1088. [CrossRef]

9. European Environmental Agency. *The Changing Faces of Europe's Coastal Areas*; Office for Official Publications of the European Communities: Bruxelles, Belgium, 2006.

10. Carter, R.W.G. *Coastal Environments*; Academic Press: San Diego, CA, USA, 1988; p. 617, ISBN 0-12-161855-2.

11. Houser, C.; Hapke, C.; Hamilton, S. Controls on coastal dune morphology, shoreline erosion and barrier island response to extreme storms. *Geomorphology* **2008**, *100*, 223–240. [CrossRef]

12. Anfuso, G.; Rangel-Buitrago, N.; Cortés-Useche, C.; Castillo, B.I.; Gracia, F. Characterization of storm events along the Gulf of Cadiz (eastern central Atlantic Ocean). *Int. J. Climatol.* **2016**, *36*, 3690–3707. [CrossRef]

13. Molina, R.; Manno, G.; Lo Re, C.; Anfuso, G.; Ciraolo, G. A Methodological Approach to Determine Sound Response Modalities to Coastal Erosion Processes in Mediterranean Andalusia (Spain). *JMSE* **2020**, *8*, 154. [CrossRef]

14. European Environement Agency. The Millennium Ecosystem Assessment. Available online: https://www.eea.europa.eu/policy-documents/the-millennium-ecosystem-assessment (accessed on 22 July 2020).

15. Mir Gual, M.; Pons, G.X.; Martín Prieto, J.A.; Rodríguez Perea, A. A critical view of the blue flag beaches in Spain using environmental variables. *Ocean Coast. Manag.* **2015**, *105*, 106–115. [CrossRef]

16. Semeoshenkova, V.; Newton, A. Overview of erosion and beach quality issues in three southern European countries: Portugal, Spain and Italy. *Ocean Coast. Manag.* **2016**, *118*, 12–21. [CrossRef]

17. Anfuso, G.; Williams, A.T.; Martínez, G.C.; Botero, C.; Hernández, J.C.; Pranzini, E. Evaluation of the scenic value of 100 beaches in Cuba: Implications for coastal tourism management. *Ocean Coast. Manag.* **2017**, *142*, 173–185. [CrossRef]

18. Dolan, R.; Davis, R.E. An intensity scale for Atlantic coast northeast storms. *J. Coast. Res.* **1992**, *8*, 840–853.

19. Masselink, G.; Pattiaratchi, C.B. Seasonal changes in beach morphology along the sheltered coastline of Perth, Western Australia. *Mar. Geol.* **2001**, *172*, 243–263. [CrossRef]

20. Donnelly, J.P.; Bryant, S.S.; Butler, J.; Dowling, J.; Fan, L.; Hausmann, N.; Westover, K.; Webb, T. 700 yr sedimentary record of intense hurricane landfalls in southern New England. *GSA Bull.* **2001**, *113*, 714–727. [CrossRef]

21. Anfuso, G.; Gracia, F.J. Morphodynamic characteristics and short-term evolution of a coastal sector in SW Spain: Implications for coastal erosion management. *J. Coast. Res.* **2005**, *21*, 1139–1153. [CrossRef]

22. Komar, P.D.; Allan, J.C. Increasing hurricane-generated wave heights along the US east coast and their climate controls. *J. Coast. Res.* **2008**, *24*, 479–488. [CrossRef]

23. Rangel-Buitrago, N.; Anfuso, G. Winter wave climate, storms and regional cycles: The SW Spanish Atlantic coast. *Int. J. Climatol.* **2013**, *33*, 2142–2156. [CrossRef]

24. Rangel-Buitrago, N.; Anfuso, G. Coastal storm characterization and morphological impacts on sandy coasts. *Earth Surf. Processes Landf.* **2011**, *36*, 1997–2010. [CrossRef]

25. Sanjaume Saumel, E.; Gracia Prieto, F.J. *Las Dunas en España*; Sociedad Española de Geomorfología: Zaragoza, Spain, 2011; ISBN 978-84-615-3780-8.

26. Anfuso, G.; Dominguez, L.; Gracia, F. Short and medium-term evolution of a coastal sector in Cadiz, SW Spain. *Catena* **2007**, *70*, 229–242. [CrossRef]

27. Duarte, C.M.; Losada, I.J.; Hendriks, I.E.; Mazarrasa, I.; Marbà, N. The role of coastal plant communities for climate change mitigation and adaptation. *Nat. Clim. Chang.* **2013**, *3*, 961–968. [CrossRef]

28. Fernández-Montblanc, T.; Duo, E.; Ciavola, P. Dune reconstruction and revegetation as a potential measure to decrease coastal erosion and flooding under extreme storm conditions. *Ocean Coast. Manag.* **2020**, *188*, 105075. [CrossRef]

29. European Union. Directive 2011/92/EU of the European Parliament and of the Council of 13 December 2011 on the Assessment of the Effects of Certain Public and Private Projects on the Environment. Available online: https://eur-lex.europa.eu/legal-content/en/TXT/?uri=CELEX:32011L0092 (accessed on 22 July 2020).

30. European Union. Directive 92/43/EEC of 21 May 1992 on the Conservation of Natural Habitats and of Wild Fauna and Flora. Available online: https://eur-lex.europa.eu/legal-content/EN/TXT/?uri=celex%3A31992L0043 (accessed on 22 July 2020).

31. European Union. Directive 2001/42/EC on the Assessment of the Effects of Certain Plans and Programmes on the Environment. Available online: https://eur-lex.europa.eu/legal-content/EN/ALL/?uri=CELEX%3A32001L0042 (accessed on 22 July 2020).

32. European Union. Directive 2009/147/EC on the Conservation of Wild Birds. Available online: https: //eur-lex.europa.eu/legal-content/EN/TXT/?uri=CELEX%3A32009L0147 (accessed on 22 July 2020).

33. Nicholls, R.J.; Wong, P.P.; Burkett, V.R.; Codignotto, J.O.; Hay, J.E.; McLean, R.F.; Ragoonaden, S.; Woodroffe, C.D. Coastal systems and low-lying areas. In *Climate Change 2007: Impacts, Adaptation and Vulnerability. Contribution of Working Group II to the Fourth Assessment Report of the Intergovernmental Panel on Climate Change*; Parry, M.L., Canziani, O.F., Palutikof, J.P., van der Linden, P.J., Hanson, C.E., Eds.; Cambridge University Press: Cambridge, UK, 2007; ISBN 9780521880107.

34. Bochev-van der Burgh, L.M.; Wijnberg, K.M.; Hulscher, S.J.M.H. Decadal-scale morphologic variability of managed coastal dunes. *Coast. Eng.* **2011**, *58*, 927–936. [CrossRef]

35. Carter, R.W.G. Near-future sea level impacts on coastal dune landscapes. *Landsc. Ecol.* **1991**, *6*, 29–39. [CrossRef]

36. Church, J.A.; Gregory, J.M.; Huybrechts, P.; Kuhn, M.; Lambeck, K.; Nhuan, M.T.; Qin, D.; Woodworth, P.L.; Anisimov, O.A.; Bryan, F.O.; et al. Changes in sea level. In *Climate Change 2001: The Scientific Basis. Contribution of Working Group I to the Third Assessment Report of the Intergovernmental Panel on Climate Change*; Houghton, J.T., Ding, Y., Griggs, D.J., Noguer, M., van der Linden, P.J., Dai, X., Maskell, K., Johnson, C.A., Eds.; Cambridge University Press: Cambridge, UK, 2001; ISBN 0521-80767-0.

37. Pye, K.; Blott, S.J. Decadal-scale variation in dune erosion and accretion rates: An investigation of the significance of changing storm tide frequency and magnitude on the Sefton coast, UK. *Geomorphology* **2008**, *102*, 652–666. [CrossRef]

38. Sterr, H. Assessment of vulnerability and adaptation to sea-level rise for the coastal zone of Germany. *J. Coast. Res.* **2008**, *24*, 380–393. [CrossRef]

39. Molina, R.; Anfuso, G.; Manno, G.; Gracia Prieto, F.J. The Mediterranean Coast of Andalusia (Spain): Medium-Term Evolution and Impacts of Coastal Structures. *Sustainability* **2019**, *11*, 3539. [CrossRef]

40. Gracia Prieto, F.J.; Sanjaume, E.; Hernández, L.; Hernández, A.I.; Flor, G.; Gómez-Serrano, M.Á. Dunas marítimas y continentales. In *Bases Ecológicas Preliminares Para la Conservación de los Tipos de Hábitat de Interés Comunitario en España*; Ministerio de Medio Ambiente, y Medio Rural y Marino: Madrid, Spain, 2009; p. 106, ISBN 978-84-491-0911-9.

41. Gómez-Zotano, J. La degradación de dunas litorales en Andalucía: Aproximación geohistórica y multiescalar. *Investig. Geogr.* **2014**, *62*, 23–39. [CrossRef]

42. Guisado, E.; Malvárez, G.C.; Navas, F. Morphodynamic environments of the Costa del Sol, Spain. *J. Coast. Res.* **2013**, *65*, 500–506. [CrossRef]

43. Prieto, A.; Ojeda, J.; Rodríguez, S.; Gracia, J.; Del Río, L. Procesos erosivos (tasas de erosión) en los deltas mediterráneos andaluces: Herramientas de análisis espacial (DSAS) y evolución temporal (servicios OGC). In *Tecnologías de la Información Geográfica en el Contexto del Cambio Global, Proceedings of the XV Congreso Nacional de Tecnologías de la Información Geográfica, Madrid, Spain, 19–21 September 2012*; Martínez Vega, J., Martín Isabel, P., Eds.; CSIC—Instituto de Economía, Geografía y Demografía: Madrid, Spain, 2012; pp. 185–193, ISBN 978-84-695-4759-5.

44. Félix, A.; Baquerizo, A.; Santiago, J.; Losada, M. Coastal zone management with stochastic multi-criteria analysis. *J. Environ. Manag.* **2012**, *112*, 252–266. [CrossRef]

45. DGPC. Dirección General de Puertos y Costas. In *Actuaciones en la Costa 1988–1990*; MOPV: Madrid, Spain, 1991; p. 307.

46. Malvárez, G.; Pollard, J.; Rodriguez, R.D. Origins, management, and measurement of stress on the coast of southern Spain. *Coast. Manag.* **2000**, *28*, 215–234. [CrossRef]

47. Malvárez, G.; Pollard, J.; Domínguez, R. The planning and practice of coastal zone management in Southern Spain. *J. Sustain. Tour.* **2003**, *11*, 204–223. [CrossRef]

48. Manno, G.; Anfuso, G.; Messina, E.; Williams, A.T.; Suffo, M.; Liguori, V. Decadal evolution of coastline armouring along the Mediterranean Andalusia littoral (South of Spain). *Ocean Coast. Manag.* **2016**, *124*, 84–99. [CrossRef]

49. Chica Ruiz, J.A.; Barragan, J.M. *Estado y Tendencia de los Servicios de los Ecosistemas Litorales de Andalucía*; Consejería de Medio Ambiente: Sevilla, Spain, 2011; p. 112.

50. Molina, R.; Manno, G.; Lo Re, C.; Anfuso, G.; Ciraolo, G. Storm Energy Flux Characterization along the Mediterranean Coast of Andalusia (Spain). *Water* **2019**, *11*, 509. [CrossRef]

51. López, M.F.P. El clima de Andalucía. In *Geografía de Andalucía*; Ariel Geografía: Barcelona, Spain, 2003; pp. 137–173, ISBN 84-344-3476-8.
52. REDIAM. Ortofotografía Regional. Available online: http://www.juntadeandalucia.es/medioambiente/site/rediam/menuitem.aedc2250f6db83cf8ca78ca731525ea0/?vgnextoid=867122ad8470f210VgnVCM1000001325e50aRCRD&lr=lang_es (accessed on 22 July 2020).
53. Hidalgo, R. *Bases Ecológicas Preliminares Para la Conservación de los Tipos de Hábitat de Interés Comunitario en España*; Ministerio de Medio Ambiente, y Medio Rural y Marino: Madrid, Spain, 2009; ISBN 978-84-491-0911-9.
54. European Commisison. The Mediterranean Region—List of Sites of Community Importance (SCI's) for the Mediterranean Biogeographical Region. Available online: https://ec.europa.eu/environment/nature/natura2000/biogeog_regions/mediterranean/index_en.htm#list_of_sites (accessed on 22 July 2020).
55. Díez-Garretas, B.; Comino, O.; Pereña, J.; Asensi, A. Spatio-temporal changes (1956–2013) of coastal ecosystems in Southern Iberian Peninsula (Spain). *Mediterr. Bot.* **2019**, *40*, 111–119. [CrossRef]
56. Pintó, J.; Martí, C.; Fraguell, R.M. Assessing current conditions of coastal dune systems of Mediterranean developed shores. *J. Coast. Res.* **2014**, *30*, 832–842. [CrossRef]
57. García-Mora, M.R.; Gallego-Fernández, J.B.; Williams, A.T.; García-Novo, F. A coastal dune vulnerability classification. A case study of the SW Iberian Peninsula. *J. Coast. Res.* **2001**, *17*, 802–811.
58. Rangel-Buitrago, N.; Anfuso, G. *Risk Assessment of Storms in Coastal Zones: Case Studies from Cartagena (Colombia) and Cadiz (Spain)*; Springer: Dordrecht, The Netherlands, 2015; p. 63. [CrossRef]
59. Rizzo, A.; Aucelli, P.P.C.; Gracia, F.J.; Anfuso, G. A novelty coastal susceptibility assessment method: Application to Valdelagrana area (SW Spain). *J. Coast. Conserv.* **2018**, *22*, 973–987. [CrossRef]
60. Manno, G.; Lo Re, C.; Ciraolo, G. Shoreline detection in a gentle slope Mediterranean beach. In Proceedings of the 5th International Short Conference on Applied Coastal Research, Aachen, Germany, 9–11 June 2011; ISBN 978-3844011326.
61. Anfuso, G.; Martínez-del-Pozo, J.Á.; Rangel-Buitrago, N. Morphological cells in the Ragusa littoral (Sicily, Italy). *J. Coast. Conserv.* **2013**, *17*, 369–377. [CrossRef]
62. Manno, G.; Lo Re, C.; Ciraolo, G. Uncertainties in shoreline position analysis: The role of run-up and tide in a gentle slope beach. *Ocean Sci.* **2017**, *13*, 661–671. [CrossRef]
63. Aybulatov, N.A.; Artyukhin, Y.V. *Geo-Ecology of the World Ocean's Shelf and Coasts*; Hydrometeo Publishing: Leningrad, Russia, 1993; p. 304.
64. Jenks, G.F.; Caspall, F.C. Error on choroplethic maps: Definition, measurement, reduction. *Ann. Assoc. Am. Geogr.* **1971**, *61*, 217–244. [CrossRef]
65. Hesp, P. Foredunes and blowouts: Initiation, geomorphology and dynamics. *Geomorphology* **2002**, *48*, 245–268. [CrossRef]
66. Ley Vega de Seoane, C.; Gallego Fernández, J.B.; Vidal Pascual, C. *Manual de Restauración de Dunas Costeras*; Ministerio de Medio Ambiente. Dirección General de Costas: Madrid, Spain, 2007; ISBN 978-84-8320-409-2.
67. Sanjaume, E.; Pardo-Pascual, J.E. Degradación de sistemas dunares. In *Las Dunas en España*; Sanjaume Saumel, E., Gracia Prieto, F.J., Eds.; Sociedad Española de Geomorfología: Zaragoza, Spain, 2011; ISBN 978-84-615-3780-8.
68. Castaño Camero, N.; Arteaga Cardineau, C.; Gómez Zotano, J. Erosión en la playa del "Saladillo-Matas Verdes" (Estepona, Málaga): Situación actual y causas potenciales. *Geotemas* **2017**, *17*, 59–62.
69. Bayo Martínez, A. Tratamiento técnico del borde litoral almeriense. In *Actas de las Jornadas sobre el Litoral de Almería: Caracterización, Ordenación y Gestión de un Espacio Geográfico Celebradas en Almería, 20 a 24 de Mayo de 1997*; Instituto de Estudios Almerienses: Almería, Spain, 1999; pp. 207–232, ISBN 84-8108-175-2.
70. Viciana Martínez-Lage, A. Las extracciones de áridos en el litoral de Almería para su utilización en la agricultura intensiva (1956–1997). In *Actas de las Jornadas sobre el Litoral de Almería: Caracterización, Ordenación y Gestión de un Espacio Geográfico Celebradas en Almería, 20 a 24 de Mayo de 1997*; Instituto de Estudios Almerienses: Almería, Spain, 1999; pp. 83–110, ISBN 84-8108-175-2.
71. Viciana Martínez-Lage, A. La costa de Almería: Desarrollo socio-económico y degradación físico-ambiental (1957–2007). *Paralelo 37* **2007**, *19*, 149–183.
72. Bardají, T.; Zazo, C.; Lario, J.; Goy, J.L.; Cabero, A.; Dabrio, C.J.; Silva, P.G. Las dunas costeras del presente y último interglaciar en Málaga, Almería y Murcia. In *Las Dunas en España*; Sanjaume Saumel, E., Gracia Prieto, F.J., Eds.; Sociedad Española de Geomorfología: Zaragoza, Spain, 2011; ISBN 978-84-615-3780-8.

73. Gracia, F.J.; Benavente, J.; Alonso, C.; Del Río, L.; Abarca, J.M.; Anfuso, G.; García de Lomas, J. Las dunas del litoral gaditano. In *Las Dunas en España*; Sanjaume Saumel, E., Gracia Prieto, F.J., Eds.; Sociedad Española de Geomorfología: Zaragoza, Spain, 2011; ISBN 978-84-615-3780-8.

74. El Mrini, A.; Anthony, E.J.; Maanan, M.; Taaouati, M.; Nachite, D. Beach-dune degradation in a Mediterranean context of strong development pressures, and the missing integrated management perspective. *Ocean Coast. Manag.* **2012**, *69*, 299–306. [CrossRef]

75. Del Río, J.L.; Malvárez, G. Impacto de la regulación del río Verde en la erosión del sistema sedimentario litoral de la ensenada de Marbella, Costa del Sol. In *XV Coloquio Ibérico de Geografía*; Universidad de Murcia: Murcia, Spain, 2016; pp. 1–10, ISBN 978-84-944193-4-8.

76. Jiménez, J.A.; Sánchez-Arcilla, A. Medium-term coastal response at the Ebro delta, Spain. *Mar. Geol.* **1993**, *114*, 105–118. [CrossRef]

77. Pranzini, E. Airborne LIDAR survey applied to the analysis of the historical evolution of the Arno River delta (Italy). *J. Coast. Res.* **2007**, *50*, 400–409.

78. Arteaga Cardineau, C.; González, J.A. Natural and human erosive factors in Liencres beach spit and dunes (Cantabria, Spain). *J. Coast. Res.* **2005**, *49*, 70–75.

79. Feagin, R.A.; Sherman, D.J.; Grant, W.E. Coastal erosion, global sea-level rise, and the loss of sand dune plant habitats. *Front. Ecol. Environ.* **2005**, *3*, 359–364. [CrossRef]

80. Saye, S.E.; Pye, K. Implications of sea level rise for coastal dune habitat conservation in Wales, UK. *J. Coast. Conserv.* **2007**, *11*, 31–52. [CrossRef]

81. Costa, S.; Coelho, C. Northwest Coast of Portugal—Past behavior and future coastal defense options. *J. Coast. Res.* **2013**, *65*, 921–926. [CrossRef]

82. Houser, C.; Ellis, J. Beach and Dune Interaction. In *Treatise on Geomorphology*; John, F., Ed.; Academic Press: Cambridge, MA, USA, 2013; Volume 10, pp. 267–288, ISBN 9780123747396.

83. De Winter, R.C.; Ruessink, B.G. Sensitivity analysis of climate change impacts on dune erosion: Case study for the Dutch Holland coast. *Clim. Chang.* **2017**, *141*, 685–701. [CrossRef]

84. Criado-Aldeanueva, F.; Del Río Vera, J.; García-Lafuente, J. Steric and mass-induced Mediterranean sea level trends from 14 years of altimetry data. *Glob. Planet. Chang.* **2008**, *60*, 563–575. [CrossRef]

85. Tsimplis, M.; Spada, G.; Marcos, M.; Flemming, N. Multi-decadal sea level trends and land movements in the Mediterranean Sea with estimates of factors perturbing tide gauge data and cumulative uncertainties. *Glob. Planet. Chang.* **2011**, *76*, 63–76. [CrossRef]

86. Puertos del Estado. *REDMAR—RED de MAReógrafos de Puertos del Estado. Resumen de Parámetros Relacionados con el Nivel del Mar y la Marea que Afectan a las Condiciones de Diseño y Explotación Portuaria*; Puertos del Estado: Madrid, Spain, 2017; p. 33.

87. Fernández-Salas, L.M.; Dabrio, C.J.; Díaz del Río, V.; Lobo, F.J.; Sanz, J.L.; Lario, J. Land-sea correlation between late Holocene coastal and infralittoral deposits in the SE Iberian Peninsula (Western Mediterranean). *Geomorphology* **2009**, *104*, 4–11. [CrossRef]

88. Malvárez, G.; Navas, F.; Guisado-Pintado, E.; Jackson, D.W.T. Morphodynamic interactions of continental shelf, beach and dunes: The Cabopino dune system in southern Mediterranean Spain. *Earth Surf. Proc. Land.* **2019**, *44*, 1647–1658. [CrossRef]

89. Kraus, N.C.; Militello, A.; Todoroff, G. Barrier breaching processes and barrier spit breach, Stone Lagoon, California. *Shore Beach* **2002**, *70*, 22.

90. Ceia, F.R.; Patrícia, J.; Marques, J.C.; Dias, J.A. Coastal vulnerability in barrier islands: The high risk areas of the Ria Formosa (Portugal) system. *Ocean Coast. Manag.* **2010**, *53*, 478–486. [CrossRef]

91. Raji, O.; Niazi, S.; Snoussi, M.; Dezileau, L.; Khouakhi, A. Vulnerability assessment of a lagoon to sea level rise and storm events: Nador lagoon (NE Morocco). *J. Coast. Res.* **2013**, *65*, 802–808. [CrossRef]

92. Hesp, P.A. Foredune formation in southeast Australia. In *Coastal Geomorphology in Australia*; Thom, B.G., Ed.; Academic Press: London, UK, 1984; pp. 69–97, ISBN 0126878803.

93. Nordstrom, K.F. *Beaches and Dunes of Developed Coasts*; Cambridge University Press: Cambridge, UK, 2000; ISBN 0-521-47013-7.

94. Martínez, M.L.; Psuty, N.P. *Coastal Dunes. Ecology and Conservation, Ecological Studies 171*; Springer: Dordrecht, The Netherlands, 2008; ISBN 978-3-540-74001-8.

95. Psuty, N.P. The coastal foredune: A morphological basis for regional coastal dune development. In *Coastal Dunes. Ecology and Conservation, Ecological Studies 171*; Martínez, M.L., Psuty, N.P., Eds.; Springer: Dordrecht, The Netherlands, 2008; ISBN 978-3-540-74001-8.
96. REDIAM. WMS, Tasas de Erosión en el Litoral Andaluz. Available online: http://www.juntadeandalucia.es/medioambiente/site/rediam/menuitem.04dc44281e5d53cf8ca78ca731525ea0/?vgnextoid=97136276467e2510VgnVCM1000001325e50aRCRD&vgnextchannel=8ca090a63670f210VgnVCM2000000624e50aRCRD&vgnextfmt=rediam&lr=lang_es (accessed on 22 July 2020).

Article

New Geomorphological and Historical Elements on Morpho-Evolutive Trends and Relative Sea-Level Changes of Naples Coast in the Last 6000 Years

Gaia Mattei [1,*], **Pietro P. C. Aucelli** [1], **Claudia Caporizzo** [1], **Angela Rizzo** [2] and **Gerardo Pappone** [1]

[1] Dipartimento di Scienze e Tecnologie, Università degli Studi di Napoli Parthenope, Centro Direzionale Is. C4, 80121 Napoli, Italy; pietro.aucelli@uniparthenope.it (P.P.C.A.); claudia.caporizzo@uniparthenope.it (C.C.); gerardo.pappone@uniparthenope.it (G.P.)

[2] REgional Models and geo-Hydrological Impacts (REMHI Division), Fondazione CMCC Centro Euro-Mediterraneo sui Cambiamenti Climatici, 73100 Lecce, Italy; angela.rizzo@cmcc.it

* Correspondence: gaia.mattei@uniparthenope.it

Received: 10 July 2020; Accepted: 17 September 2020; Published: 22 September 2020

Abstract: This research aims to present new data regarding the relative sea-level variations and related morpho-evolutive trends of Naples coast since the mid-Holocene, by interpreting several geomorphological and historical elements. The geomorphological analysis, which was applied to the emerged and submerged sector between Chiaia plain and Pizzofalcone promontory, took into account a dataset that is mainly composed of: measurements from direct surveys; bibliographic data from geological studies; historical sources; ancient pictures and maps; high-resolution digital terrain model (DTM) from Lidar; and, geo-acoustic and optical data from marine surveys off Castel dell' Ovo carried out by using an USV (Unmanned Surface Vehicle). The GIS analysis of those data combined with iconographic researches allowed for reconstructing the high-resolution geomorphological map and three new palaeoenvironmental scenarios of the study area during the Holocene, deriving from the evaluation of the relative sea-level changes and vertical ground movements of volcano-tectonic origin affecting the coastal sector in the same period. In particular, three different relative sea-level stands were identified, dated around 6.5, 4.5, and 2.0 ky BP, respectively at +7, −5, and −3 m MSL, due to the precise mapping of several paleo-shore platforms that were ordered based on the altimetry and dated thanks to archaeological and geological interpretations.

Keywords: coastal landscape evolution; geomorphological analysis; palaeo-shore platform; relative sea-level changes; sea-level proxy; vertical ground movements; campi flegrei volcanic area

---

1. Introduction

The susceptibility of coastal towns and metropolis to the negative effects of the ongoing sea-level rise [1–5] depends on a series of factors, among which the behaviour of the land in terms of vertical ground movements (VGMs). In this regard, remote sensing methods can resolve ongoing vertical motions at sub-mm/year precision, which provides very precise and useful data [6–9].

However, in tectonically active areas, the chronological observation window has to be enlarged, by also considering the local vertical ground movements that occurred in the last centuries or millennia. Said movements can be identified and measured by dating a variety of possible proxies [10–16] indicating past positions of the relative sea levels (RSL). The obtained data are then corrected by subtracting the component due to eustatic and glacio-hydro-isostatic processes [10,12,17–21].

Among the proxies traditionally used to determine past RSL positions are those of geomorphological nature, such as coastal notches [22], marine terrace and shore platforms [23–25], or sedimentary facies [26–32]. Shore platforms, in particular, are initially created by wave quarrying and abrasion activities, but they can also be modelled by bio-erosion and weathering, particularly where tidal exposure is significant. The dimensions of shore platforms, under stable sea-level conditions, depend on the duration of the sea-level stand and they are intimately related to the intensity of wave processes, hardness, and structure of the rocks forming the platform, and the height of the sea cliff to be worn back [25,33,34]. Relict shore platforms also have a relevant role in the present study, since Naples, the second most populated conurbation of Italy, has a hilly topography and its coast alternates promontories and narrow coastal plains.

On the other hand, on a Mediterranean scale, the geomorphological analysis of these coastal landforms along with the interdisciplinary interpretation of the main sea-level indicators carried out in several recent studies [5,10,11] allowed for reconstructing the glacio- and hydro-isostatic influence on the Holocene sea-level changes. In particular, in Mediterranean regions that were affected by minimal tectonic influence, the glacio-isostatic adjustment (GIA) was the major driver of the decametric RSL rise in the last 8000 years. On the contrary, in geologically complex sectors, as the case of the mid-eastern Tyrrhenian coast in which the study area falls, complex volcano-tectonic behaviours that are primarily influenced the RSLs variations and related coastal changes during the Holocene [35]. Nevertheless, coastal vertical movements can be also fundamental for calibrating earth rheology models and ice sheet reconstructions in seismic active sectors, as in the case of North-Eastern Aegean Sea, which is mainly influenced by the activity of the North Anatolian Fault [20].

In the last ten years, the knowledge about the ancient vertical ground movements and related RSLs variation along the Naples coasts was deeply improved thanks to several geoarchaeological studies carried out during the excavation of a new coastal line of the subways systems. These studies analyzed complex stratigraphic records of former littoral deposits, well-dated by the presence of pottery fragments and the remains of ancient coastal structures [36–41].

In particular, during the Late-Holocene, the Neapolitan coastal sector has been affected by a subsiding trend, both along the high rocky coast of Posillipo and in the adjacent low coast of Chiaia and Municipio up to the limits of the Sebeto Plain [36,37,42–46]. In the case of Posillipo promontory (western periphery of Naples), the relative sea-level rise that was caused by this subsidence resulted in a retreating trend of the coastline. On the contrary, despite the subsidence, the coastal plains of Naples experienced a progradation of hundreds of meters thanks to the strong sediment supply that is ensured by both the longshore drift and torrential discharge from the hilly hinterland, probably favoured by anthropic impact [37–40].

A sector for which no specific study has been carried out till now is the Pizzofacone promontory. It was the starting point of our investigation, which was later extended to the nearby coastal plain of Chiaia and the seafloor between the said promontory and the islet of Castel dell'Ovo, formerly called Megaride. (see Figure 1 for location).

In this paper, new data are presented regarding the relic landforms that charcterize the western part of Naples coasts. In particular, we present the results of a geomorphological interpretation of high-resolution topographic models of the onshore and nearshore stripes, obtained from both direct and indirect measurements. For landforms that were masked by modern constructions, it was also essential for the study of documental data from historical sources. The aim was to gather elements for reconstructing the late Holocene evolution of the coastal landscape and gain new additional data about the local history of vertical ground movements during the Holocene.

2. Geological and Geomorphological Background

The city of Naples is located along the Tyrrhenian coasts, in the western sector of Campi Flegrei volcanic area [41]. This 60 ky old active volcano is known worldwide for the VGMs that have affected its territory in the last millennia and they have strongly influenced the Holocene evolution of its coasts [47–54].

The Campi Flegrei volcanic area is a poly-calderic system made of structural depressions that cover an area of about 230 km^2 and it is mainly shaped by three super-eruptions. The older one was the Campanian Ignimbrite (CI) eruption that occurred 40 ky BP [55,56]. After this complex event, the northern part of the just-formed caldera was invaded by the sea. The second eruption, which led to the formation of the Masseria del Monte Tuff, occurred 29.3 ky BP [57]. It preceded the Neapolitan Yellow Tuff (NYT Deino et al. [58]) eruption occurred 15 ky BP that contributed to the formation of the younger caldera [59,60].

After 15 ky BP, the volcanic activity of Campi Flegrei was characterized by three eruptive epochs, E1 between 15 and 10.6 ky BP, E2 between 9.6 and 9.1 ky BP, and E3 between 5.5 and 3.5 ky BP [61]. During each epoch, alternating magma/hydrothermal fluid inflation and deflation processes controlled the morphological evolution of this volcanic area.

Further, vertical ground movements (VGM) of meter-scale occurred in the periods preceding and following each eruption, which produced rapid relative sea-level variations along the whole coastal sector (Isaia et al. [62] and reference therein).

During the periods of volcanic quiescence dividing each eruptive epoch, the Campi Flegrei coasts were modelled by the wave action, recording erosional and depositional traces of these relative sea-level stands. The better-preserved evidence of this alternation (between VGMs and quiescence) is recorded in the sedimentary succession of La Starza hanging marine terrace that shows a sequence of depositional events that outline the interplay between inflation and deflation episodes (62 and reference therein). In this sequence, the three eruptive epochs (E1,15–10.6 ky, E2,9.6–9.1 ky, and E3, 5.5–3.8 ky, [61]) appear to be coupled with prevalent uplift and possible minor deflating episodes. The major uplift, greater than 100 m, occurred before the beginning of the E3 epoch activity. On the contrary, prevailing subsidence occurred during quiescent periods, from 8.59 to 5.86 ky, and generally after 3.8 ky BP and before the Monte Nuovo eruption (1538 AD), the last volcanic event that deeply modified the morphology of the area. This eruption, which led to the formation of Monte Nuovo tuff cone [63–66], represents the only example showing important ground uplift (up to 15 m), coupled with an eruptive activity (1538CE; [67,68]).

Several effects of the CF volcano-tectonic activity were recorded in Naples city that is located in the peripheral sector of this volcano. Indeed, its present landscape is the result of the mantling of preexisting volcanic edifices [36] by the NYT, and pyroclastic units that are related to the three eruptive epochs. Important evidence of this intense volcano-tectonic activity is also represented by several fault scarps, mainly SW-NE and NW-SE oriented (e.g., [36,37,69]).

The morphology of Naples coastal sector (Figure 1) is characterized by landforms due to endogenous dynamics [37,70], superimposed torrential dissections, and erosive traces of wave action [36].

The resulting coastal landscape is made of alternating small bays hosting both narrow coastal plains and cliffed promontories, such as that of Pizzofalcone (Figure 1).

Depositional sequences of pyroclastic deposits, reworked by alluvial and/or colluvial processes, are recorded in the main coastal plains of Chiaia and Municipio and they were deeply studied in the last years, thanks to the underground excavations [37–40]. The present elevation of well-dated shoreface deposits demonstrates that an overall subsidence affected these areas in the last 4000 years [37], even if a prograding/aggrading trend mostly prevailed because of sedimentary inputs coming from the hillslopes, probably favoured by anthropic forcing.

The present morphology of the cliffed promontories led to suppose that complex morphogenesis, which are related to the interplay between volcano-tectonic activity and marine processes, modelled these landforms since the Holocene transgression.

In the case of Pendino terrace, Cinque et al. [36] supposed that it formed during the early Holocene, because the depositional volcanic sequence presents at the top the fallout of "Soccavo 4" eruption (belonging to E1 eruptive epoch), and it also includes the fallout deposits of the Agnano-Monte Spina eruption (belonging to E3 eruptive epoch). Furthermore, other minor hanging terraces can be detected along the slope of Pizzofalcone promontory and in Chiaia plain. Regardless, presently, not enough chronological and morphogenetic data have been acquired about these landforms, even if they can be considered to be crucial to understanding the coastal evolution of this sector, when considering that they represent clear evidence of ancient sea-level stands.

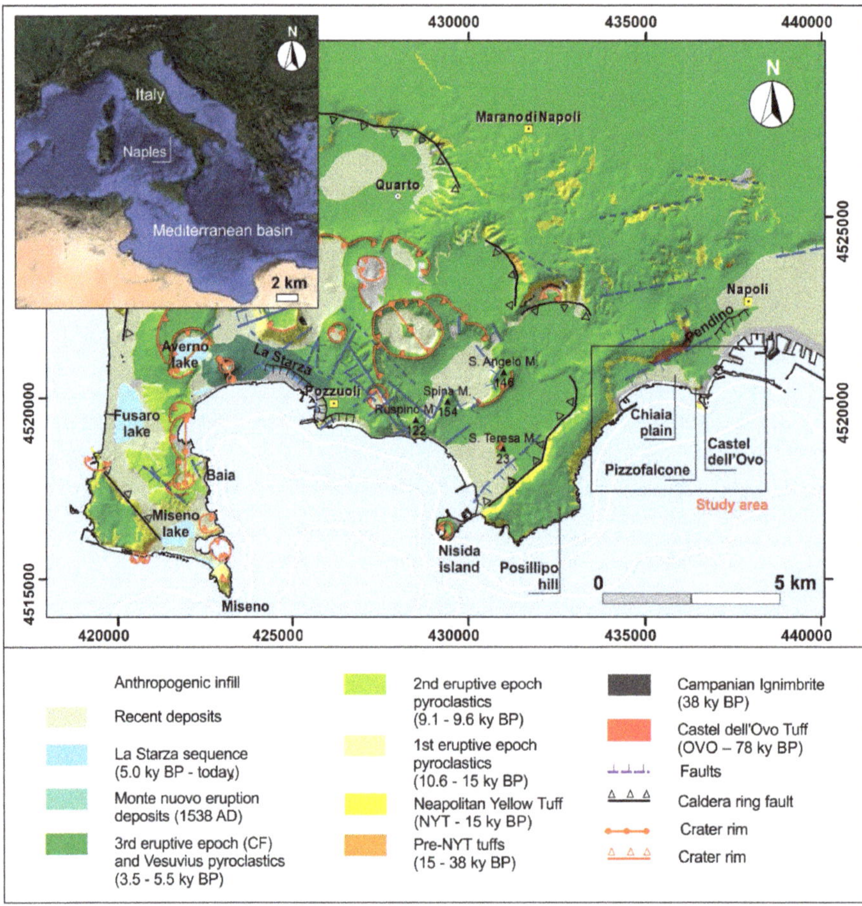

Figure 1. Geological sketch of Naples and Campi Flegrei (after Isaia et al. [71]; DiVito et al. [68]).

3. Materials and Methods

In this study, several direct and indirect surveys were carried out in the emerged and submerged sectors to obtain a detailed geomorphological characterization of the whole study area (black square in Figure 1). In particular, several direct surveys were carried out in the emerged coastal sector between Chiaia coastal plain and Castel dell'Ovo. Instead, the underwater sector off Castel dell'Ovo was surveyed using a USV (Unmanned Surface Vehicle) that was equipped with morpho-acoustic and optical sensors [72–75].

By overlaying these new data with historical sources, ancient pictures and maps, bibliographic, geological, and coring data, High-resolution digital terrain model (DTM) from Lidar and bathymetric data, the knowledge on the Holocene morpho-evolutive trend of Naples coasts were improved (Figure 2). On the other hand, the analysis of iconographic sources is a visualizing approach to geomorphological interpretations that can be considered to be highly efficient in urban contexts where the original landforms were masked by anthropogenic interventions [76,77].

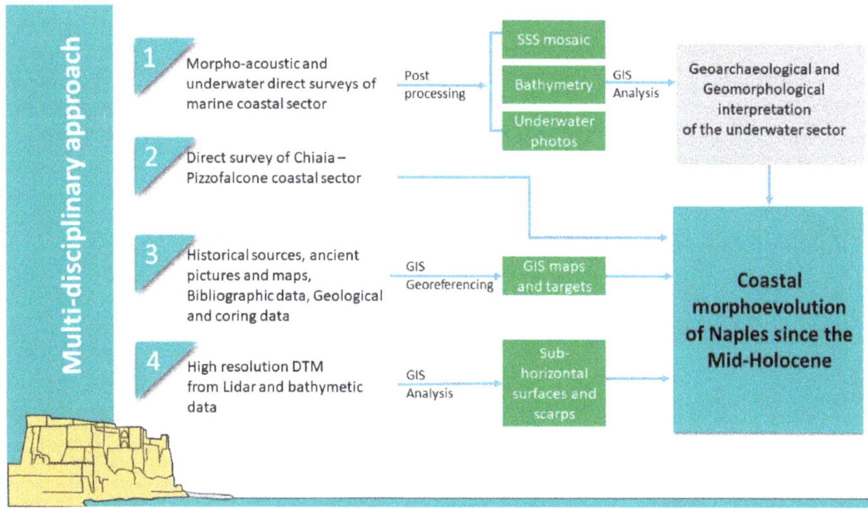

Figure 2. Flowchart describing the methodological approach used in this research.

3.1. Data and Sources

As described in the following sections, the new data from morpho-acoustic surveys (SSS ad Bathymetry) were GIS-processed together with data from direct underwater surveys to interpret the underwater landforms (point 1 in Figure 2), i.e., submerged paleo-shore platforms sculptured in tuff and precisely mapped in the following geomorphological map of the study area.

The geological units presented in the map were GIS-reconstructed by integrating data from [71] with those that were obtained by reinterpreting several stratigraphic records (see Figure 4 for location) contained in the 1967 Geological Report of Naples Municipality, and measurements taken during direct inspections carried out in the whole coastal sector (point 2 in Figure 2, Table 1).

Table 1. List of data and sources used in this research.

Type	Source	Date	Resolution (m)
Recent photos	New data		-
SSS mosaic	New data	2019	0.5 × 0.5
Bathymetric data	New data	2019	0.1 × 0.1
DTM from LIDAR	Ministry of Environment	2013	
Isobaths	CARG project	2013	1 line/meter
Geological units	[68,71]	2016	-
Coring n. 66	Naples Municipality Report	1967	-
Painting by Consalvo Carelli	Jpg—free source	1836	-
n. 2 1800' photos	Jpg—free source	1800	-
Painting by William Marlow	Jpg—free source	1790	
The painting by Lancelot-Théodore Turpin de Crissé	Jpg—free source	1819	
Painting by Anton Pitlo	Jpg—free source	1820	
1800' painting of Chiaia	Jpg—free source	1800	
Map of Naples by Lafrery	Jpg—free source	1566	

Besides, the GIS project was populated with several bibliographic data from previous studies, georeferenced historical maps, data interpreted from ancient pictures, and high-resolution DTM from Lidar (Table 1, Figure 2). The result was the reconstruction of the original shape of some ancient emerged shore-platforms, used here as sea-level markers (points 2,3,4 in Figure 2) and deeply described in the Results section.

3.2. Morpho-Acoustic and Underwater Survey

The underwater sector was surveyed with direct and indirect methods.

The indirect survey mainly consisted of an integrated marine survey (Figure 3) that was carried out using a prototype of unmanned surface vessel (USV). MicroVeGA is a technological project of USV pursued by the GEAC (Geologia degli Ambienti Costieri) research group of Parthenope University, intending to carry out high-resolution surveys of the submerged landscape, particularly in the presence of archaeological structures that can be intended as witnesses of this ancient coastal landscape [54,72–75].

Figure 3. MicroVeGA survey in the marine area off Castel dell' Ovo (photos taken by the authors during the surveys).

All of the data are broadcast in real-time both to a base station and to all operators involved in the research (geophysicist, archaeologist, geomorphologist, etc.).

The main task is the acquisition of data that are related to the morphology of the seabed, in order to reconstruct the underwater three-dimensional landscape, using geophysical (Single Beam Echo Sounder and Side Scan Sonar) and optical (underwater cameras for photogrammetry) instruments [78–85].

The onboard instrumentation mainly consists of a 200 KHz digital echosounder; a 450 KHz digital side-scan sonar; and, a high definition underwater photographic system.

The Single-Beam Echo Sounders (SBES) is an Ohmex with 200 KHz acquisition frequency and 60 m as maximum measured depth, which is therefore optimized for coastal bathymetric measurements.

The Side Scan Sonar (SSS) Tritech StarFish 450 C is optimized for coastal waters (450 KHz CHIRP transmission) and engineered for installation on drones. Under optimal conditions, the instrument has a resolution of 0.02 m. During the survey, the slant range of the instrument was 50 m with a 50% overlap between lines, in order to obtain a total coverage and reconstruct the morphology of both the target and the seabed [70].

The photographic system that was installed on-board of MicroVeGA consists of two Xiaomi YI Action cameras and a GoPro Hero 3 [74].

The main underwater targets that were detected during the indirect survey were surveyed by a team of specialized scuba divers (two geomorphologists and an archaeologist), assisted by two surveyors on a support boat with GIS-GPS cartographic station, which were useful for detecting the georeferenced targets. For each target, a direct measurement of its submersion was performed using a graduated level staff, in order to better constrain the archaeological interpretations and evaluate the present submersion of both natural landforms and main archaeological structural elements.

3.3. Geomorphological and GIS Analysis

The geomorphological analysis that was applied to the emerged and submerged sectors of the study area allowed for discriminating between natural/anthropic landform shapes. Among the main morphological elements interpreted as evidence of ancient landscapes, several terraced surfaces were detected by overlaying photo-interpretation, onsite surveys, analysis of high-resolution DTMs, and the interpretation of historical paintings and/or maps.

These landforms were interpreted according to the following criteria:

1. morpho-structural analysis of the coastal sector by in situ surveys;
2. morphometric analysis of extension, slope, and borders by spatial-analyzing the high-resolution DTMs;
3. altimetric ordering and connectability between terrace relicts;
4. analysis of historical sources, painting, and maps describing the landforms before the anthropic modifications; and,
5. definition of the primary process forming the surface (erosion, deposition, etc.).

A specific procedure was applied to the collected data in order to obtain the above-mentioned interpretations. In the first instance, an onshore-offshore DTM was built by interpolating the LIDAR (from the Ministry of Environment, 0–200 m MSL) and bathymetric data (from the CARG project, 0–20 m MSL) with a Topo to Raster interpolator (1 × 1 m grid).

Secondly, the slope map, which was obtained from the abovementioned DTM, was reclassified in three classes based on the slope value: sub-horizontal surfaces between 0 and 5%; gently sloping surface between 5 and 15%; steep slopes greater than 15%.

The sub-horizontal surfaces thus identified were interpreted as terraced surfaces, if sculptured in rock and matching with criteria at points 1 and 2. In the case of emerged terraced surfaces, the geological characterization was obtained from direct surveys and the analysis of both coring data and historical painting. Instead, in the case of the submerged ones, the classification of their geological and geomorphological characters was obtained from the analysis of the SSS signal and direct underwater surveys [45].

Finally, the terraced surfaces were interpreted as paleo-shore platforms, according to [25], and then classified in three orders, depending on their altimetry and dating, as described in the Results section.

3.4. Shore Platforms as Sea-Level Indicators

The paleo-shore platforms that were detected in this study were interpreted as sea-level index points (SLIPs), by detecting the inner margin of each landform [2]. The inner margin forms at the same level of the mean higher high water (MHHW), i.e., the average of the higher high water height of each tidal day observed over a Tidal Datum Epoch (National Oceanic and Atmospheric Administration-NOAA). Consequently, the RSL that is associated with this indicator is:

$$\text{SLIP}_i = \text{E}_i - \text{MHHW}$$

where E_i is the present elevation of the inner margin of paleo-shore platform i.

When considering that a micro-tidal range characterizes the study area [86], the present elevation of their inner margin was corrected for the value of MHHW that, in the Gulf of Naples, is 0.45 m (data from Rete Ondametrica Nazionale). About the local wave climate, the Gulf is characterized by frequent marine stormy events during winter and autumn with wave height values up to 4.8 m, mainly approaching from 180° to 210°. In the same way, seasonal data demonstrated that the study area is also strongly exposed to prevailing waves whose approaching directions range from S to SW [87–94].

If the inner margin of a shore platform is not precisely detectable, the medium-altitude of these landforms can be considered a marine limiting point (*sensu* Vacchi et al. [10]).

4. Results

The geology of the study area is characterized by a bedrock made of NYT, only outcropping along the steepest hillslopes, and a cover of Holocene pyroclastic and volcanoclastic units occurring with a variable thickness on terraces, footslopes, and coastal plains.

In the plains and, in particular, in Chiaia coastal plain, the bedrock is covered by alluvial deposits related to the climatic instability that occurred during the transition from humid to arid conditions in the late Holocene [37]. Indeed, in the near high rocky coast sector, the NYT crops out along the paleo-sea cliffs still visible at Pizzofalcone promontory and Castel dell' Ovo islet.

The trans-disciplinary analysis of documental data from historical sources and high-resolution topographic data from direct and indirect measurements of the onshore-offshore sector of Naples produced several relevant results concerning the Holocene evolution of this coastal sector.

Firstly, the relicts of three orders of paleo-shore platforms were detected along the coastal sector and intended as evidence of ancient sea level stations. Precisely, these forms represented the starting point for the geomorphological interpretations that were carried out in this study, which produced three main results: the high-resolution geomorphological map of the Naples coastal sector (Figure 4); the morpho-evolution model since the mid-Holocene (see Figure 14); and, the relative sea-level variations and related vertical grounds movements occurred in the area in the same period (see Figure 15).

The first result, i.e the geomorphological map of the study in Figure 4, was obtained by overlaying the bibliographic data with those derived from our geological and geomorphological analysis, as described in the Methods section.

In the map, the three orders of the paleo-shore platforms and the Greek–Roman littoral deposits were mapped and they represented the key strength of the Holocene evolutive model proposed in this study for the Naples coastal sector. In addition, the emerged scarps were differed in natural, when preserved their original shape; anthropically modified, when the primary morphogenetic factor was of natural origin, but the shape was modified by anthropogenic activities; anthropogenic scarps, when the primary morphogenetic factor was the human intervention.

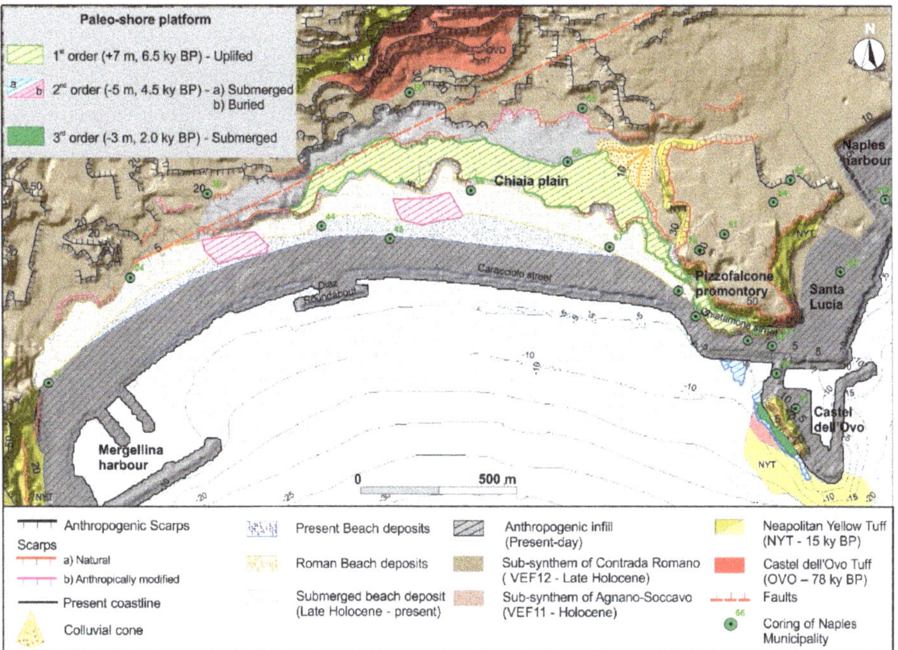

Figure 4. Geomorphological map of the study area: geological units were reconstructed by integrating data from [71] with those obtained by reinterpreting several stratigraphic records contained in the 1967 Geological Report of Naples Municipality (numbered in green in the map), and measurements taken during direct inspections carried out in the whole coastal sector.

4.1. First-Order Paleo-Shore Platforms

The first order (i.e., highest) of platforms is an emerged landform that extends from near the point of Pizzofalcone promontory to the Chiaia plain. This terrace has never been reported previously, probably because the strong urbanization affecting the area since the late 19th century made it difficult to perceive all of the topographic details of the area and—more relevant—to discriminate the natural component from the anthropic one.

Fortunately, this coastal sector appears depicted in several ancient painting and this helped us to understand the local geomorphological situation before the complete urbanization.

By overlaying the results from field surveys, high-resolution DTM analysis (Figure 5), and the study of ancient maps and photos, it became evident the old age of the 4°-inclined terrace that occurs a few meters above sea level, below the cliff that borders the Pizzofalcone hill. Nowadays, its inner margin runs at about 9 ± 1 m MSL.

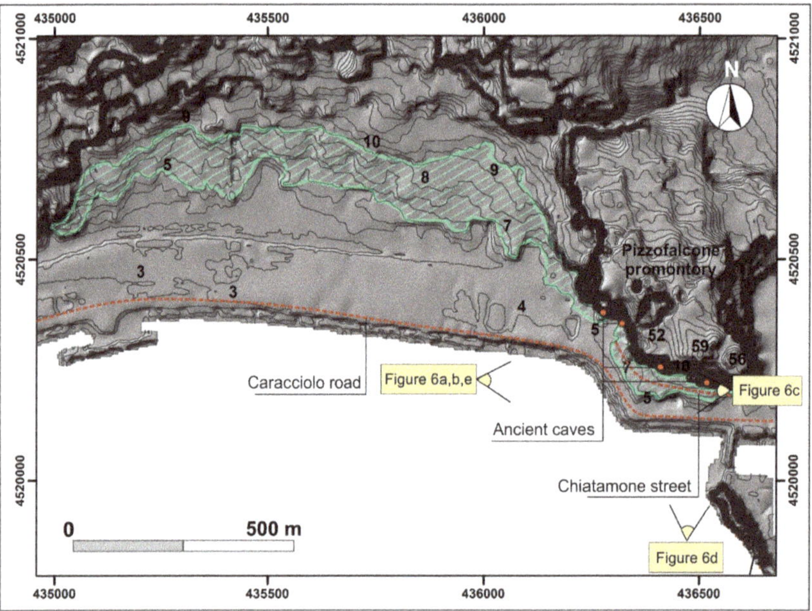

Figure 5. High-resolution DTM of the study area (1 × 1 m) with the first order paleo-shore platform highlighted with inclined green lines, contour lines in grey, elevation spots in black, and the roads described in the main text in dashed red.

Nowadays, the outer edge of the terrace is poorly visible as it is partially buried. The active sea cliff in NYT, which appears in ancient depictions (Figure 6), was buried by landfills in 1869 when a new coastal road (Via Caracciolo) was also constructed.

(a)

Figure 6. *Cont.*

(b)

(c)

Figure 6. *Cont.*

(**d**)

(**e**)

Figure 6. Five images showing ancient views from the south of Pizzofalcone promontory, see Figure 5 for point of view. The black arrows indicate the active sea cliff in NYT, detected downslope of the first order platform (red arrows). (**a**) Painting by Consalvo Carelli 1836, view from SW; (**b**) 1800′ photo, view from SW; (**c**) 1800′ photo, view from SE; (**d**) painting by Anton Sminck Pitloo 1820, view from S; and, (**e**) painting by William Marlow 1790, view from SW.

Measurements that were taken on more paintings of the 18th and 19th centuries allowed for estimating in 6–7 m the height of the ancient sea cliff (Figure 6).

The fact that the platform at issue is cut in the same cemented tuff forming the whole promontory (the NYT; see Section 2) was also ascertained by some direct inspections in the basements of modern buildings that are dug into the platform itself.

It was impossible to base the interpretation of the terrace on sedimentological and/or palaeo-ecological characters of associate deposits because of the urbanization of the area and the consequent lack of outcrops. However, it was interpreted as an ancient abrasion platform by excluding other possible interpretations in virtue of the morphological properties of the terrace itself and the scarp above it (Figure 7).

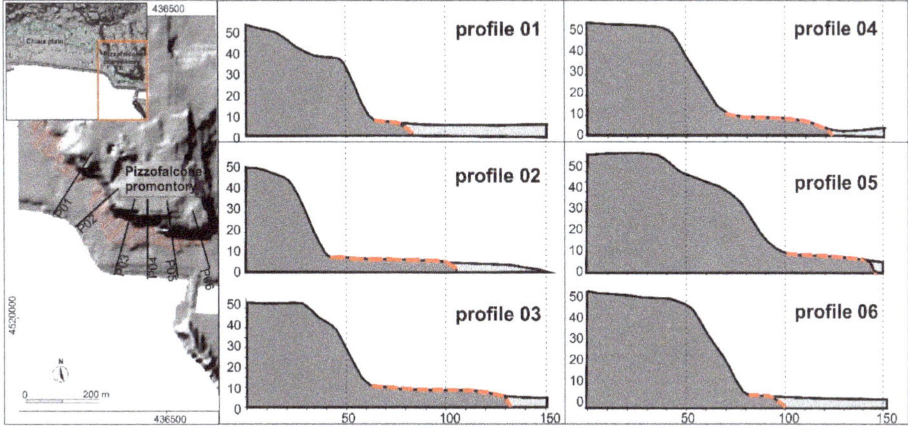

Figure 7. Six topographic profiles across the first-order platform (x and y axis are in metres): the black line is the present profile; the dotted red line is the original profile of the terrace and the light-grey area is the artificial anthropic infill built in the last 150 years. The green area in the inset map defines the perimeter of the 1st order platform and the red box represent the area in which the six profiles were reconstructed.

In particular, the hypothesis of a structural origin was discarded because of the angular unconformity between the terraced surface and the attitude of the bedrock strata. On the other hand, the terrace at issue cannot be ascribed to fault duplication of Pizzofalcone hill flat top, because the course of the scarp above the terrace is too sinuous (Figures 6e and 7) to be interpreted as a fault scarp.

Finally, the alternative interpretation of the terrace as an ancient tuff quarry floor, which is shaped by human activities, implies that the scarp above the terrace should be interpreted as the corresponding quarry front. However, this hypothesis can be excluded when considering the lack of any geometrical regularity—typical of quarry fronts—in the projections and recesses characterising the cliff (Figure 7), whose original shape is visible on paintings that predate the masking by buildings, barbicans, and retaining walls (Figures 6e and 8).

Figure 8. Pizzofalcone promontory. The highest reach of the paleo-sea cliff marking the inner margin of the cliff is cut and disappears under retaining walls and buildings (photo taken by the authors during the surveys).

Against the "quarry hypothesis", there are also (i) the great areal extent of the landform (about 13,000 m^2) and (ii) the excessive height of the hypothetical quarry front (up to 50 m, Figures 7 and 8). Two arguments that acquire particular value by considering that the quarry at issue should be dated to the Greek–Roman antiquity. One of the artificial caves marking the inner edge of the terrace at issue hosted a sanctuary of Serapis during the Greek–Roman period (Figure 9), as demonstrated by both archaeological evidence and narrations by the old authors [95,96].

Figure 9. The painting by Lancelot-Théodore Turpin de Crissé "View of a Villa, Pizzofalcone, Naples" (1819), National Gallery of Art, New York.

Moreover, the antiquity of the terrace is also suggested by the traditional toponym Chiatamone, which derives from the Greek word Platamon, meaning—among others—'flat and large stone' (in Homer, 6th century BC) and Flat rock emerging from the sea (in Aratus, 3rd century BC).

Taking into account all of the above-mentioned evidence, the platform at 7–10 m MSL (hereinafter "Chiatamone terrace") is interpreted as a substantially natural landform. More exactly, it is interpreted as an abrasion platform cut in the NYT. Its development occurred side-by-side with the retreat (and height increase) of the sea cliff bordering Pizzofalcone promontory. This platform—whose age will be discussed below—arose above sea level not later than the 2nd century BC, allowing for the extraction of tuff blocks in caves at the base of the former sea cliff. Later on, the caves (Figure 9) were used for religious purposes, as demonstrated by an ancient marble inscription that was dedicated to Serapis [96].

The Chiatamone terrace becomes less and less topographically expressed moving from the point of Pizzofalcone promontory north-eastward (Figures 7 and 8). Nevertheless, the analysis of historical painting led to recovering pieces of evidence destroyed by the strong urbanization. In particular, in the foreground of a painting by Pitloo (1820, Figure 6d), a flattened relict in NYT of this platform is visible.

In the adjacent central part of the Chaia plain, the relative sea-level stand that allowed for the sculpturing of the Chiatamone terrace is probably testified by a depositional coastal terrace. Notwithstanding the presence of a late cover of primary and reworked pyroclastics and the complete urbanization of the area, such a terrace may be recognized in the strip of minimum topographic gradient that occurs between 7 and 11 m MSL.

This geomorphological element appears clearly on the DTM of Figure 5 and it is followed downslope by a scarplet a couple of meters high, which is also evident in some paintings of the 19th century, such as the two in Figure 10, where the Lafrery (1566) map, also show this evidence.

(**a**)

(**b**)

Figure 10. *Cont.*

(c)

Figure 10. (**a**) Painting by Anton Sminck Pitlo 1820, view from S; (**b**) 1800′ painting, view from S; and, (**c**) Antony Lafrery (1566), map of Naples with the evidence of the uplifted terrace at Chiaia plain in the central part of the map.

The terrace width ranges from 40 to about 230 m, while its inclination does not exceed 4.2°. Indeed, historical beach deposits [37] buried the footslope of the downslope paleo-sea cliff.

Unfortunately, our scrutiny of pre-existing drilling data did not discover any well-log reporting also palaeo-environmental information for the sediments that form the terrace at issue. However, the log of a well-drilled near the inner margin of the terrace (n. 66, see Figure 4 for location) shows the top of the NYT at −24 m MSL, followed by beds of pyroclastic materials up to −17.5 m MSL and then by alternations of sands and gravelly sands that reach up to 3.5 m MSL. They are tentatively interpreted as shoreface deposits, probably truncated by a phase of subaerial erosion. The said littoral sands are covered by 4.5 m of primary and reworked pyroclastics that could be related to Agnano Monte Spina and Astroni eruptions (4.5 and 4 ky BP; [61], already detected at the footslope of the scarplet [37].

Other stratigraphic indications come from a study that was conducted a few years ago during the excavation of a new subway rail station [37]. It regarded a site located immediately downslope of the scarplet delimiting seaward the terrace at issue. That study allowed for detecting—at elevations ranging from −4.8 to −0.5 m MSL—an abrasion platform that was cut in Agnano Monte Spina and Astroni tephras that were buried by a sequence of littoral sediments testifying palaeo-shorelines migrations between 4.0 ky BP and the 16th century AD [37].

Around the age of this first-order terrace, it postdates the NYT eruption (15 ky BP) and predates the second-order of shore platforms (4.0 ky BP, see Section 3.2).

The data emerging from other studies in the coastal sector east of Pizzofalcone promontory also corroborate this assumption. There, the studies conducted during the excavations for the new subway stations Municipio, Università, Duomo, and Piazza Garibaldi [38–41,97] showed that the tectonic behaviour of the area has been dominated by subsidence with a rate between −2.5 and −1.2 mm/year [40] since at least the Early Bronze Age. As the first order terrace rests at an elevation (7–10 m MSL), demonstrating a subsequent uplift, it must be older than, at least, the 4.0 ky BP.

The residual uncertainty (15 to 4.0 ky BP) can be further reduced by discarding those millennia, during which the rate of post-glacial sea-level rise was too fast to permit the formation of sub-horizontal wave-cut platforms on rocks of appreciable resistance, such as the NYT. Therefore, we believe that the most probable age of our first order terrace does not exceed 6.5 ky (Figure 11).

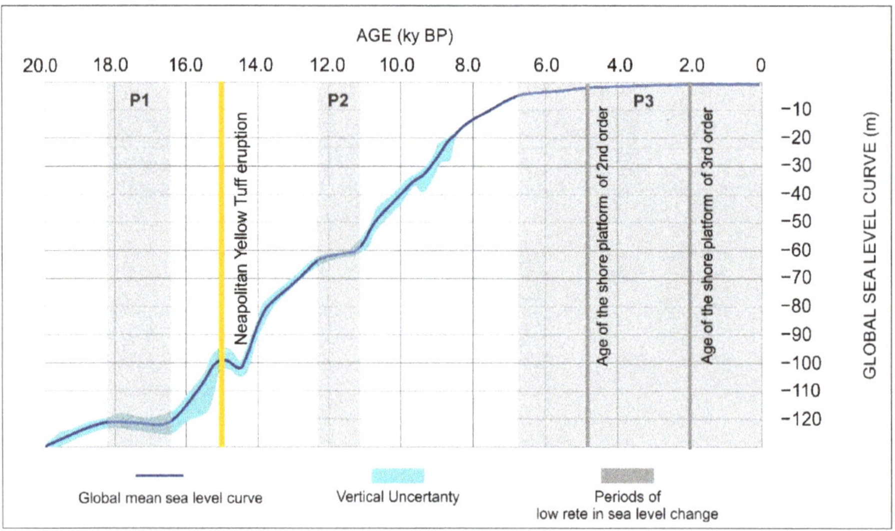

Figure 11. Global eustatic sea-level curve in the last glacial-interglacial circle, after Benjamin et al. [12].

Consequently, an uplift could have been responsible for the lifting of the first order platform some 15 m above the relative sea level (−4.5 m MSL) that was obtained from the Early Bronze Age second order shore-platforms.

Anyway, when considering that in the Gulf of Naples, the average retreat rate of sea cliffs in tuff is around 0.04 m/year [14,98], the time that is required to abrade the Chiatamone terrace (up to 60 m wide) can be estimated in about 1500 years, under conditions of sea-level stand or low rate in sea-level rise.

4.2. Second and Third Order Paleo-Shore Platforms

The second-order paleo-shore platforms (Figure 12a) was precisely mapped at −4.5/−5 m MSL in the nearshore sector of Castel dell' Ovo islet during the morpho-acoustic survey (Figure 12a,b). The analysis of the acoustic signal, along with several direct surveys, allowed for defining that this landform is cut in NYT.

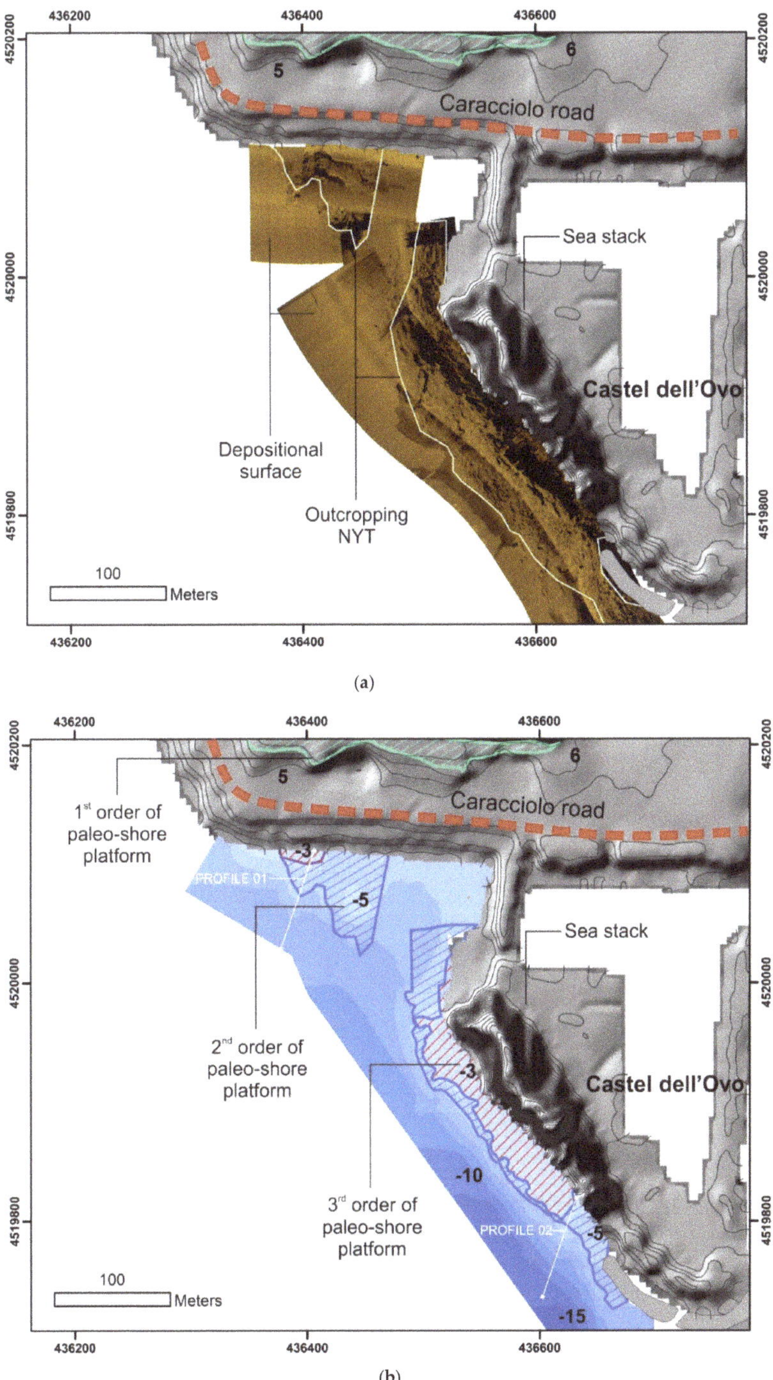

Figure 12. (**a**) Side Scan Sonar (SSS) map of the nearshore sector of Castel dell' Ovo islet; (**b**) Bathymetric map of the nearshore sector of Castel dell' Ovo islet with the precise mapping of the second (blue dashed lines) and third (red dashed lines) order submerged paleo-shore platforms (after Pappone et al. [45]).

The southern part of this platform has a very narrow shape in SW-NE direction (max 15 m) and it extends for 350 m in the NW-SE direction, with a medium slope of 2.5°. Instead, in the northernmost sector, overlooking the promontory of Pizzofalcone, the platform reaches about 100 m of extension seaward (Figure 12b). The two relicts of this paleo-shore platform are separated by a depositional surface, as demonstrated by the backscattering signal analysis [45].

When considering that the main storms and prevailing waves in the area coming from the S and SW sectors [87–94], it is possible to assume that the underwater channel separating the two landforms was shaped by the erosive effects of the wave action that also induced the formation of the sea-stack, where Castel dell' Ovo is now located.

In the close Chiaia plain, a buried landform at the same elevation range was detected during the geoarchaeological excavations that were related to the subway construction [37]. This buried landform is cut into Agnano Monte Spina and Astroni tephras (4.5 and 4.0 ky BP, respectively) and it is covered by archaeological remains, leading to suppose that it was shaped between the 4.0 ky BP and 600 BC.

It is reasonable to assume that submerged landform at −5 m MSL near Castel dell' Ovo was shaped in the same period.

The RSL in this period was deduced at −4 m MSL by measuring the present elevation of the inner margin in a well-preserved part, as described in the method section.

The spatial analysis of the morpho-acoustic data also allowed the detection of the third order paleo-shore platform at −3 m MSL (Figure 12).

This landform extended seaward 25–30 m in the nearshore sector close to Castel dell' Ovo and has a slope up to 2.2° (Profile 02 in Figure 13). Indeed, only a small witness is still visible in the northern area, as the platform was buried by an anthropogenic infill during modern times (Profile 01 in Figure 13).

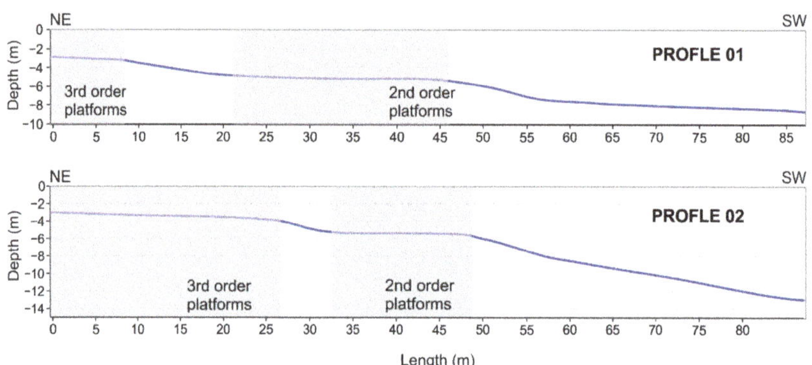

Figure 13. NE–SW profiles of the two relicts of the submerged paleo-shore platforms; for their location see Figure 11b.

The dating of this landform was possible thanks to the geoarchaeological study that was carried out by Pappone et al. [45] on the submerged archaeological remains laid on this platform. During direct and indirect surveys, several submerged targets were detected on a complex structure sculpted in tuff and extending along the outer margin of the third order of platform. The archaeological interpretation of these targets led to suppose that the structure was an ancient fish tank related to the Lucullus villa built on the Castel dell' Ovo islet during the 1st century BC [99]. The fundamental structural element that was used to evaluate the relative seal level (RSL) in this period was a well-preserved walkway, nowadays positioned at −1.8 m MSL. When considering that the functional clearance of this element with respect the ancient high tide (MHW, *sensu* Shennan, [24]) has been evaluated in 0.2 m by several authors (Aucelli et al. [53] and reference therein), a RSL at −2.2 m during the first century BC was deduced.

Moreover, this paleo-sea level is perfectly compatible with that obtained from the present elevation of the inner margin of the third order paleo-shore platforms measured in the best-preserved part.

Consequently, we can deduce that this shore platform, what was probably widened by human activity, was active during the Roman period.

5. Discussion

The interpretation of new data coming from direct and indirect surveys carried out in the coastal sector between Chiaia plain and Pizzofalcone promontory, corroborated by the analysis of bibliographic data and historical pictures, led to the detection of emerged and submerged terraces, testifying three Holocene episodes of relative sea-level stand, at +8 m, −4.5 m, and −2.2 m MSL, respectively.

On the promontory of Pizzofalcone and in the proximity of the islet Castel dell' Ovo, the three orders of paleo-shore platforms are represented by wave-cut platforms, while, in the adjacent bays (Chiaia plain and Piazza Municipio), also by sedimentary bodies.

The geomorphological evolution of the study area during the Holocene can be summarized, as follows (Figure 14).

During the Holocene, maximum marine ingression in the area occurred 6000 years ago, and the decrease of the sea-level rise rate favoured a sea cliff retreat in the whole coastal sector-characterized at that time by active sea cliffs—and the formation of the first order paleo-shore platform nowadays uplifted at 7–9 m MSL. In the same period, the sea stack nowadays hosting Castel dell' Ovo probably formed (Figure 14a).

Later on (between 4.0 ky BP and 2.2 ky BP), after an uplift of a metric entity, the RSL stood at about −4 m MSL, which was long enough to allow the wave action to partially dismantle that raised platform and form a new one below it (2nd order platform; Figure 14b).

The relicts of these platforms are nowadays buried by recent littoral sands in Chiaia plain and sculptured in tuff in the submerged sector of Castel dell' Ovo. Other erosional traces of the palaeo-sea level at −4.5/−5 m MSL were detected in the nearby Posillipo coastal sector [43], where they mainly consist of abandoned sea cliffs in NYT. In this sector, several port-like landing structures dated at the 1st century BC were built on these platforms, demonstrating that during the Roman period these platforms were already completely submerged [43,44].

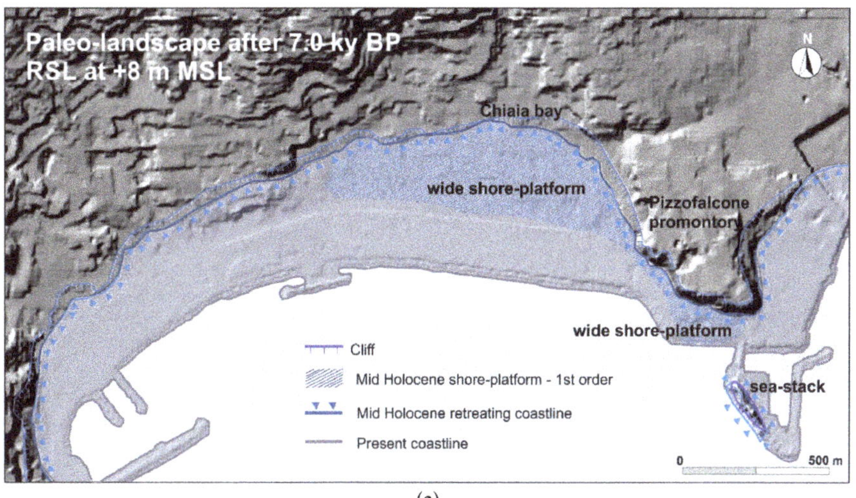

(a)

Figure 14. *Cont.*

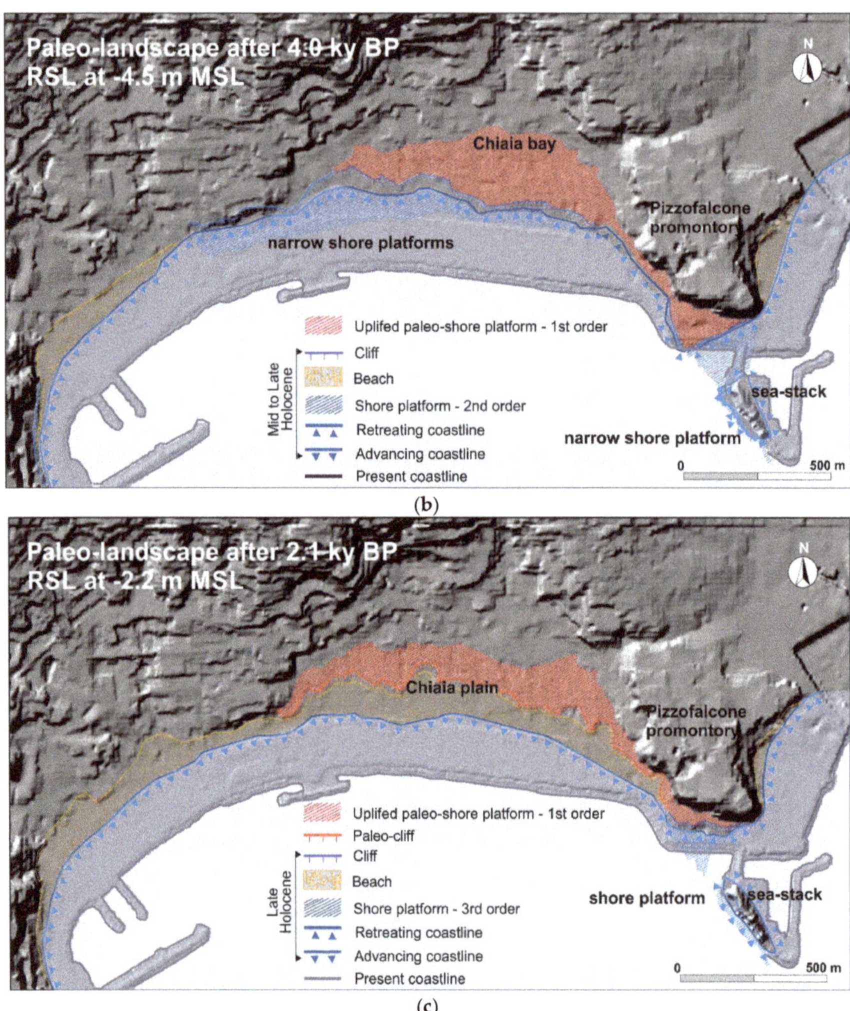

Figure 14. Morpho-evolution of Naples coastal sector since the mid-Holocene (**a**) Paleo-landscape after the Holocene marine transgression (6.5 ky BP); (**b**) Paleo-landscape after the volcano-tectonic uplift occurred 5.0–4.5 ky BP; and, (**c**) Paleo-landscape during the Roman period.

Since that time, a subsiding trend—which was also recorded in La Starza sequence [62]—affected the area bringing the relative sea-level up to −2.2 m during the Roman age. In this period, a low rate in sea level change favoured the formation of the third order paleo-shore platforms in the Castel dell' Ovo sector at −3 m MSL, as a result of sea cliff retreat. The effect was different inside the adjacent bay of Chiaia [37]. Being a coastal reach with a positive sedimentary budget, a prograding trend prevailed, as demonstrated by the littoral sands dated thanks to pottery remains of Greek–Roman (Figure 14c).

The natural evolution of the area ended in 1800 with the construction of coastal gardens and streets.

A new curve of Mid-Holocene relative sea-level variation was proposed for the Naples coastal sector as a result of our geomorphological interpretations (Figure 15).

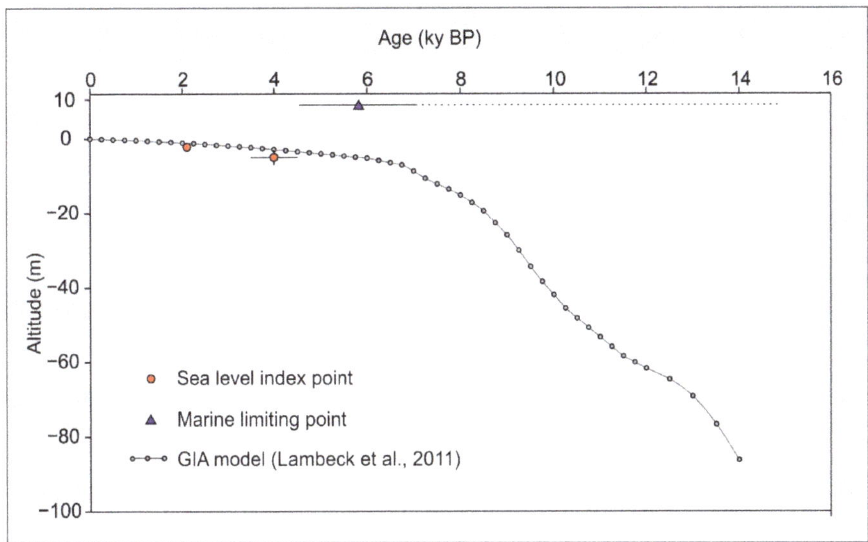

Figure 15. Relative sea-level curve for Naples in the last 6000 years. Black lines indicate the bi-directional uncertainty, but in the case of the marine limiting point related to the first order paleo-shore platform, the temporal uncertainty between 15 and 7 ky BP is a dotted line, because our interpretations led to discarding those millennia during which the rate of post-glacial sea-level rise was too fast to permit the formation of sub-horizontal wave-cut platforms.

Some novelties emerged by comparing our data with the framework of knowledge that was acquired during the underground excavations [36–41]. In the first instance, the geoarchaeological studies that were carried out in Chiaia and Municipio coastal plains precisely reconstructed the landscape modifications and the relative sea-level variations occurred in the last 4000 years, but no specific data have been acquired until now on the coastal palaeo-morphologies detectable along the interposed Pizzofalcone promontory, which separates the two coastal plains. In particular, such studies mainly measured the differential rate of a subsiding trend that occurred in the same period, but any evidence had been highlighted on the mid-Holocene uplift of a metric entity described in this study. On the other hand, the overlay between this result, previous studies, and several morpho-acoustic surveys of the underwater sector allowed for reconstructing the main morpho-evolutive trends of Naples coasts in the last 6000 years.

By comparing our study at a wider Mediterranean scale, it demonstrates the efficiency of an integrated interpretation of geomorphological and historical elements in terms of former sea-level indicators in tectonically complex sectors, as various authors similarly proposed for the Greek coasts, another Mediterranean coastal area. This sector suffered a subsiding trend during the Holocene with some episodes of uplift related to seismic activities [20,100–105]. In particular, Crete Island and the Perachora Peninsula recorded several geomorphological and archaeological traces of a nine-metre coseismic uplift that occurred in the historical period and were related to the tectonic activity of North-Eastern Aegean Sea [104,106]. Even in this case, historical sources allowed for interpreting this event, leading to obtaining some substantial information on the effect that the 8.5-magnitude earthquake occurred in 365 AD had the fasten landscape changes also occurred in the surrounding areas [104,105].

This comparison between Greek and Neapolitan coasts demonstrates the high reliability of multidisciplinary studies in interpreting past geological events that strongly changed coastal landscapes, thus inducing fast RSL variations.

6. Conclusions

In this study, some novelties were proposed for the Late-Holocene morpho-evolution of Naples coast. The main result obtained is the unveiling of a raised paleo-shore platform, ranging between 7 and 9 m MSL. nowadays almost undetectable due to the massive urbanization. No absolute chronological constraints were found; nevertheless, some regional geological dynamics and eustatic considerations led to suppose that it formed between 7 ky BP and 4.5 ky BP.

The mentioned strong urbanization of the area with the consequent lack of outcrops, and the absence of recent stratigraphic records on the raised platform, prevented from obtaining direct evidence on sedimentological and/or palaeo-ecological characters of deposits that are associated to this sea-level stand. Consequently, landform was interpreted as a paleo-shore platform by interpreting the morphological properties of the platform itself and the scarp above it, obtained from the interplay between historical sources, direct inspection of the few outcrops still visible, and a morphometric analysis of the high-resolution DTMs, as discussed in the Results section. Although that platform is not yet precisely dated and the details of its original morphology are obscured by the heavy urbanization of the area, that landform is very important because it is the first appeared evidence that the Naples area also experienced some uplift during the Holocene.

In the underwater domain of the study area, we identified two orders (second and third) of wave-cut platforms at −4.5 and −3 m MSL, respectively, by integrating the high-resolution morpho-acoustic and direct surveys. The second order was shaped between the 4.0 ky BP and 600 BC, when considering that, in the Chiaia plain, it is cut into Agnano Monte Spina and Astroni tephras (4.5 and 4.0 ky BP, respectively) and covered by archaeological remains. The third order, separated from the previous one by a 1.5-m high scarp, was dated at the Roman period due to the presence of the remnants of a fish tank.

By comparing our measurements of ancient relative sea levels with the Glacio-Hydro-Isostatic Adjustment (GIA) model that was proposed by Lambeck et al. [19], the vertical ground movements of the area during the Holocene can be reconstructed, as follows:

(T1) period of substantial stability between 7.0 and 5.0 ky BP during which the first order platforms formed;

(T2) an uplift of 10 to 13 m occurred shortly before the formation of the second order platforms about 5.0 ky BP;

(T3) short period of stability during which the second order platform was abraded; and,

(T4) period of prevailing subsidence covering the last 4 ky, which accelerated before the second century BC, considering that the RSL changed from −4.5 to −2.2 m MSL. This trend totalled about 2 m of lowering and was interrupted by a brief pause during the second century BC, when the third order platform formed, and the fish tank of the maritime villa at Castel dell' Ovo was constructed on it.

The subsiding movements at issue can be interpreted as the background regional trend due to the extensional tectonics of the whole Campania Plain and Gulf of Naples [106,107]. On the contrary, the uplift characterizing the period T2 is interpreted as a volcano-tectonic event that is probably related to the Campi Flegrei volcanic complex. According to this attribution, there is the fact that—about 5 ky BP—the area around Pozzuoli (central part of Campi Flegrei) experienced a phase of strong and fast volcano-tectonic uplift that raised the La Starza marine terrace. There, the displacement was in the order of tens of meters (up to about 100 m according to Isaia et al. [62], while Naples could have been more slightly affected due to its peripheral position.

In conclusion, this research, which belongs to a wider framework of studies concerning the Holocene landscape evolution of the Gulf of Naples [42–45,53,54,79,108,109], on one hand, improved the knowledge on the coastal landscape evolution of Naples and, on the other hand, provided new data on the VGM affecting this area and referable to the CF recent activity.

Author Contributions: Conceptualization, G.M., P.P.C.A. and G.P.; methodology, G.M., P.P.C.A. and C.C.; software, G.M. and C.C.; validation G.M., P.P.C.A. and G.P.; formal analysis, G.M., P.P.C.A. and G.P.; investigation, G.M. and C.C.; resources, P.P.C.A. and G.P.; data curation, G.M., A.R. and C.C.; writing—original draft preparation, G.M., P.P.C.A., G.P., A.R. and C.C.; writing—review and editing, G.M., P.P.C.A., G.P., A.R. and C.C.; visualization, G.M.; supervision, G.M., P.P.C.A. and G.P.; funding acquisition, P.P.C.A., G.P. All authors have read and agreed to the published version of the manuscript.

Funding: This research was financially supported by research found of Parthenope University granted to P.P.C. Aucelli and G. Pappone. This research was also funded by Distretto ad alta tecnologia per i beni culturali DATABENC, PON 03PE_00164 Rete Intelligente dei Parchi Archeologici (RIPA-PAUN), and the APC was funded by the same fund.

Acknowledgments: The author sincerely thank Francesco Peluso, Ferdinando Sposito and Alberto Greco for their unevaluable contribution during the marine surveys; Luigi Maglio, President of the Italian Institute of Castles, Campania Region, Castel dell'Ovo, for his scientific contribution in the historical researches. In this paper, the Ministry of the Environment kindly provided the coastal LIDAR data in March 2013, and the Soil Defense Service of the Campania Region kindly provided the Multibeam data.

Conflicts of Interest: The authors declare no conflict of interest.

References

1. Pappone, G.; Aucelli, P.P.C.; Alberico, I.; Amato, V.; Antonioli, F.; Cesarano, M.; Di Paola, G.; Pelosi, N. Relative sea-level rise and marine erosion and inundation in the Sele river coastal plain (Southern Italy): Scenarios for the next century. *Rend. Lincei* **2012**, *23*, 121–129. [CrossRef]

2. Rovere, A.; Stocchi, P.; Vacchi, M. Eustatic and relative sea level changes. *Curr. Clim. Chang. Rep.* **2016**, *2*, 221–231.

3. Stocchi, P.; Vacchi, M.; Lorscheid, T.; de Boer, B.; Simms, A.R.; van de Wal, R.S.W.; Vermeersen, B.L.A.; Pappalardo, M.; Rovere, A. MIS 5e relative sea-level changes in the Mediterranean Sea: Contribution of isostatic disequilibrium. *Quat. Sci. Rev.* **2018**, *185*, 122–134. [CrossRef]

4. Khan, N.S.; Horton, B.P.; Engelhart, S.; Rovere, A.; Vacchi, M.; Ashe, E.L.; Törnqvist, T.E.; Dutton, A.; Hijma, M.P.; Shennan, I. Inception of a global atlas of sea levels since the Last Glacial Maximum. *Quat. Sci. Rev.* **2019**, *220*, 359–371. [CrossRef]

5. Mann, T.; Bender, M.; Lorscheid, T.; Stocchi, P.; Vacchi, M.; Switzer, A.; Rovere, A. Relative sea-level data from the SEAMIS database compared to ICE-5G model predictions of glacial isostatic adjustment. *Data Brief* **2019**, *27*, 104600. [CrossRef]

6. Prampolini, M.; Savini, A.; Foglini, F.; Soldati, M. Seven Good Reasons for Integrating Terrestrial and Marine Spatial Datasets in Changing Environments. *Water* **2020**, *12*, 2221. [CrossRef]

7. Casella, E.; Drechsel, J.; Winter, C.; Benninghoff, M.; Rovere, A. Accuracy of sand beach topography surveying by drones and photogrammetry. *Geo-Mar. Lett.* **2020**, *40*, 1–14. [CrossRef]

8. Fuhrmann, T.; Garthwaite, M.C. Resolving three-dimensional surface motion with InSAR: Constraints from multi-geometry data fusion. *Remote Sens.* **2019**, *11*, 241. [CrossRef]

9. Di Paola, G.; Alberico, I.; Aucelli, P.P.C.; Matano, F.; Rizzo, A.; Vilardo, G. Coastal subsidence detected by Synthetic Aperture Radar interferometry and its effects coupled with future sea-level rise: The case of the Sele Plain (Southern Italy). *J. Flood Risk Manag.* **2018**, *11*, 191–206. [CrossRef]

10. Vacchi, M.; Marriner, N.; Morhange, C.; Spada, G.; Fontana, A.; Rovere, A. Multiproxy assessment of. Holocene relative sea-level changes in the western Mediterranean: Sea-level variability and improvements. in the definition of the isostatic signal. *Earth Sci. Rev.* **2016**, *155*, 172–197. [CrossRef]

11. Vacchi, M.; Ghilardi, M.; Melis, R.T.; Spada, G.; Giaime, M.; Marriner, N.; Lorscheid, T.; Morhange, C.; Burjachs, F.; Rovere, A. New relative sea-level insights into the isostatic history of the Western. Mediterranean. *Quat. Sci. Rev.* **2018**, *201*, 396–408. [CrossRef]

12. Benjamin, J.; Rovere, A.; Fontana, A.; Furlani, S.; Vacchi, M.; Inglis, R.H.; Galili, E.; Antonioli, F.; Sivan, D.; Miko, S.; et al. Late Quaternary sea-level changes and early human societies in the central and eastern, Mediterranean Basin: An interdisciplinary review. *Quat. Int.* **2017**, *449*, 29–57. [CrossRef]

13. Vecchio, A.; Anzidei, M.; Serpelloni, E.; Florindo, F. Natural Variability and Vertical Land Motion Contributions in the Mediterranean Sea-Level Records over the Last Two Centuries and Projections for 2100. *Water* **2019**, *11*, 1480. [CrossRef]

14. Aucelli, P.; Cinque, A.; Mattei, G.; Pappone, G. Historical sea level changes and effects on the coasts of.Sorrento Peninsula (Gulf of Naples): New constrains from recent geoarchaeological investigations. *Palaeogeogr. Palaeoclim. Palaeoecol.* **2016**, *463*, 112–125. [CrossRef]

15. Amato, V.; Aucelli, P.P.C.; Mattei, G.; Pennetta, M.; Rizzo, A.; Rosskopf, C.M.; Schiattarella, M.A. Geodatabase of Late Pleistocene-Holocene palaeo sea-level markers in the Gulf of Naples. *Alp. Mediterr. Quat.* **2018**, *31*, 5–9.

16. Lorscheid, T.; Stocchi, P.; Casella, E.; Gómez-Pujol, L.; Vacchi, M.; Mann, T.; Rovere, A. Paleo sea-level changes and relative sea-level indicators: Precise measurements, indicative meaning and glacial isostatic adjustment perspectives from Mallorca (Western Mediterranean). *Palaeogeogr. Palaeoclim. Palaeoecol.* **2017**, *473*, 94–107. [CrossRef]

17. Peltier, W.R. Global glacial isostasy and the surface of the ice-age Earth: The ICE-5 G (VM2) model and GRACE. *Annu. Rev. Earth Planet. Sci.* **2004**, *32*, 111–149. [CrossRef]

18. Spada, G.; Stocchi, P. SELEN: A Fortran 90 program for solving the "sea-level equation". *Comput. Geosci.* **2007**, *33*, 538–562. [CrossRef]

19. Lambeck, K.; Antonioli, F.; Anzidei, M.; Ferranti, L.; Leoni, G.; Scicchitano, G.; Silenzi, S. Sea level change along the Italian coasts during Holocene and prediction for the future. *Quat. Int.* **2011**, *232*, 250–257. [CrossRef]

20. Vacchi, M.; Rovere, A.; Zouros, N.; Desruelles, S.; Caron, V.; Firpo, M. Spatial distribution of sea-level markers on Lesvos Island (NE Aegean Sea): Evidence of differential relative sea-level changes and the neotectonic implications. *Geomorphology* **2012**, *159*, 50–62. [CrossRef]

21. Vacchi, M.; Ghilardi, M.; Spada, G.; Currás, A.; Robresco, S. New insights into the sea-level evolution in Corsica (NW Mediterranean) since the late Neolithic. *J. Archaeol. Sci. Rep.* **2017**, *12*, 782–793. [CrossRef]

22. Carobene, L. Marine Notches and Sea-Cave Bioerosional Grooves in Microtidal Areas: Examples from the Tyrrhenian and Ligurian Coasts-Italy. *J. Coast Res.* **2015**, *31*, 536–556. [CrossRef]

23. Anzidei, M.; Lambeck, K.; Antonioli, F.; Furlani, S.; Mastronuzzi, G.; Serpelloni, E.; Vannucci, G. Coastal structure, sea-level changes and vertical motion of the land in the Mediterranean. In *Sedimentary Coastal Zones from High to Low Latitudes: Similarities and Differences Special Publications 388*; Martini, I.P., Ed.; The Geological Society: London, UK, 2014. [CrossRef]

24. Shennan, I. Handbook of sea-level research: Framing research questions. In *Handbook of Sea-Level Research*; Shennan, I., Long, A.J., Horton, B.P., Eds.; John Wiley & Sons: Oxford, UK, 2015; pp. 3–25.

25. Rovere, A.; Raymo, M.E.; Vacchi, M.; Lorscheid, T.; Stocchi, P.; Gómez-Pujol, L.; Harris, D.L.; Casella, E.; O'Leary, M.J.; Hearty, P.J. The analysis of Last Interglacial (MIS 5e) relative sea-level indicators: Reconstructing sea-level in a warmer world. *Earth-Sci. Rev.* **2016**, *159*, 404–427. [CrossRef]

26. Amato, V.; Aucelli, P.P.C.; Cinque, A.; D'Argenio, B.; Di Donato, V.; Pappone, G.; Petrosino, P.; Rosskopf, C.M.; Russo Ermolli, E. Holocene palaeo-geographical evolution of the Sele river coastal plain (Southern Italy): New morpho-sedimentary data from the Paestum area. *Il Quat.* **2011**, *24*, 5–7.

27. Pappone, G.; Alberico, I.; Amato, V.; Aucelli, P.P.C.; Di Paola, G. Recent evolution and the present-day conditions of the Campanian Coastal plains (South Italy): The case history of the Sele River Coastal plain. *WIT Trans. Ecol. Env.* **2011**, *149*, 15–27.

28. Alberico, I.; Amato, V.; Aucelli, P.P.C.; Di Paola, G.; Pappone, G.; Rosskopf, C.M. Historical and recent changes of the Sele River coastal plain (Southern Italy): Natural variations and human pressures. *Rend. Lincei* **2012**, *23*, 3–12. [CrossRef]

29. Vacchi, M.; Rovere, A.; Chatzipetros, A.; Zouros, N.; Firpo, M. An updated database of Holocene relative sea level changes in NE Aegean Sea. *Quat. Int.* **2014**, *328*, 301–310. [CrossRef]

30. Engelhart, S.E.; Vacchi, M.; Horton, B.P.; Nelson, A.R.; Kopp, R.E. A sea-level database for the Pacific coast of central North America. *Quat. Sci. Rev.* **2015**, *113*, 78–92. [CrossRef]

31. Seeliger, M.; Pint, A.; Frenzel, P.; Weisenseel, P.K.; Erkul, E.; Wilken, D.; Wunderlich, T.; Başaran, S.; Bücherl, H.; Herbrecht, M.; et al. Using a multi-proxy approach to detect and date a buried part of the Hellenistic City Wall of Ainos (NW Turkey). *Geosciences* **2018**, *8*, 357. [CrossRef]

32. Ronchi, L.; Fontana, A.; Correggiari, A.; Remia, A. Anatomy of a transgressive tidal inlet reconstructed through high-resolution seismic profiling. *Geomorphology* **2019**, *343*, 65–80. [CrossRef]

33. Cinque, A.; De Pippo, T.; Romano, P. Coastal slope terracing and relative sea-level changes: Deductions based on computer simulations. *Earth Surf. Proc. Land.* **1995**, *20*, 87–103. [CrossRef]

34. Kanyaya, J.I.; Trenhaile, A.S. Tidal wetting and drying on shore platforms: An experimental assessment. *Geomorphology* **2005**, *70*, 129–146. [CrossRef]

35. Ascione, A.; Aucelli, P.P.C.; Cinque, A.; Di Paola, G.; Mattei, G.; Ruello, M.; Russo Ermolli, E.; Santangelo, N.; Valente, E. Geomorphology of Naples and the Campi Flegrei: Human and natural landscapes in a restless land. *J. Maps* **2020**, 1–11. [CrossRef]

36. Cinque, A.; Irollo, G.; Romano, P.; Ruello, M.R.; Amato, L.; Giampaola, D. Ground movements and sea level changes in urban areas: 5000 years of geological and archaeological record from Naples (Southern Italy). *Quat. Int.* **2011**, *232*, 45–55. [CrossRef]

37. Romano, P.; Di Vito, M.A.; Giampaola, D.; Cinque, A.; Bartoli, C.; Boenzi, G.; Detta, F.; Di Marco, M.; Giglio, M.; Iodice, S.; et al. Intersection of exogenous, endogenous and anthropogenic factors in the Holocene landscape: A study of the Naples coastline during the last 6000 years. *Quat. Int.* **2013**, *303*, 107–119. [CrossRef]

38. Russo Ermolli, E.; Romano, P.; Ruello, M.R.; Lumaga, M.R.B. The natural and cultural landscape of Naples (southern Italy) during the Graeco-Roman and Late Antique periods. *J. Archaeol. Sci.* **2014**, *42*, 399–411. [CrossRef]

39. Di Donato, V.; Ruello, M.R.; Liuzza, V.; Carsana, V.; Giampaola, D.; Di Vito, M.A.; Morhange, C.; Cinque, A.; Russo Ermolli, E. Development and decline of the ancient harbor of Neapolis. *Geoarcheaology* **2018**, *33*, 542–557. [CrossRef]

40. Vacchi, M.; Russo Ermolli, E.; Morhange, C.; Ruello, M.; Di Donato, V.; Di Vito, M.; Giampaola, D.; Carsana, V.; Liuzza, V.; Cinque, A.; et al. Millennial variability of rates of sea-level rise in the ancient harbour of Naples (Italy, western Mediterranean Sea). *Quat. Res.* **2019**, *93*, 284–298. [CrossRef]

41. Delile, H.; Goiran, J.P.; Blichert-Toft, J.; Arnaud-Godet, F.; Romano, P.; Bravard, J.P. A geochemical and sedimentological perspective of the life cycle of Neapolis harbor (Naples, southern Italy). *Quat. Sci. Rev.* **2016**, *150*, 84–97. [CrossRef]

42. Aucelli, P.P.C.; Cinque, A.; Mattei, G.; Pappone, G. Late Holocene landscape evolution of the gulf of Naples (Italy) inferred from geoarchaeological data. *J. Maps* **2017**, *13*, 300–310. [CrossRef]

43. Aucelli, P.P.C.; Cinque, A.; Mattei, G.; Pappone, G.; Stefanile, M. Coastal landscape evolution of Naples (Southern Italy) since the Roman period from archaeological and geomorphological data at Palazzo degli Spiriti site. *Quat. Int.* **2018**, *483*, 23–38. [CrossRef]

44. Aucelli, P.P.C.; Cinque, A.; Mattei, G.; Pappone, G.; Rizzo, A. Studying relative sea level change and correlative adaptation of coastal structures on submerged Roman time ruins nearby Naples (southern Italy). *Quat. Int.* **2019**, *501*, 328–348. [CrossRef]

45. Pappone, G.; Aucelli, P.P.C.; Mattei, G.; Peluso, F.; Stefanile, M.; Carola, A. A Detailed Reconstruction of the Roman Landscape and the Submerged Archaeological Structure at "Castel dell'Ovo islet" (Naples, Southern Italy). *Geosciences* **2019**, *9*, 170. [CrossRef]

46. Carsana, V.; Febbraro, S.; Giampaola, D.; Guastaferro, C.; Irollo, G.; Ruello, M.R. Evoluzione del paesaggio costiero tra *Parthenope Neapolis*. *Méditerranée Revue Géographique Pays Méditerranéens J. Mediterr. Geogr.* **2009**, *112*, 14–22. [CrossRef]

47. Aucelli, P.P.C.; Cinque, A.; Mattei, G.; Pappone, G.; Stefanile, M. First results on the coastal changes related to local sea level variations along the Puteoli sector (Campi Flegrei, Italy) during the historical times. *Alp. Mediterr. Quat.* **2018**, *31*, 13–16.

48. Pappalardo, U.; Russo, F. Il Bradisismo dei Campi Flegrei (Campania): Dati Geomorfologici ed Evidenze. Archeologiche. In *Forma Maris, Atti Della Rassegna Internazionale di Archeologia Subacquea (Pozzuoli 1998)*; Gianfrotta, P.A., Maniscalco, F., Eds.; Massa Editore: Pozzuoli, Italy, 2001.

49. Morhange, C.; Marriner, N.; Laborel, J.; Todesco, M.; Oberlin, C. Rapid sea-level movements and noneruptive crustal deformation in the phlegrean Fields caldera, Italy. *Geology* **2006**, *34*, 93–96. [CrossRef]

50. Bellucci, F.; Woo, J.; Kilburn, C.R.; Rolandi, G. Ground deformation at Campi Flegrei, Italy: Implications for hazard assessment. *Geol. Soc. Lond. Spec. Publ.* **2006**, *269*, 141–157. [CrossRef]

51. Del Gaudio, C.; Aquino, I.; Ricciardi, G.P.; Ricco, C.; Scandone, R. Unrest episodes at Campi Flegrei: A reconstruction of vertical ground movements during 1905–2009. *J. Volcanol. Geotherm. Res.* **2010**, *195*, 48–56. [CrossRef]

52. Passaro, S.; Barra, M.; Saggiomo, R.; Di Giacomo, S.; Leotta, A.; Uhlen, H.; Mazzola, S. Multi-resolution morphobathymetric survey results at the Pozzuoli-Baia underwater archaeological site (Naples, Italy). *J. Archaeol. Sci.* **2013**, *40*, 1268–1278. [CrossRef]

53. Aucelli, P.P.C.; Mattei, G.; Caporizzo, C.; Cinque, A.; Troisi, S.; Peluso, F.; Stefanile, M.; Pappone, G. Ancient Coastal Changes Due to Ground Movements and Human Interventions in the Roman Portus Julius (Pozzuoli Gulf, Italy): Results from Photogrammetric and Direct Surveys. *Water* **2020**, *12*, 658. [CrossRef]

54. Stefanile, M.; Mattei, G.; Troisi, S.; Aucelli, P.P.C.; Pappone, G.; Peluso, F. The Pilae of Nisida. Some geological and archaeological observations. *Archaeol. Marit. Mediterr. Int. J. Underw. Archaeol.* **2018**, *15*, 81–100.

55. Giaccio, B.; Isaia, R.; Fedele, F.; Di Canzio, E.; Hoffecker, J.; Ronchitelli, A.; Sinitsyn, A.; Anikovich, M.; Lisitsyn, S.; Popov, V.V. The Campanian Ignimbrite and Codola tephra layers: Two temporal/stratigraphic markers for the Early Upper Palaeolithic in southern Italy and eastern Europe. *J. Volcanol. Geoth. Res.* **2008**, *177*, 208–226. [CrossRef]

56. Costa, A.; Folch, A.; Macedonio, G.; Giaccio, B.; Isaia, R.; Smith, V. Quantifying volcanic ash dispersal and impact of the Campanian Ignimbrite super-eruption. *Geophys. Res.* **2012**, *39*, L10310. [CrossRef]

57. Albert, P.G.; Giaccio, B.; Isaia, R.; Costa, A.; Niespolo, E.M.; Nomade, S.; Pereira, A.; Renne, P.R.; Hinchliffe, A.; Mark, D.F.; et al. Evidence for a large-magnitude eruption from Campi Flegrei caldera (Italy) at 29 ka. *Geology* **2019**, *47*, 595–599. [CrossRef]

58. Deino, A.L.; Orsi, G.; Piochi, M.; de Vita, S. The age of the Neapolitan Yellow Tuff caldera-forming eruption (Campi Flegrei caldera-Italy) assessed by 40Ar/39Ar dating method. *J. Volcanol. Geoth. Res.* **2004**, *185*, 48–56. [CrossRef]

59. Orsi, G.; D'Antonio, M.; de Vita, S.; Gallo, G. The Neapolitan Yellow Tuff, a large-magnitude trachytic phreatoplinian eruption: Eruptive dynamics, magma withdrawal and caldera collapse. *J. Volcanol. Geotherm. Res.* **1992**, *53*, 275–287. [CrossRef]

60. Sacchi, M.; Pepe, F.; Corradino, M.; Insinga, D.D.; Molisso, F.; Lubritto, C. The Neapolitan Yellow Tuff caldera offshore the Campi Flegrei: Stratal architecture and kinematic reconstruction during the last 15 ky. *Mar. Geol.* **2014**, *354*, 15–33. [CrossRef]

61. Smith, V.C.; Isaia, R.; Pearce, N.J.C. Tephrostratigraphy and glass compositions of post-15 kyr Campi. Flegrei eruptions: Implications for eruption history and chronostratigraphic markers. *Quat. Sci. Rev.* **2011**, *30*, 3638–3660. [CrossRef]

62. Isaia, R.; Vitale, S.; Marturano, A.; Aiello, G.; Barra, D.; Ciarcia, S.; Iannuzzi, E.; Tramparulo, F. High-resolution geological investigations to reconstruct the long-term ground movements in the last 15 kyr at Campi Flegrei caldera (southern Italy). *J. Volcanol. Geotherm. Res.* **2019**, *385*, 143–158. [CrossRef]

63. Dvorak, J.J.; Mastrolorenzo, G. The mechanisms of recent vertical crustal movements in Campi Flegrei caldera, southern Italy. *Spec. Pap. Geol. Soc. Am.* **1991**, *263*, 49.

64. Todesco, M.; Costa, A.; Comastri, A.; Colleoni, F.; Spada, G.; Quareni, F. Vertical ground displacement at Campi Flegrei (Italy) in the fifth century: Rapid subsidence driven by pore pressure drop. *Geophys. Res. Lett.* **2014**, *41*, 1471–1478. [CrossRef]

65. Di Vito, M.; Lirer, L.; Mastrolorenzo, G.; Rolandi, G. The 1538 Monte Nuovo eruption (Campi Flegrei, Italy). *Bull. Volcanol.* **1987**, *49*, 608–615. [CrossRef]

66. D'Antonio, M.; Civetta, L.; Orsi, G.; Pappalardo, L.; Piochi, M.; Caradente, A.; de Vita, S.; Di Vito, M.A.; Isaia, R. The present state of the magmatic system of the Campi Flegrei caldera based on a reconstruction of its behavior in the past 12 Ka. *J. Volcanol. Geotherm. Res.* **1999**, *2–4*, 247–268. [CrossRef]

67. Guidoboni, E.; Ciuccarelli, C. The Campi Flegrei caldera: Historical revision and new data on seismic crises, bradyseisms, the Monte Nuovo eruption and ensuing earthquakes (twelfth century 1582 AD). *Bull. Volcanol.* **2011**, *73*, 655–677. [CrossRef]

68. Di Vito, A.; Acocella, V.; Aiello, G.; Barra, D.; Battaglia, M.; Carandente, A.; Del Gaudio, C.; de Vita, S.; Ricciardi, G.P.; Ricco, C.; et al. Magma transfer at Campi Flegrei caldera (Italy) before the 1538 AD eruption. *Sci. Rep.* **2016**, *6*, 32245. [CrossRef]

69. Orsi, G.; Di Vito, M.A.; Isaia, R. Volcanic hazard assessment at the restless Campi Flegrei caldera. *Bull. Volcanol.* **2004**, *66*, 514–530. [CrossRef]

70. Aucelli, P.; Brancaccio, L.; Cinque, A. Vesuvius and Campi Flegrei volcanoes. Activity, landforms and impact on settlements. In *Landscapes and Landforms of Italy*; Soldati, M., Marchetti, M., Eds.; Springer International Publishing: Cham, Switzerland, 2017; pp. 389–398. [CrossRef]

71. Isaia, R.; Iannuzzi, E.; Sbrana, A.; Marianelli, P. Note Illustrative Della Carta Geologica d'Italia Alla Scala 1: 50.000. 2016; Foglio 446-447 Napoli (aree emerse). Available online: https://www.isprambiente.gov.it/Media/carg/note_illustrative/446_447_Napoli.pdf (accessed on 18 September 2020).

72. Giordano, F.; Mattei, G.; Parente, C.; Peluso, F.; Santamaria, R. MicroVeGA (micro vessel for geodetics application): A marine drone for the acquisition of bathymetric data for GIS applications. The international archives of photogrammetry. *ISPRS* **2015**, *40*, 123.

73. Giordano, F.; Mattei, G.; Parente, C.; Peluso, F.; Santamaria, R. Integrating sensors into a marine drone for bathymetric 3D surveys in shallow waters. *Sensors* **2016**, *16*, 41. [CrossRef]

74. Mattei, G.; Troisi, S.; Aucelli, P.; Pappone, G.; Peluso, F.; Stefanile, M. Sensing the Submerged Landscape of Nisida Roman Harbour in the Gulf of Naples from Integrated Measurements on a USV. *Water* **2018**, *10*, 1686. [CrossRef]

75. Mattei, G.; Troisi, S.; Aucelli, P.P.C.; Pappone, G.; Peluso, F.; Stefanile, M. Multiscale Reconstruction of Natural and Archaeological Underwater Landscape by Optical and aCoustic Sensors. In Proceedings of the 2018 IEEE International Workshop on Metrology for the Sea; Learning to Measure Sea Health Parameters (MetroSea), Bari, Italy, 8–10 October 2018; IEEE: Piscataway, NJ, USA, 2018; pp. 46–49.

76. Tooth, S.; Viles, H.A.; Dickinson, A.; Dixon, S.J.; Falcini, A.; Griffiths, H.M.; Hawkins, H.; Lloyd-Jones, J.; Whalley, B. Visualising geomorphology: Improving communication of data and concepts through engagement with the arts. *Earth Surf. Process. Landf.* **2016**, *41*, 1793–1796. [CrossRef]

77. Reynard, E.; Pica, A.; Coratza, P. Urban geomorphological heritage. An overview. *Quaest. Geogr.* **2017**, *36*, 7–20. [CrossRef]

78. Mattei, G.; Giordano, F. Integrated geophysical research of Bourbonic shipwrecks sunk in the Gulf of Naples in 1799. *J. Archaeol. Sci. Rep.* **2015**, *1*, 64–72. [CrossRef]

79. Aucelli, P.; Cinque, A.; Giordano, F.; Mattei, G. A geoarchaeological survey of the marine extension of the Roman archaeological site Villa del Pezzolo, Vico Equense, on the Sorrento Peninsula, Italy. *Geoarchaeology* **2016**, *31*, 244–252. [CrossRef]

80. Fontana, A.; Ronchi, L.; Rossato, S.; Mozzi, P. Lidar-derived dems for geoarchaeological investigations in alluvial and coastal plains. *Alp. Mediterr. Quat.* **2018**, *31*, 209–212.

81. Trobec, A.; Busetti, M.; Zgur, F.; Baradello, L.; Babich, A.; Cova, A.; Gordini, E.; Romeo, R.; Tomini, I.; Poglajen, S.; et al. Thickness of marine Holocene sediment in the Gulf of Trieste (northern Adriatic Sea). *Earth Syst. Sci. Data* **2018**, *10*, 1077. [CrossRef]

82. Fediuk, A.; Wilken, D.; Wunderlich, T.; Rabbel, W.; Seeliger, M.; Laufer, E.; Pirson, F. Marine seismic investigation of the ancient Kane harbour bay, Turkey. *Quat. Int.* **2019**, *511*, 43–50. [CrossRef]

83. Novak, A.; Šmuc, A.; Poglajen, S.; Vrabec, M. Linking the high-resolution acoustic and sedimentary facies of a transgressed Late Quaternary alluvial plain (Gulf of Trieste, northern Adriatic). *Mar. Geol.* **2020**, *419*, 106061. [CrossRef]

84. Novak, A.; Šmuc, A.; Poglajen, S.; Celarc, B.; Vrabec, M. Sound Velocity in a Thin Shallowly Submerged Terrestrial-Marine Quaternary Succession (Northern Adriatic Sea). *Water* **2020**, *12*, 560. [CrossRef]

85. Troisi, S.; Del Pizzo, S.; Gaglione, S.; Miccio, A.; Testa, R.L. 3D models comparison of complex shell in underwater and dry environments. *Int. Arch. Photogramm. Remote Sens. Spat. Inf. Sci.* **2015**, *40*, 215–222. [CrossRef]

86. Buonocore, B.; Sansone, E.; Zambardino, G. Rilievi ondametrici nel Golfo di Napoli. *Ann. IUN* **2003**, *67*, 203–211.

87. Benassai, G.; de Maio, A.; Sansone, E. Altezze e periodi delle onde significative nel Golfo di Napoli dall'aprile 1986 al giugno 1987. *Ann. IUN* **1994**, *61*, 3–9.

88. Menna, M.; Mercatini, A.; Uttieri, M.; Buonocore, B.; Zambianchi, E. Wintertime transport processes in the Gulf of Naples investigated by HF radar measurements of surface currents. *Riv. Nuovo. Cim.* **2007**, *30*, 605–622.

89. Uttieri, M.; Cianelli, D.; Buongiorno Nardelli, B.; Buonocore, B.; Falco, P.; Colella, S.; Zambianchi, E. Multiplatform observation of the surface circulation in the Gulf of Naples (Southern Tyrrhenian Sea). *Ocean Dyn.* **2011**, *61*, 779–796. [CrossRef]

90. de Ruggiero, P.; Napolitano, E.; Iacono, R.; Pierini, S. A high-resolution modelling study of the circulation along the Campania coastal system, with a special focus on the Gulf of Naples. *Cont. Shelf. Res.* **2016**, *122*, 85–101. [CrossRef]

91. Falco, P.; Buonocore, B.; Cianelli, D.; De Luca, L.; Giordano, A.; Iermano, I.; Kalampokis, A.; Saviano, S.; Uttieri, M.; Zambardino, G.; et al. Dynamics and sea state in the Gulf of Naples: Potential use of high-frequency radar data in an operational oceanographic context. *J. Oper. Oceanogr.* **2016**, *9*, 33–45. [CrossRef]

92. de Ruggiero, P.; Napolitano, E.; Iacono, R.; Pierini, S.; Spezie, G. A baroclinic coastal trapped wave event in the Gulf of Naples (Tyrrhenian Sea). *Ocean Dyn.* **2018**, *8*, 1683–1694. [CrossRef]

93. Saviano, S.; Kalampokis, A.; Zambianchi, E.; Uttieri, M. A year-long assessment of wave measurements retrieved from an HF radar network in the Gulf of Naples (Tyrrhenian Sea, Western Mediterranean Sea). *J. Oper. Oceanogr.* **2019**, *12*, 1–15. [CrossRef]

94. Saviano, S.; Cianelli, D.; Zambianchi, E.; Conversano, F.; Uttieri, M. Multiple year characterization of the wave field in the Gulf of Naples by multi-platform measurements. In Proceedings of the International Workshop on Metrology for the Sea Genoa, Genoa, Italy, 3–5 October 2019; pp. 3–5.

95. Carletti, N. *Topografia Universale Della Citta'di Napoli in Campagna Felice Enote Enciclopediche Storiografe*; Stamperia Raimondiana: Napoli, Italy, 1776.

96. Dall'Osso, L. Napoli trogloditica e preellenica. *Napoli Nobilissima* **1906**, *15*, 33–51.

97. Amato, L.; Carsana, V.; Cinque, A.; Di Donato, V.; Giampaola, D.; Guastaferro, C.; Irollo, G.; Morhange, C.; Romano, P.; Ruello, M.R.; et al. Ricostruzioni morfoevolutive nel territorio di Napoli: l'evoluzione tardo pleistocenica-olocenica e le linee di riva di epoca storica. *Mediterranee* **2009**, *112*, 23–31. [CrossRef]

98. Esposito, G.; Salvini, R.; Matano, F.; Sacchi, M.; Troise, C. Evaluation of geomorphic changes and retreat rates of a coastal pyroclastic cliff in the Campi Flegrei volcanic district, southern Italy. *J. Coast. Conserv.* **2018**, *22*, 957–972. [CrossRef]

99. Maglio, L. *Castel dell'Ovo Dalle Origini Al Secolo XX, Quaderno 1*, 2nd ed.; AF Istituto Italiano dei Castelli: Sezione Campania, Italy, 2015; pp. 1–64.

100. Pavlopoulos, K.; Vasilios, K.; Katerina, T. Relative sea-level changes in Aegean coastal areas during Holocene: A geoarchaeological view. *J. Earth Sci.* **2010**, *21*, 244–246.

101. Pirazzoli, P.A.; Stiros, S.C.; Arnold, M.; Laborel, J.; Laborel-Deguen, F.; Papageorgiou, S. Episodic uplift deduced from Holocene shorelines in the Perachora Peninsula, Corinth area, Greece. *Tectonophysics* **1994**, *229*, 201–209. [CrossRef]

102. Poulos, S.E.; Ghionis, G.; Maroukian, H. Sea-level rise trends in the Attico–Cycladic region (Aegean Sea) during the last 5000 years. *Geomorphology* **2009**, *107*, 10–17. [CrossRef]

103. Stiros, S.C. The 8.5+ magnitude, AD365 earthquake in Crete: Coastal uplift, topography changes, archaeological and historical signature. *Quat. Int.* **2010**, *216*, 54–63. [CrossRef]

104. Stiros, S.C. Identification of earthquakes from archaeological data: Methodology, criteria and limitations. *Archaeoseismology* **1996**, *7*, 129–152.

105. Stiros, S.C.; Pirazzoli, P.A.; Laborel, J.; Laborel-Deguen, F. The 1953 earthquake in Cephalonia (Western Hellenic Arc): Coastal uplift and halotectonic faulting. *Geophys. J. Int.* **1994**, *117*, 834–849. [CrossRef]

106. Auger, E.; Gasparini, P.; Virieux, J.; Zollo, A. Seismic evidence of an extended magmatic sill under Mt. Vesuvius. *Science* **2001**, *294*, 1510–1512. [CrossRef]

107. Fedi, M.; Cella, F.; D'Antonio, M.; Florio, G.; Paoletti, V.; Morra, V. Gravity modeling finds a large magma body in the deep crust below the Gulf of Naples, Italy. *Sci. Rep.* **2018**, *8*, 8229. [CrossRef]

108. Mattei, G.; Rizzo, A.; Anfuso, G.; Aucelli, P.P.C.; Gracia, F.J. Enhancing the protection of archaeological sites as an integrated coastal management strategy: The case of the Posillipo Hill (Naples, Italy). *Rendiconti Lincei Scienze Fisiche Naturali* **2020**, *31*, 1–14. [CrossRef]

109. Mattei, G.; Rizzo, A.; Anfuso, G.; Aucelli, P.P.C.; Gracia, F.J. A tool for evaluating the archaeological heritage vulnerability to coastal processes: The case study of Naples Gulf (southern Italy). *Ocean Coast. Manag.* **2019**, *179*, 104876. [CrossRef]

 *water*

Article

Key Aesthetic Appeal Concepts of Coastal Dunes and Forests on the Example of the Curonian Spit (Lithuania)

Arvydas Urbis [1], Ramūnas Povilanskas [1,*], Rasa Šimanauskienė [2] and Julius Taminskas [2]

[1] Department of Natural Sciences, Klaipeda University, Herkaus Manto str. 84, 92294 Klaipeda, Lithuania; arvydasurbisku@gmail.com
[2] Laboratory of Climate and Water Research, Nature Research Centre, Akademijos str. 2, 08412 Vilnius, Lithuania; rasa.simanauskiene@gamtc.lt (R.Š.); julius.taminskas@gamtc.lt (J.T.)
* Correspondence: ramunas.povilanskas@gmail.com; Tel.: +370-615-71-711

Received: 12 May 2019; Accepted: 4 June 2019; Published: 7 June 2019

Abstract: The main objective of the study was to elicit key concepts determining the aesthetic appeal of coastal dunes and forests using the example of the Curonian Spit (Lithuania). The mixed approach included three methods: (1) paired comparison survey of 45 coastal landscapes, (2) semi-structured interviews with local inhabitants, and (3) eliciting the key aesthetic appeal concepts by a panel of experts using the Delphi technique. The results of the paired comparison survey show that the most aesthetically appealing landscapes of the Curonian Spit are: (1) white mobile dunes, (2) white dunes with grey (grassland) dunes in the background, and (3) grey dunes with white dunes in the background. The local inhabitants considered the concept of visual coherence as the best, explaining the aesthetic appeal of the dune and the forest landscapes on the spit. The experts of the Delphi survey considered that the concepts of stewardship, naturalness, imageability, and visual scale best define the scenic appeal. The appeal of the least attractive landscapes, in their opinion, was shaped by the concepts of naturalness, disturbance, and complexity. We conclude that the notions of visitors, local inhabitants and experts differ on the aesthetic appeal concepts of coastal dunes and forests, suggesting potential management conflicts.

Keywords: aesthetic appeal; Baltic Sea; coastal dunes; Curonian Spit; Delphi technique; paired comparison survey; psychophysical approach

---

1. Introduction

Aesthetic appeal of landscapes for tourism is among essential intangible ecosystem services [1]. In particular, scenery plays an ever-increasing role in the provision of ecosystem services in coastal areas with high conservation value and recreational appeal [2]. The results of numerous recent surveys of beach attractiveness to visitors around the world show that nearly all beaches with the highest scenic value are located in protected natural coastal areas in Europe [3–8] and worldwide [9–16]. However, regarding the attractiveness of coastal areas for tourism, the challenge of accommodating nature conservation requirements with tourist interests, including their quest for scenic beauty, arises. This challenge is difficult yet necessary to meet [17]. It is difficult because, very often, matching the ecosystem services, the conservation regulations, and the landscape perceptions by the visitors can be problematic [18]. It is necessary because recreation demands must be balanced with conservation needs for natural resources and ecosystem services [19].

This study aimed to explore the psychophysical relationship between the visitors' perceptions and the key aesthetic appeal concepts of coastal dunes and forests on the Curonian Spit, a UNESCO world heritage coastal cultural landscape (Figures 1 and 2).

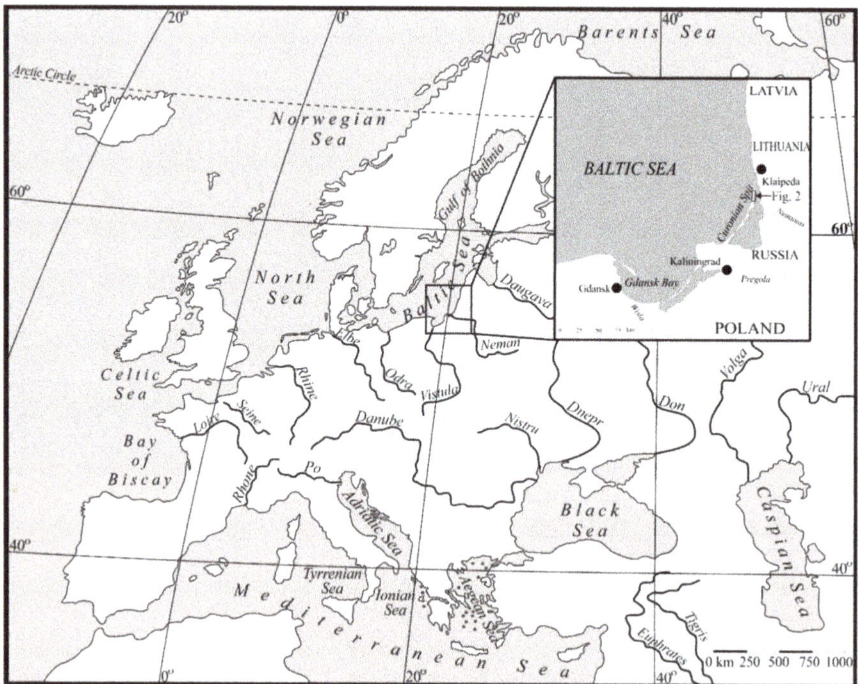

Figure 1. Location of the Curonian Spit on the southeast Baltic Sea coast.

In the study, we assessed a full range of scenes representing coastal dune and forest landscapes of the Curonian Spit—from white (open) mobile dunes and grey (grassland) dunes to Scots pine wood plantations and natural Black alder forests. Our study lies within this broad and rapidly advancing field of eliciting a comprehensive assessment of coastal non-use (intangible) ecosystem values for tourism. It is essential for integrated management of the complex coastal territories which are, as in our case, a transboundary world heritage cultural landscape, a national park, and a popular seaside resort.

We hypothesize that open white and grey dunes are an overlooked asset for sustainable tourism development on the Curonian Spit. The scenic appeal of the dunes is one of the main reasons why tourists visit this unique peninsula. However, despite their immense aesthetic value, the dunes are mostly inaccessible for lay visitors and are overgrown with a natural mixed forest of Scots pine and Silver birch because of excessively strict conservation policies. Hence, the main conclusion of our study was that imageability and visual scale of open and semi-open dune landscapes support the notion of the importance of the "tourist gaze" [20] for the aesthetic appeal of coastal areas as tourist destinations.

Considering the assessment of aesthetic appeal of landscapes, we faced a dilemma of choosing between visitor-based, i.e., "subjective", and expert-based, i.e., "objective" approaches. It is noted in [21] that when investigating visual landscape aesthetics and preferences, a subjective approach is more meaningful and can be adopted, especially concerning the issues involving public interests. On the contrary, as argued in [8], perception varies from person to person, and while there is nothing wrong with a subjective view of the scenery, from a management viewpoint, objectivity should be a must.

This methodological divide predetermines the answer to the critical question of whether the scenic value is a real property inherent in a landscape or a subjective impression based on individual cognitive processes [22]. In other words, the former approach presupposes that scenic beauty is inherent in an original setting [15,23], whereas the latter assumes that scenic beauty is "in the eyes of the beholder" [24,25].

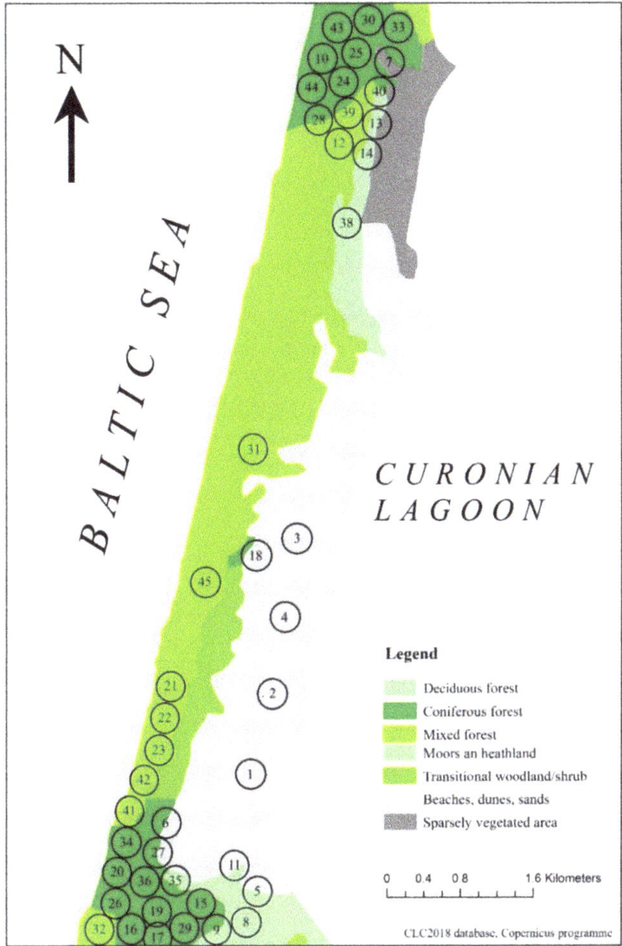

Figure 2. Locations of the sites for taking photos (the numbers refer to the habitats from Table 3).

Daniel et al. [26] were the first to separate aesthetic appraisal into the detailed inventory and the public preference approaches. Zube et al. [25] provided the most comprehensive classification of the approaches in landscape aesthetics, identifying four paradigms of landscape perception research (Table 1). Daniel and Vining [27] refined the classification, and Uzzell [28] extended it to include socio-cultural dimensions. The psychophysical paradigm, an approach grounded in environmental psychology, is a quantitative holistic methodology integrating both "objectivist" and "subjectivist" aesthetic appraisal approaches [29,30]. Psychophysics is defined as a segment of measurement theory and procedure, which attempts to relate environmental stimuli to human sensations, perceptions, and judgments [31,32].

The psychophysical paradigm relies on the assumption that specific, measurable parameters contribute to the perception of scenic beauty and therefore can be predicted [33]. It emphasizes the relationship between the occurrence of environmental events (the physical world) and the perceptual response (the psychological world) of the observer [34].

In other words, from the psychophysical perspective, the aesthetic appeal of a landscape results from an interaction between the physical features of the environment and the perceptual and the judgmental processes of a human observer. It means that aesthetic appeal is neither "inherent in the landscape" nor purely "in the eye of the beholder". The psychophysical paradigm is, therefore, particularly suited for modeling landscape preferences based on geographical data of the physical landscape [35].

Table 1. Paradigms of landscape perception research.

Authors	Paradigms			
	Objectivist (Physical)	Subjectivist (Psychological)		
Lothian [22], Zube et al. [25]	Expert	Psychophysical	Cognitive	Experiential
Daniel and Vining [27] Uzzell [28]	Ecological/Formal aesthetic	Psychophysical	Psychological	Phenomenological

In psychophysical landscape preference surveys, the respondents are asked to evaluate the attractiveness of a scene, typically with a single response concerning aesthetic preference or appeal [36]. The relationships of interest are those between the physical features of landscapes, such as topography, vegetation, and water, and the psychological responses of the viewers expressing judgments of preference or attributing scores for attractiveness. Multiple linear regression is the most commonly used technique to determine relationships [37]. Many studies do not specify any quantitative functions but describe the characteristics of preferred and less-preferred landscapes in relative terms [27,36,38]. Extensive psychophysical aesthetic valuation studies require a full range of scenes to be selected to represent all of the physical characteristics used as predictors of aesthetic values [39]. They are sensitive to subtle landscape variations and consistently provide robust aesthetic quality assessments regarding changes in both landscapes and observers [27,32].

Another advantage of the psychophysical approach towards aesthetic valuation is that it can accommodate evolutionary and cultural interpretations of landscape aesthetics. The evolutionary interpretation advocates for a persistent cross-cultural preference for certain landscape types or the environmental context in which the transactions between individuals and landscapes occur. It stems from an intrinsic standard of beauty developed by humans in the hunter-gatherer phase of their evolutionary development [21,40–43]. It can be especially pertinent for studying the aesthetic appeal of coastal landscapes, since sheltered coasts, along with savannah-type landscapes, have been the habitats satisfying biological needs of early humans to survive and to spread worldwide [44].

Adepts of the cultural interpretation of landscape aesthetics argue that the appreciation of natural landscapes relies on a dynamic, poly-sensual, and active interaction with the environment and an understanding of its ecosystem functions and services [29]. In other words, the perception or the experience of the landscape is dependent on cultural background and personal attitudes, beliefs, and ideas of each observer focusing either on perceived functions of the observer [45] or on cultural conventions regarding landscapes [46]. Human factors affecting aesthetic preference include socio-cultural groupings, education, personality, profession, and involvement [47]. There are many pieces of evidence that individual preferences change in response to the context in which they are sought depending on an individual's qualifications, their understanding of the cultural background of a landscape, and their experiential connection with it [48,49].

An integrated approach to accommodate both interpretations was developed summarizing the aesthetic quality research outputs of the last half-century, covering both theories and identifying four hierarchical levels of aesthetic abstraction ranging from nine abstract concepts to concrete descriptors of the physical landscape: concepts → dimensions → landscape attributes → indicators [50]. The "concept" and the "dimension" level are both abstract conceptual levels, whereas landscape attributes and indicators are the aspects of a physical landscape.

A concept is an umbrella term that includes several visual dimensions [50]. These dimensions define different aspects of the concepts at a general level. Dimensions are predetermined by physical attributes of the landscape, while the landscape attributes are expressed using visual indicators. The indicators represent the basic level at which the physical landscape attributes are measured to allow different landscapes to be compared or to identify any change in the same landscape over time. Such a systematic rendering is consistent with the psychophysical approach towards aesthetic valuation, reflecting the interdependence and the multi-dimensionality among human attitudes and landscape perceptions [51–54].

2. Materials and Methods

The 32.6-km long Great Curonian Dune Ridge of 40 m to 60 m high mobile dunes is the second longest coastal mobile dune ridge in Europe [55]. It is protected as a strict nature reserve within Kurshskaya Kosa national park on the Russian part of the spit (est. 1987) and Kuršių Nerija national park on the Lithuanian part (est. 1991). Due to its unique blend of nature and culture values, in 2000, the whole Curonian Spit was included in the UNESCO World Heritage List as a transboundary cultural landscape of outstanding international importance [56].

The unique landscape features of the Curonian Spit as we see them today have resulted from an intensive advance of mobile dunes from the 17th to the 18th century and their stabilization in the 19th century. In that period, ancient forested parabolic dunes, which have prevailed since Holocene, were destroyed by the sand intensively washed ashore from the Baltic Sea and were replaced by mobile barchans [57]. Much of the necessary primary work on dune stabilization of the Curonian Spit ended by the beginning of the 20th century [58]. These efforts stopped the rapid evolution of dune landscapes on the spit, thereby creating a rather specific landscape mosaic where the natural landscapes of the mobile dunes and the relics of the parabolic dunes—overgrown with natural mixed forest—are found along with various forest landscapes [59].

The Curonian Spit is a special place among the Lithuanian coastal areas, since it bears salient societal connotations such as "unique landscape", "place of outstanding natural beauty", "the greatest natural treasure of Lithuania", and "national pride". The objective of the Kuršių nerija national park is to preserve the Grand Curonian Dune Ridge, mobile bare dunes with marram grass ("white dunes"), fixed dunes with herbaceous vegetation ("grey dunes"), fossil dune and soil relics. Although forests are not even mentioned among the nature conservation targets of the National Park, they form an indispensable part of the local landscape. A forest plantation symbolizes the victory of the human order over the dune wilderness. Care for the forest plantations is ideologically motivated as the responsibility for the future generations [56], and foresters are cherished as wardens of the care for the coastal environment [60].

On an international scale, mobile dunes are valued as very diverse and variable environments and as multifunctional systems with great importance for our society [61]. Human activities such as forestation, recreation, or urbanization are perceived as disorder and left outside the reversed and the ecologically delimited boundaries of the concept of dunes. Forests and other human interventions into dunes are aliens to the intrinsic natural order of the very dynamic mobile dune area. As dynamic systems, mobile dunes are considered to be highly fragile, and their natural order is very vulnerable. The appeal is apparent; it is a challenge to ecologists, planners, engineers, and politicians to stop further deterioration and destruction of these fragile environments [61].

Along with Silver birch and Black alder forests, other forest landscapes under consideration comprise Norway spruce, Scots pine, and Mugo pine plantations that were planted to stabilize mobile dunes in the 19th and the 20th centuries. As a side effect of the maintenance of these plantations (which play an essential recreational role), forest vegetation proliferates into the white and the grey dune areas, thus accelerating the succession of vegetation and the natural forestation [62]. This process results in a loss of open dune landscapes. Hence, some visitors perceive the forested dune landscape of the Curonian Spit as a natural habitat, while others view it as a monument to the arduous toil of foresters who stopped the menacing sand drift.

The third group of visitors appreciate the beauty of open dunes and regard forest plantations as a redundant alien intrusion to be erased [56].

The variety of approaches taken to apply visual methods for examining people's perceptions of destinations range from the highly interventionist, wherein researchers choose the visual images to be studied, to the highly participatory, where the research subjects themselves collect photographic representations of those images, which are subsequently analyzed by the researchers [63]. Hitherto, practical application of the psychophysical approach for eliciting the preferences for and the attractiveness of coastal landscapes founded in the landscape's physical attributes was limited. Due to the complexity of the psychophysical approach using photographs as visual stimuli, coastal research case studies focusing on aesthetic appeal still mostly rely on the objectivist paradigm [64].

However, a mixed approach to the psychophysical interpretation of aesthetic appeal combining both quantitative (paired comparison survey) and qualitative methods (semi-structured interviews and the Delphi technique) can deliver results that contribute substantially to the development of a comprehensive methodology for the aesthetic appraisal of coastal landscapes as tourist destinations. The rationale for employing the mixed approach for interdisciplinary coastal management studies is explained in-depth in our previous study [18]. This innovative approach advances research in scenic aesthetics perception and appraisal and is the first study of this kind worldwide (Figure 3). The study employed a combination of quantitative and qualitative methods requiring a quantitative paired comparison survey, whose results were interpreted using the Delphi technique and face-to-face semi-structured interviews.

We applied the method of paired comparisons for the relative ranking of dune and forest landscapes of the Curonian Spit using an approach drawn from the literature [22,65,66]. This response method consists of the systematic pairing of objects or stimuli. As each pair is presented, the observer makes a judgment indicating which member of the pair has a higher value of some attribute. In this way, aesthetic values of several landscapes can be elicited by presenting pairs of landscape scenes to the observer. On each presentation, the observer would indicate which scene is perceived to be more attractive regarding scenery. The advantage of this method is that it presents a relatively simple task for the respondent [19].

The photographer used a digital camera Nikon N5005 wide zoom with a 28–200 mm focal length Tamron lens with an approximately 75° degree field of view. It was fixed on a high-angle Gitzo tripod. It is commonly considered that photographs taken with a 50 mm lens are closer to the experience of the landscape; yet, work undertaken by Wherrett [32] found the differences between 35 mm and 50 mm focal lengths negligible. Images were obtained at a high resolution (1536×1024 pixels). All photographs were taken on several beautiful days in summer by the same photographer.

To avoid a potential problem of photographic representations recording a limited field of view [67], we used panoramic images presenting a viewshed within an angle of approximately 150° (double width wide zoom photos). The foreground (20 m × 20 m) was the spatial dimension used for the photographs of the forested dunes, and the near view (next 20 m × 20 m) was the spatial dimension for the photographs of the open dune landscapes. It was of an appropriate size for habitat definition and heterogeneity to be evident on a photograph measuring 30 cm × 15 cm while allowing for vegetation detail to be seen [65,68].

Finally, in the screening phase, a team of professional judges made a unanimous group decision on the ultimate selection of the photographs, ensuring that all possible combinations of the surveyed landscape attributes (stages of succession, heterogeneity, and naturalness of scenery) were indeed represented by the images selected [32,68]. The final set comprised monochromatic double photos representing all dune landscape types and land features on the Curonian Spit from different viewing positions. As the visual conditions are quite diverse, and the level of patchiness of the landscapes is quite high in the dunes, we used three photographs for every analyzed landscape type [19].

Although some references on photograph-based landscape simulations indicate that lack of color makes a difference in simulation validity [69], applying larger samples of landscape photographs in the questionnaires in field conditions requires substantial funds for printing. This limitation rendered it possible to use only monochrome photographs in our survey.

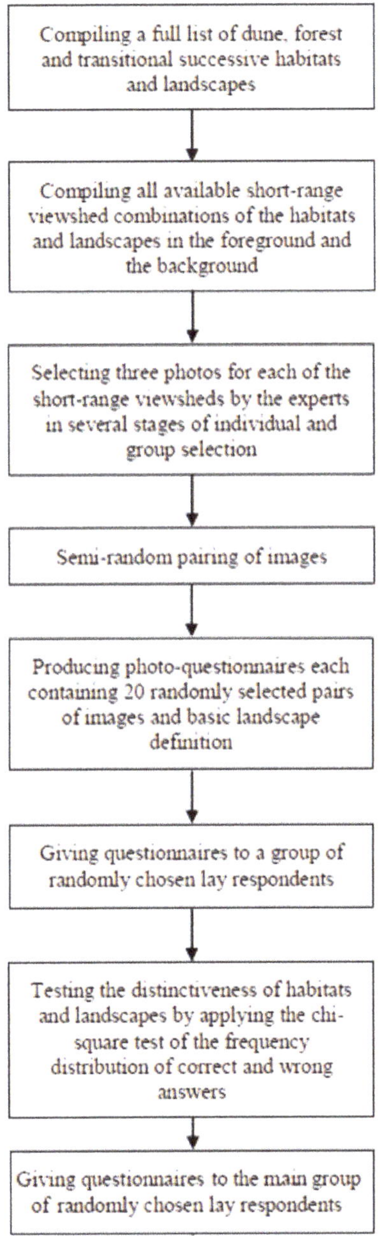

Figure 3. Research flow chart.

We relied on the argument that monochrome photography supposedly captures both the historical ambience and the closeness of tourists to the "other" nature, landscape, heritage, or local culture [19,70]. This argument is coherent with the psychophysical approach on which our study relied.

Before the survey, we compiled a full list of dune, forest, and successive transitional habitats of the Curonian Spit using a pre-existing national database of NATURA 2000 habitats. Then, we compiled all short-range viewshed combinations of the habitats in the foreground and the near view. As a result, the complete list of the short-range viewsheds of the Curonian Spit landscapes under scrutiny comprised 45 combinations of dunes, brushwood, and forest habitats in the foreground and the near view. Other potential short-range viewsheds of dunes, brushwood, and forests are either absent on the spit or are so rare that visitors rarely encounter them in practice.

A pre-test of the survey instrument was conducted using a Chi squared-test to identify whether the difference among different landscape type images and the equivalence of the three different images representing the same landscape type was statistically significant. As the main result of the pre-testing, we proved ($p < 0.00001$) that lay visitors of the Curonian Spit indeed distinguish different landscapes and their combinations portrayed in monochrome photographs [19]. Therefore, the 135 images representing 45 forest and dune landscapes of the Curonian Spit as well as their intermediate succession and management stages served as visual stimuli for the main paired comparison study.

The respondents had a chance to see any ten of all possible pairs of the objects (N * (N − 1)/2 = 990 if N = 45 sorted in random order. The objects were repeated 15 times with each of three photos representing each landscape with equal chances to be paired with any of the photos representing other landscapes. It yielded a final sample of 14,850 pairs presented to 1485 randomly selected visitors of the Curonian Spit. Every respondent had equal chances to receive any of the 14,850 pairs of large monochrome images for the assessment provided in ten randomly sorted pairs of photographs in a flipper album. The limit for a single evaluator was kept as low as ten pairs of photographs to be examined, considering this number to be the highest possible for an attentive and dedicated judgment by an individual [19,21,65], thus preventing visual and mental fatigue of the evaluators.

The paired comparison survey took place in July and August 2017 at the ferry pier upon leaving the Curonian Spit for the continent. While the ferry pier was convenient for the interviewer, it also was the best location because the visitors had completed their stay on the spit and had enough time while waiting for the ferry. An additional advantage was that upon departure, information was accumulated from the whole vacation experience [71], since a vacation provides an episodic memory containing personal experiences related to a particular time and place [72]. The surveys took place throughout the day, both on weekdays and weekends. The average time used by a visitor in responding to the questionnaire, including socio-demographic questions, was approximately 15 min. Respondents were assured verbally that the information they provided would be used only for research purposes and that their responses were anonymous and confidential.

The population in the study comprised all domestic summer visitors to the Lithuanian part of Curonian Spit, currently amounting to 387,000 according to the data from the toll-point (Table 2). The sample ($n = 1485$) included randomly selected adult domestic visitors to the Curonian Spit (0.38% of the total visitors population). We assessed two principal socio-demographic factors (age and gender) for their effects on landscape preference. The results show that the sample of respondents was not significantly different from all domestic visitors to the Lithuanian part of the Curonian Spit during the two-month study period of the peak tourism season (July and August) in terms of average age and gender ($p < 0.05$). We also found that age and gender did not have any significant effects on landscape preferences, which was in line with other similar studies [32].

Interpretation of the survey results started from the ranking of the surveyed landscapes relative to each other regarding their aesthetic appeal. The ranking was done using a simple ranking technique. It involved the calculation of the proportion of times each photo was judged as "more attractive" than any other photo. These proportions were then transformed to normalized values that represented the judgments on aesthetic values of various landscapes. For the interpretation of the ranking results, we conducted twelve interviews with local inhabitants of the Curonian Spit in April 2018 and February 2019. The interviews were in-depth, open-ended, and semi-structured around the concepts defining the aesthetic appeal of the Curonian Spit [50,73,74].

The interviewees included officials, park rangers, resort staff, and inhabitants of the Curonian Spit (Table 2). The sample was evenly distributed between women (50%) and men (50%).

Table 2. The sample characteristics and the survey methods of quantitative and qualitative surveys.

Sample Characteristics and Survey Methods	Quantitative Survey	Qualitative Survey
Number of respondents	1485	12
Representativity	Representative sample of domestic summer visitors	Representative sample of local stakeholders
Survey method applied	Paired comparison survey	Semi-structured face-to-face interviews
Survey months	July, August	April, February
Average duration	~15 min.	25–30 min.
Number of landscape types	45	45
Number of photographs	135	135
Additional questions asked	Socio-demographic questions (age, gender, residence, visit frequency)	Socio-demographic questions and opinion about the presented landscape appeal concepts

The mean age of the respondents was 43 years. The kick-off question for the interviews was "What are, in your opinion, the key elements of scenic beauty and conservation values of the spit?"; next, the stakeholders were introduced to the different concepts of landscape appeal and shown different landscape types used in the quantitative survey. They were not asked to compare but rather to tell which concepts best described the most appealing landscape types.

The concepts defining the aesthetic appeal of the five most attractive and the five least attractive landscapes of the Curonian Spit were elicited using the Delphi technique. The technique uses a series of iterative rounds interspersed by controlled feedback for the formulation of consensus judgment within a group of experts in a relevant field who do not interact directly with each other. For our study, three rounds of expert judgment were performed using an approach drawn from previous Delphi studies in landscape management and heritage tourism found in the literature [75–81]. This Delphi study was completed in April–September 2018 using the e-mail communication to relate a panel of ten experts in coastal tourism, landscape architecture, and dune geomorphology.

The visual concepts of landscape aesthetics defined by Tveit et al. [50] include:

1. *Stewardship* (the presence of a sense of order and care contributing to perceived accordance with an "ideal" situation and reflecting care for the landscape through management).
2. *Coherence* (a reflection of the unity of a scene enhanced through repeating patterns of color and texture, and a reflection of the correspondence between land use and natural conditions).
3. *Disturbance* (lack of contextual fit and coherence where elements deviate from the context due to temporary and permanent interventions occurring in the landscape).
4. *Historicity* (historical continuity reflecting the visual presence of different time layers and historical richness related to the amount, the condition, and the diversity of cultural elements).
5. *Visual scale* (perceptual units reflecting the experience of landscape visibility and openness).
6. *Imageability* (qualities of a landscape present in totality or through elements; landmarks and unique features, both natural and cultural, making the landscape create a strong visual image in the observer and making landscapes distinguishable and memorable).
7. *Complexity* (diversity and richness of landscape elements and features and their interspersion).
8. *Naturalness* (closeness to a preconceived natural state).
9. *Ephemera* (elements and land-cover types changing with season and weather).

In the scoping round, the selected experts were introduced to the conceptual framework of the hierarchical levels of aesthetic abstraction ranging from the abstract concepts to the concrete descriptors of the landscape [50]. They were asked to attribute the most suitable primary and secondary concepts from the offered nine options to the landscape types of the Curonian Spit using the photos from the paired comparison survey and to explain their choice. The first judgments were collected, summarized, and sent back to the panelists for further evaluation.

They included the responses of other panelists so that participants could read the other opinions and adjust their own opinions.

In the next two iterative convergence rounds, the revised judgments with the feedback from other panelists were sent back again to the participants. The Delphi technique aims to achieve a higher quality of response on expert issues than a single round questionnaire could achieve [75]. As we fed back the results from the previous rounds, there was a tendency among the panelists to converge their opinions around the emerging key concepts and look for a consensus.

The objective to narrow the judgments of the respondents was successfully achieved, and a partial consensus of 80% (162 of 200) of the judgments by the experts regarding the primary and the secondary concepts defining the aesthetic appeal of the ten landscape types was reached after the third round.

3. Results

The main results of the study are summarized in Table 3 and explained in Figure 4. They show that the five most aesthetically appealing landscapes of the Curonian Spit are: (1) white mobile dunes (Figure 5), (2) white dunes with grey dunes in the background (Figure 6), (3) grey dunes with white dunes in the background (Figure 7), (4) mature Scots pine wood with dry grassland in the background (Figure 8), and (5) mature Scots pine wood (Figure 9). Meanwhile, the five least attractive landscapes of the Curonian Spit are: (41) middle-aged Scots pine woods with Juniper, (42) mature Black alder forest with middle-aged Black alder in the background, (43) young Silver birch stands (Figure 10), (44) young Black alder stands, and (45) deciduous forest clearings (Figure 11).

The results of our research also show that summer visitors of the Curonian Spit prefer open landscapes with a clear visual scale (white and grey dunes and edges of mature Scots pine woods with dunes and grassland). They also prefer artificially nurtured landscapes (Scots pine plantations) to ecologically richer landscapes of mixed young stands and brushwood. Also, all local inhabitants of the Curonian Spit value the landscapes that are denoted by the concept of coherence (Figure 4).

Coherence is a reflection of the correspondence between land use and natural conditions [50]. The key dimensions defining coherence are harmony, unity/holistic, land-use suitability, and readability of a landscape, while the main physical landscape attributes are pattern and structure [73]. In the case of the spit, coherence designates familiarity, comprehensiveness, and readability of landscapes both as a recreational space and a fragile heritage. It is enhanced by anything that helps to organize the patterns of light, size, texture, or other elements into a few significant units [82].

The most coherent landscape of the spit, i.e., which best "hangs together", according to the interviewed local inhabitants comprises mobile dunes with some patches and edges of forest vegetation (up to 50%) and occasional clumps of trees. As a whole, it presents a complex yet comprehensible scene [83].

The coherence of the plantations with an open background is notable for its similarity to the savanna environments of our speciation [42]. It offers the visual relationship between the landscape components that simultaneously afford prospect (open views from which hazards can be spotted) with refuge (protected settings that prevent one from being seen) [40]. Coherence enhances people's ability to orient themselves, which is dependent on the legibility of the landscape [84]. A coherent landscape presents itself as a structured wholeness—as a unity—which is experienced aesthetically as a pleasing harmonic entirety, a capability "to tell its story" or to deliver orientation patterns [85].

On the contrary, in the opinion of the Delphi expert panel, another two primary concepts define the aesthetic appeal of the most and the least attractive of the 45 dune and forest landscapes on the Curonian Spit, namely, naturalness (both in favorable and unfavorable terms) and stewardship (only in favorable terms) (Table 3). The secondary concepts defining the aesthetic appeal are the concepts of imageability and visual scale defining the aesthetic appeal in favorable terms, while the concepts of disturbance and complexity define the aesthetic appeal in unfavorable terms.

Table 3. Results of the paired comparison and the Delphi survey of the Curonian Spit landscapes.

Rank	Normalized Rating	Landscape Type	Key Concepts [50]	
			Primary Concept	Secondary Concept
1.	1.00	White (open) mobile dunes (Figure 5)	Naturalness	Imageability
2.	0.92	White dunes with grey dunes (Figure 6)	Naturalness	Visual scale
3.	0.91	Grey dunes with white dunes (Figure 7)	Naturalness	Visual scale
4.	0.82	Mature Scots pine with grey dunes (Figure 8)	Stewardship	Visual scale
5.	0.79	Mature Scots pine (Figure 9)	Stewardship	Visual scale
6.	0.76	Young Scots pine with Grey dunes	n.a.	n.a.
7.	0.76	Grey dunes	n.a.	n.a.
8.	0.74	Middle-aged Silver birch	n.a.	n.a.
9.	0.73	Mature Norway spruce	n.a.	n.a.
10.	0.69	Middle-aged Scots pine with Mugo pine	n.a.	n.a.
11.	0.69	Mature and middle-aged Scots pine	n.a.	n.a.
12.	0.68	Forest glade in mature Scots pine stands	n.a.	n.a.
13.	0.68	Mature Silver birch	n.a.	n.a.
14.	0.68	Mature Silver birch with dry grasslands	n.a.	n.a.
15.	0.67	Mature Silver birch with Scots pine	n.a.	n.a.
16.	0.66	Dry grasslands	n.a.	n.a.
17.	0.62	Middle-aged and young Scots pine	n.a.	n.a.
18.	0.59	Grey dunes with middle-aged Scots pine	n.a.	n.a.
19.	0.59	Young Scots pine stands with Mugo pine	n.a.	n.a.
20.	0.59	Young and mature Norway spruce	n.a.	n.a.
21.	0.59	Middle-aged Scots pine	n.a.	n.a.
22.	0.59	Middle-aged Black alder	n.a.	n.a.
23.	0.59	Middle-aged Norway spruce	n.a.	n.a.
24.	0.59	Mature and middle-aged Norway spruce	n.a.	n.a.
25.	0.58	Juniper brushwood	n.a.	n.a.
26.	0.58	Young and middle-aged Scots pine	n.a.	n.a.
27.	0.58	Mature Silver birch with Norway spruce	n.a.	n.a.
28.	0.57	Grey dunes with young Scots pine	n.a.	n.a.
29.	0.57	Young Scots pine	n.a.	n.a.
30.	0.56	Mugo pine with middle-aged Scots pine	n.a.	n.a.
31.	0.52	Young Norway spruce	n.a.	n.a.
32.	0.52	Mature Black alder forest	n.a.	n.a.
33.	0.49	Dry grasslands with Mugo pine	n.a.	n.a.
34.	0.48	Middle-aged Scots pine with Grey dunes	n.a.	n.a.
35.	0.47	Juniper with middle-aged Scots pine	n.a.	n.a.
36.	0.46	Willow stands with young Scots pine	n.a.	n.a.
37.	0.45	Grey dunes with Willow stands	n.a.	n.a.
38.	0.45	Mugo pine plantation	n.a.	n.a.
39.	0.42	Young and mature Silver birch	n.a.	n.a.
40.	0.42	Willow stands with Grey dunes	n.a.	n.a.
41.	0.39	Middle-aged Scots pine with Juniper	Naturalness	Disturbance
42.	0.39	Mature and middle-aged Black alder	Naturalness	Disturbance
43.	0.35	Young Silver birch (Figure 10)	Naturalness	Disturbance
44.	0.35	Young Black alder	Naturalness	Complexity
45.	0.34	Deciduous forest clearing (Figure 11)	Naturalness	Complexity

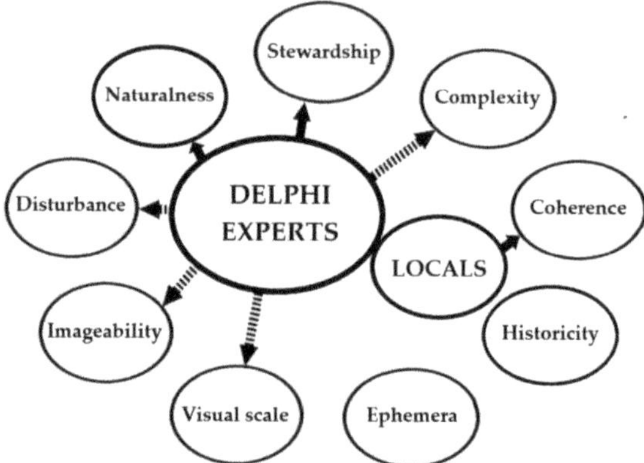

Figure 4. Key aesthetic appeal concepts of coastal dunes and forests of the Curonian Spit.

Figure 5. White (open) mobile dunes of the Curonian Spit (photo by Arvydas Urbis).

Figure 6. White dunes with grey (grassland) dunes in the background (photo by Arvydas Urbis).

Figure 7. Grey dunes with white dunes in the background (photo by Arvydas Urbis).

Figure 8. Mature Scots pine wood with dry grasslands in the background (photo by Arvydas Urbis).

Figure 9. Mature Scots pine wood (photo by Arvydas Urbis).

Figure 10. Young Silver birch stands (photo by Arvydas Urbis).

Figure 11. Deciduous forest clearing with undergrowth (photo by Arvydas Urbis).

The concept of *naturalness* indicates how close a landscape is to a perceived natural state, where perceived naturalness can be different from ecological naturalness [50]. The main dimensions of naturalness are intactness, wilderness, naturalness, and ecological robustness. As the results of the aesthetic valuation studies in the Netherlands show, opinions of different social groups differ most concerning the appreciation of natural, unmanaged coastal landscapes, such as marshes and dunes [35].

Indeed, concerning the "naturalness" concept, mobile dunes are widely acknowledged as a "natural" coastal landscape in terms of aesthetics [14] but also ecological terms [86].

Additionally, forests of the Curonian Spit might look entirely natural for lay visitors, particularly the old stands of Scots pine and Norway spruce with dense understorey vegetation. However, different from a standard assertion that the level of the succession is an expression of naturalness [84], the succession on the Curonian Spit does not indicate the degree of naturalness. The proliferation of forest to mobile dunes is a natural phenomenon. However, the national park rangers heavily regulate it by forest management activities. Instead, a more significant indicator is a varied edge between the forest and the dune habitats, which may be perceived as more natural compared to a straight edge [45].

The concept of *stewardship* reflects human care for the landscape through active and careful management as the presence of a sense of order and care contributing to perceived accordance with an "ideal" situation [50]. Sevenant and Antrop [87] found that stewardship or maintenance as a concept describing scenic beauty is correlated by viewers with the term "well maintained" as a cognitive item. In order to analyze the presence of care in the landscape, "cues of care" are used [46]. These cues are familiar to visitors and tell them whether the landscape is well-nurtured or not.

In the case of the Curonian Spit, tidy and well-kept Scots pine plantations are the most obvious example of dedicated human care for the landscape [88]. The visitors of the national park regularly mention the sense of order, the sense of care, and the upkeep as essential items determining landscape preference [18]. On the spit, large monocultural forest plantations indicate a lower level of naturalness than small forest patches in the mobile dunes. This observation is different from the notion that a landscape in which woodland consists of small, fragmented patches rather than one sizeable woodland patch may be interpreted as less natural [73].

The secondary concept of *imageability* designates qualities of a landscape present in totality or through elements, landmarks, and unique features—both natural and cultural—making the landscape create a sharp visual image in the observer and making landscapes distinguishable and memorable [50]. Imageability applies to qualities that are special for a landscape and hence make the place distinguishable from other places [89]. The landscape of the mobile dunes of the Curonian Spit is one of the most iconic landscapes in Lithuania due to its exotic features, which are unique for this country, and due to its liminality and sublimity [90].

Visual scale as a concept comprises perceptual units that reflect the experience of landscape visibility and openness. Weinstoerffer and Girardin [91] use it as an indicator defined by the ease with which an observer can obtain an extensive view over the landscape. Sevenant and Antrop [87] also found that visual scale as a concept describing scenic beauty is correlated by viewers with the term "vast" as a cognitive item. The vast landscapes of white and grey dunes of the Curonian Spit together with the dry grassland in the background of mature pine woods are appreciated for their beauty and salient societal connotations [56,90].

The concept of *disturbance* is defined as a lack of contextual fit and coherence where elements deviate from the context [50]. Other dimensions related to disturbance are intrusion, alteration, and impact understood both in terms of scenery and of landscape ecology [92]. It is a broadly interpreted concept, as the context of introduced elements determines the pattern and the scale of the impact [50].

For instance, in the worldwide survey of coastal scenery, the disturbance is defined as an audial factor rather than a visual intrusion, causing discomfort for beach visitors [3,13]. For the natural and the semi-natural landscapes of the Curonian Spit, the perceived disturbance is of natural origin epitomized by a mixture of dense forest stands with brushwood with low appeal for tourists in terms of both aesthetic delight and visitation comfort.

The concept of *complexity* is similar to the heterogeneity or the patchiness of the landscape structure correlated with landscape diversity [65]. The main dimensions are diversity, variation, the complexity of patterns, and shapes. Also, it can be divided into two properties (or dimensions), diversity and edge [93]. Diversity refers to the abundance and the evenness of land-cover classes in the view, and edge refers to the number of ecotones dividing up habitat types. However, in the case of the Curonian

Spit, complexity is interpreted as a proxy concept for landscape fuzziness in terms of attractiveness for tourists according to comments provided by the Delphi study panelists.

4. Discussion

The results of this study are coherent with the findings of other similar studies from various parts of the world. Experiences developed through tourist–landscape interactions are independent of any historical knowledge or cultural awareness of the landscape itself. They are dependent upon the cultural and the knowledge background of the viewer [94]. While comparing the landscape perception of an expert and lay people, Vouligny et al. [95] found that those two groups use different visual criteria, since an expert does not experience the landscape continuously.

Therefore, to capture the value of ordinary landscapes (i.e., landscapes that do not stand out in any particular way), a combination of expert and lay people approaches is necessary [95]. Our case study proves that this is also true for outstanding landscapes as well. As results of a similar case study from Australia show, people tend to give high importance to those attractions which are known as tourist icons for the region [96]. Likewise, a survey of forested landscapes in east Texas has shown that spatial configuration, especially openness, has an essential impact on scenic beauty [97].

Another important feature that matches the dune landscapes of the Curonian Spit with other coastal landscapes, such as the coastal landscapes of the North Sea or the Baltic Sea or the riparian landscapes of vast river valleys, is their widely appreciated aesthetic appeal [98,99]. Hence, a broad agreement is that the attention recently paid to ecosystem services, which includes cultural services such as spiritual and aesthetic experiences, encourages the consideration of scenic landscape values in land management policies [100]. However, the specific perceptions of landscape values may widely differ from one society to another. These cultural characteristics play a crucial role in how people act on the landscape; different cultures see the physical environment in different ways [101].

5. Conclusions

Three conclusions regarding the key aspects defining the aesthetic appeal of coastal landscapes for tourism could be drawn from the results of the study. First, the findings of our study support the notion that the psychophysical approach towards aesthetic appraisal using proxy visual stimuli is appropriate for the assessment of the aesthetic appeal of coastal landscapes eliciting realistic and reasonable rankings. The psychophysical approach towards valuation of the aesthetic appeal of coastal dunes and forests of the Curonian Spit embraces both the evolutionary and the cultural interpretations of landscape aesthetics in terms of "ordering of nature" [56]. It takes into account the interdependence and the multi-dimensionality among human attitudes and landscape perceptions.

The second conclusion is that it is essential to collect a representative sample of both lay and professional evaluators in order to elicit realistic ranking values of aesthetic appeal and to set a proper framework for their interpretation. As we have seen from the results of the survey and their interpretation, ecological robustness and aesthetical appeal may mean different things to different stakeholders. For instance, the locals of the Curonian Spit are well aware of the World Heritage status awarded to the Curonian Spit by UNESCO. However, there is no clear understanding or consensus of what this means. Many locals relate the UNESCO status to the beauty of the local nature and to keeping the Curonian Spit, as a seaside resort, tidy and clean [18].

Finally, it is necessary to realize that the aesthetic appeal of a landscape or a scene for tourism is transient and contingent on various landscape history and legacy effects resulting in variations of how different landscapes are valued [1,102]. The values, particularly those of coastal landscapes, change with changes in perceptions of the users towards ecosystems and their functions [103,104]. Environmental attitudes, knowledge, and experiences might influence the aesthetic appreciation of iconic and symbolic landscapes [105].

However, it is evident that imageability and visual scale of open and semi-open dune landscapes in contrast to the "messiness" of disturbance and complexity of ecologically more diverse wet forests and clearings support the notion of the importance of the "tourist gaze" [20] for the aesthetic appeal of coastal areas as tourist destinations. With both key concepts of naturalness and stewardship related to the highest aesthetic value scores of the dune and the pine wood landscapes, we cannot confirm an essential cognitive opposition between these two concepts [50,73]. The main practical implication of the study is that the proposed method can help the National Park authorities in regulating uses and activities in landscapes that are under protection without conducting new surveys each time when they have to decide about the landscape management priorities.

Author Contributions: Conceptualization, A.U. and R.P.; methodology, A.U. and R.P.; validation, A.U.; formal analysis, A.U. and J.T.; investigation, A.U.; data curation, A.U.; writing—original draft preparation, A.U.; writing—review and editing, R.P.; visualization, A.U. and R.Š.; supervision, R.Š. and J.T.; project administration, R.Š.; funding acquisition, R.P.

Funding: Lithuanian Research Council funded this research, grant number NKPDOKT15019.

References

1. Bürgi, M.; Silbernagel, J.; Wu, J.; Kienast, F. Linking ecosystem services with landscape history. *Landsc. Ecol.* **2015**, *30*, 11–20. [CrossRef]

2. Robert, S. Assessing the visual landscape potential of coastal territories for spatial planning. A case study in the French Mediterranean. *Land Use Policy* **2018**, *72*, 138–151. [CrossRef]

3. Anfuso, G.; Williams, A.T.; Rangel-Buitrago, N. Examples of Class Divisions and Country Synopsis for Coastal Scenic Evaluations. In *Coastal Scenery: Evaluation and Management*; Rangel-Buitrago, N., Ed.; Springer: Cham, Switzerland, 2018; pp. 143–210.

4. Armaitienė, A.; Boldyrev, V.L.; Povilanskas, R.; Taminskas, J. Integrated shoreline management and tourism development on the cross-border World Heritage Site: A case study from the Curonian Spit (Lithuania/Russia). *J. Coast. Conserv.* **2007**, *11*, 13–22. [CrossRef]

5. Ergin, A.; Williams, A.T.; Micallef, A. Coastal scenery: Appreciation and evaluation. *J. Coast. Res.* **2006**, *22*, 958–964. [CrossRef]

6. Iglesias, B.; Anfuso, G.; Uterga, A.; Arenas, P.; Williams, A.T. Scenic value of the Basque Country and Catalonia coasts (Spain): Impacts of tourist occupation. *J. Coast. Conserv.* **2018**, *22*, 247–261. [CrossRef]

7. Mooser, A.; Anfuso, G.; Mestanza, C.; Williams, A.T. Management Implications for the Most Attractive Scenic Sites along the Andalusia Coast (SW Spain). *Sustainability* **2018**, *10*, 1328. [CrossRef]

8. Phillips, M.R.; Edwards, A.M.; Williams, A.T. An incremental scenic assessment of the Glamorgan Heritage Coast, UK. *Geogr. J.* **2010**, *176*, 291–303. [CrossRef]

9. Anfuso, G.; Williams, A.T.; Cabrera Hernández, J.A.; Pranzini, E. Coastal scenic assessment and tourism management in western Cuba. *Tour. Manag.* **2014**, *42*, 307–320. [CrossRef]

10. Anfuso, G.; Williams, A.T.; Casas Martínez, G.; Botero, C.M.; Cabrera Hernández, J.A.; Pranzini, E. Evaluation of the scenic value of 100 beaches in Cuba: Implications for coastal tourism management. *Ocean Coast. Manag.* **2017**, *142*, 173–185. [CrossRef]

11. Da Costa, C.S.; Portz, L.C.; Anfuso, G.; Camboim Rockett, G.; Guimarães Barboza, E. Coastal scenic evaluation at Santa Catarina (Brazil): Implications for coastal management. *Ocean Coast. Manag.* **2018**, *160*, 146–157. [CrossRef]

12. De Araújo, M.C.B.; Da Costa, M.F. Environmental quality indicators for recreational beaches classification. *J. Coast. Res.* **2008**, *24*, 1439–1449. [CrossRef]

13. Ergin, A. Coastal Scenery Assessment by Means of a Fuzzy Logic Approach. In *Coastal Scenery: Evaluation and Management*; Rangel-Buitrago, N., Ed.; Springer: Cham, Switzerland, 2018; pp. 67–106.

14. Pranzini, E.; Williams, A.T.; Rangel-Buitrago, N. Coastal Scenery Assessment: Definitions and Typology. In *Coastal Scenery: Evaluation and Management*; Rangel-Buitrago, N., Ed.; Springer: Cham, Switzerland, 2018; pp. 107–142.

15. Williams, A.T. The Concept of Scenic Beauty in a Landscape. In *Coastal Scenery: Evaluation and Management*; Rangel-Buitrago, N., Ed.; Springer: Cham, Switzerland, 2018; pp. 17–42.

16. Williams, A.T. Some Scenic Evaluation Techniques. In *Coastal Scenery: Evaluation and Management*; Rangel-Buitrago, N., Ed.; Springer: Cham, Switzerland, 2018; pp. 43–66.

17. Atauri, J.A.; Bravo, M.A.; Ruiz, A. Visitors' Landscape Preferences as a Tool for Management of Recreational Use in Natural Areas: A case study in Sierra de Guadarrama (Madrid, Spain). *Landsc. Res.* **2000**, *25*, 49–62. [CrossRef]

18. Povilanskas, R.; Armaitienė, A.; Dyack, B.; Jurkus, E. Islands of prescription and islands of negotiation. *J. Destin. Market. Manag.* **2016**, *5*, 260–274. [CrossRef]

19. Povilanskas, R.; Baziukė, D.; Dučinskas, K.; Urbis, A. Can visitors visually distinguish successive coastal landscapes? A case study from the Curonian Spit (Lithuania). *Ocean Coast. Manag.* **2016**, *119*, 109–118. [CrossRef]

20. Urry, J. *The Tourist Gaze: Leisure and Travel in Contemporary Societies*, 2nd ed.; Sage Publications: London, UK; Thousand Oaks, FL, USA; New Delhi, India, 2002; 176p.

21. Huang, S.-C.L. Visitor responses to the changing character of the visual landscape as an agrarian area becomes a tourist destination: Yilan County, Taiwan. *J. Sustain. Tour.* **2013**, *21*, 154–171. [CrossRef]

22. Lothian, A. Landscape and the philosophy of aesthetics: Is landscape quality inherent in the landscape or in the eye of the beholder? *Landsc. Urban Plan.* **1999**, *44*, 177–198. [CrossRef]

23. Smardon, R.C.; Appleyard, D.; Sheppard, S.R.J.; Newman, S. *Prototype: A Visual Impact Assessment Manual*; Syracuse State University: New York, NY, USA, 1979; 88p.

24. Jacobsen, J.K.S. Use of Landscape Perception Methods in Tourism Studies: A Review of Photo-Based Research Approaches. *Tour. Geogr.* **2007**, *9*, 234–253. [CrossRef]

25. Zube, E.H.; Sell, J.L.; Taylor, J.G. Landscape perception—Research, application and theory. *Landsc. Plan.* **1982**, *9*, 1–33. [CrossRef]

26. Daniel, T.C.; Arthur, L.M.; Boster, R.S. Scenic assessment: An overview. *Landsc. Plan.* **1977**, *4*, 109–129.

27. Daniel, T.C.; Vining, J. Methodological issues in the assessment of landscape quality. In *Behavior and the Natural Environment*; Altman, I., Wohwill, J., Eds.; Plenum Press: New York, NY, USA, 1983; pp. 39–83.

28. Uzzell, D. Environmental psychological perspectives on landscape. *Landsc. Res.* **1991**, *16*, 3–10. [CrossRef]

29. Jorgensen, A. Beyond the view: Future directions in landscape aesthetics research. *Landsc. Urban Plan.* **2011**, *100*, 353–355. [CrossRef]

30. Ribe, R.G. Is Scenic Beauty a Proxy for Acceptable Management? The Influence of Environmental Attitudes on Landscape Perceptions. *Environ. Behav.* **2002**, *34*, 757–780. [CrossRef]

31. Hull, R.B.; Buhyoff, G.J.; Daniel, T.C. Measurement of scenic beauty: The law of comparative judgment and scenic beauty estimation procedures. *For. Sci.* **1984**, *30*, 1084–1096.

32. Wherrett, J.R. Creating Landscape Preference Models Using Internet Survey Techniques. *Landsc. Res.* **2000**, *25*, 79–96. [CrossRef]

33. Rolloff, D.B. Scenic Quality at Crater Lake National Park: Visitor Perceptions of Natural and Human Influence. Ph.D. Thesis, Oregon State University, Corvallis, OR, USA, 1998.

34. Vining, J.; Stevens, J.J. The assessment of landscape quality: Major methodological considerations. In *Foundations for Visual Project Analysis*; Smardon, R.C., Palmer, J.F., Felleman, J.P., Eds.; John Wiley: New York, NY, USA, 1986; pp. 167–186.

35. De Vries, S.; de Groot, M.; Boers, J. Eyesores in sight: Quantifying the impact of man-made elements on the scenic beauty of Dutch landscapes. *Landsc. Urban Plan.* **2012**, *105*, 118–127. [CrossRef]

36. Karjalainen, E.; Tyrväinen, L. Visualization in forest landscape preference research: A Finnish perspective. *Landsc. Urban Plan.* **2002**, *59*, 13–28. [CrossRef]

37. Buhyoff, G.J.; Miller, P.A.; Roach, J.W.; Zhou, D.; Fuller, L.G. An AI methodology for landscape visual assessments. *AI Appl.* **1994**, *8*, 1–13.

38. Petrova, E.G.; Mironov, Y.V.; Aoki, Y.; Matsushima, H.; Ebine, S.; Furuya, K.; Petrova, A.; Takayama, N.; Ueda, H. Comparing the visual perception and aesthetic evaluation of natural landscapes in Russia and Japan: Cultural and environmental factors. *Prog. Earth Planet. Sci.* **2015**, *2*, 6. [CrossRef]

39. Hull, R.B.; Revell, G.R.B. Issues in sampling landscapes for visual quality assessments. *Landsc. Urban Plan.* **1989**, *17*, 323–330. [CrossRef]

40. Appleton, J. *The Experience of Landscape*, 2nd ed.; John Wiley: New York, NY, USA, 1996; 282p.

41. Kaltenborn, B.P.; Bjerke, T. Associations between Landscape Preferences and Place Attachment: A study in Røros, Southern Norway. *Landsc. Res.* **2002**, *27*, 381–396. [CrossRef]

42. Kaplan, R.; Kaplan, S. *The Experience of Nature: A Psychological Perspective*; Cambridge University Press: New York, NY, USA, 1989; 340p.

43. Zube, E.H. Perceived land use patterns and landscape values. *Landsc. Ecol.* **1987**, *1*, 37–45. [CrossRef]

44. Wells, S. *The Journey of Man: A Genetic Odyssey*; Princeton University Press: Princeton, NJ, USA; Oxford, UK, 2002; 240p.

45. Bell, S. *Landscape: Pattern, Perception and Process*, 2nd ed.; Taylor and Francis: London, UK; New York, NY, USA, 2013; 352p.

46. Nassauer, J.I. Culture and changing landscape structure. *Landsc. Ecol.* **1995**, *10*, 229–237. [CrossRef]

47. Buhyoff, G.J.; Wellman, D.J.; Koch, N.E.; Gauthier, L.; Hultman, S.G. Landscape preference metrics: An international comparison. *J. Environ. Manag.* **1983**, *16*, 181–190.

48. De la Fuente de Val, G.; Mühlhauser, H.S. Visual quality: An examination of a South American Mediterranean landscape, Andean foothills east of Santiago (Chile). *Urban For. Urban Green.* **2014**, *13*, 261–271. [CrossRef]

49. Lewis, J.L. Challenges of interdisciplinarity for forest management and landscape perception research. In *From Landscape Research to Landscape Planning: Aspects of Integration, Education and Application*; Tress, B., Tress, G., Fry, G., Opdam, P., Eds.; Springer: Dordrecht, The Netherlands, 2006; pp. 83–94.

50. Tveit, M.; Ode, Å.; Fry, G. Key Concepts in a Framework for Analysing Visual Landscape Character. *Landsc. Res.* **2006**, *31*, 229–255. [CrossRef]

51. Daniel, T.C. Whither scenic beauty? Visual landscape quality assessment in the 21st century. *Landsc. Urban Plan.* **2001**, *54*, 267–281. [CrossRef]

52. Gobster, P.H.; Westphal, L.M. The human dimensions of urban greenways: Planning for recreation and related experiences. *Landsc. Urban Plan.* **2004**, *68*, 147–165. [CrossRef]

53. Jorgensen, A.; Hitchmough, J.; Calvert, T. Woodland spaces and edges: Their impact on perception of safety and preference. *Landsc. Urban Plan.* **2002**, *60*, 135–150. [CrossRef]

54. Krause, C.L. Our visual landscape: Managing the landscape under special consideration of visual aspects. *Landsc. Urban Plan.* **2001**, *54*, 239–254. [CrossRef]

55. Povilanskas, R.; Chubarenko, B.V. Interaction between the drifting dunes of the Curonian Barrier Spit and the Curonian Lagoon. *Baltica* **2000**, *13*, 8–14.

56. Povilanskas, R. *Landscape Management on the Curonian Spit: A Cross-border Perspective*; EUCC Publishers: Klaipėda, Lithuania, 2004; 242p.

57. Povilanskas, R. Spatial diversity of modern geomorphological processes on a Holocene Dune Ridge on the Curonian Spit in the South–East Baltic. *Baltica* **2009**, *22*, 77–88.

58. Povilanskas, R.; Baghdasarian, H.; Arakelyan, S.; Satkūnas, J.; Taminskas, J. Secular Morphodynamic Trends of the Holocene Dune Ridge on the Curonian Spit (Lithuania/Russia). *J. Coast. Res.* **2009**, *25*, 209–215.

59. Povilanskas, R.; Satkūnas, J.; Taminskas, J. Results of cartometric investigations of dune morphodynamics on the Curonian Spit. *Geologija* **2006**, *53*, 22–27.

60. Gudelis, V. *The Coast and the Offshore of Lithuania*; Academia: Vilnius, Lithuania, 1998; 442p.

61. Van der Meulen, F.; Jungerius, P.D.; Visser, J. Editorial. In *Perspectives in Coastal Dune Management*; Van der Meulen, F., Jungerius, P.D., Visser, J., Eds.; SPB Academic Publishing: The Hague, The Netherlands, 1989; pp. 1–7.

62. Povilanskas, R.; Riepšas, E.; Armaitienė, A.; Dučinskas, K.; Taminskas, J. Mobile Dune Types of the Curonian Spit and Factors of Their Development. *Balt. For.* **2011**, *17*, 215–226.

63. Garrod, B. Exploring Place Perception: A Photo-based Analysis. *Ann. Tour. Res.* **2008**, *35*, 381–401. [CrossRef]

64. Rangel-Buitrago, N.; Williams, A.T.; Ergin, A.; Anfuso, G.; Micallef, A.; Pranzini, E. Coastal Scenery: An Introduction. In *Coastal Scenery: Evaluation and Management*; Rangel-Buitrago, N., Ed.; Springer: Cham, Switzerland, 2018; pp. 1–17.

65. De la Fuente de Val, G.; Atauri, J.A.; de Lucio, J.V. Relationship between landscape visual attributes and spatial pattern indices: A test study in Mediterranean-climate landscapes. *Landsc. Urban Plan.* **2006**, *77*, 393–407. [CrossRef]

66. Múgica, M.; de Lucio, J.V. The role of on-site experience on landscape preferences. A case study at Doñana National Park (Spain). *J. Environ. Manag.* **1996**, *47*, 229–239. [CrossRef]

67. Palmer, J.F.; Hoffman, R.E. Rating reliability and representation validity in scenic landscape assessments. *Landsc. Urban Plan.* **2001**, *54*, 149–161. [CrossRef]

68. Urbis, A.; Povilanskas, R.; Newton, A. Valuation of aesthetic ecosystem services of protected coastal dunes and forests. *Ocean Coast Manag.* **2019**, *179*. in press. [CrossRef]

69. Stamps, A.E. Mystery, complexity, legibility and coherence: A meta-analysis. *J. Environ. Psychol.* **2004**, *24*, 1–16. [CrossRef]

70. Mowforth, M.; Munt, I. *Tourism and Sustainability: Development, Globalisation and New Tourism in the Third World*, 3rd ed.; Routledge: London, UK, 2009; 456p.

71. Poria, Y.; Butler, R.; Airey, D. The core of heritage tourism. *Ann. Tour. Res.* **2003**, *30*, 238–254. [CrossRef]

72. Eysenck, M. Memory. In *Psychology: An Integrated Approach*; Eysenck, M., Ed.; Longman: Essex, UK, 1998; pp. 167–204.

73. Ode, Å.; Fry, G.; Tveit, M.S.; Messager, P.; Miller, D. Indicators of perceived naturalness as drivers of landscape preference. *J. Environ. Manag.* **2009**, *90*, 375–383. [CrossRef] [PubMed]

74. Creswell, J.W. *Research Design: Qualitative, Quantitative, and Mixed Methods Approaches*, 4th ed.; Sage: Los Angeles, CA, USA, 2014; 273p.

75. Garrod, B.; Fyall, A. Managing Heritage Tourism. *Ann. Tour. Res.* **2000**, *27*, 682–708. [CrossRef]

76. Hsu, C.C.; Sandford, B.A. The Delphi technique: Making sense of consensus. *Pract. Assess. Res. Eval.* **2007**, *12*, 1–8.

77. La Sala, P.; Conto, F.; Conte, A.; Fiore, M. Cultural Heritage in Mediterranean Countries: The Case of an IPA Adriatic Cross Border Cooperation Project. *Int. J. Eur. Med. Stud.* **2016**, *9*, 31–50.

78. Lupp, G.; Konold, W.; Bastian, O. Landscape management and landscape changes towards more naturalness and wilderness: Effects on scenic qualities—The case of the Muritz National Park in Germany. *J. Nat. Conserv.* **2013**, *21*, 10–21. [CrossRef]

79. Monavari, S.M.; Khorasani, N.; Mirsaeed, S.S.G. Delphi-based Strategic Planning for Tourism Management—A Case Study. *Pol. J. Environ. Stud.* **2013**, *22*, 465–473.

80. Olszewska, A.A.; Marques, P.F.; Ryan, R.L.; Barbosa, F. What makes a landscape contemplative? *Environ. Plan. B Urban Anal. City Sci.* **2018**, *45*, 7–25. [CrossRef]

81. Tan, W.J.; Yang, C.F.; Château, P.A.; Lee, M.T.; Chang, Y.C. Integrated coastal-zone management for sustainable tourism using a decision support system based on system dynamics: A case study of Cijin, Kaohsiung, Taiwan. *Ocean Coast Manag.* **2018**, *153*, 131–139. [CrossRef]

82. Palmer, J.F. Reliability of Rating Visible Landscape Qualities. *Landsc. J.* **2000**, *19*, 166–178. [CrossRef]

83. Parsons, R.; Daniel, T.C. Good looking: In defense of scenic landscape aesthetics. *Landsc. Urban Plan.* **2002**, *60*, 43–56. [CrossRef]

84. Van Mansvelt, J.D.; Kuiper, J. Criteria for the humanity realm: Psychology and physiognomy and cultural heritage. In *Checklist for Sustainable Landscape Management*; van Mansvelt, J.D., van der Lubbe, M.J., Eds.; Elsevier Science: Amsterdam, The Netherlands, 1999; pp. 116–134.

85. Nohl, W. Sustainable landscape use and aesthetic perception-preliminary reflections on future landscape aesthetics. *Landsc. Urban Plan.* **2001**, *54*, 223–237. [CrossRef]

86. Doody, J.P. *Sand Dune Conservation, Management and Restoration*; Springer: Dordrecht, The Netherlands, 2013; 304p.

87. Sevenant, M.; Antrop, M. Cognitive attributes and aesthetic preferences in assessment and differentiation of landscapes. *J. Environ. Manag.* **2009**, *90*, 2889–2899. [CrossRef] [PubMed]

88. Riepšas, E. *Recreational Forestry*; Aleksandras Stulginskis University Publishers: Kaunas, Lithuania, 2012; 256p.

89. Green, R. Meaning and form in community perception of town character. *J. Environ. Psychol.* **1999**, *19*, 311–329. [CrossRef]

90. Povilanskas, R.; Armaitienė, A. Marketing of coastal barrier spits as liminal spaces of creativity. *Procedia Soc. Behav. Sci.* **2014**, *148*, 397–403. [CrossRef]

91. Weinstoerffer, J.; Girardin, P. Assessment of the contribution of land use pattern and intensity to landscape quality: Use of a landscape indicator. *Ecol. Model.* **2000**, *130*, 95–109. [CrossRef]

92. Wu, J. Landscape of culture and culture of landscape: Does landscape ecology need culture? *Landsc. Ecol.* **2010**, *25*, 1147–1150. [CrossRef]

93. Germino, M.J.; Reiners, W.A.; Blasko, B.J.; McLeod, D.; Bastian, C.T. Estimating visual properties of Rocky Mountain landscapes using GIS. *Landsc. Urban Plan.* **2001**, *53*, 71–84. [CrossRef]
94. Chhetri, P.; Arrowsmith, C.; Jackson, M. Determining hiking experiences in nature-based tourist destinations. *Tour. Manag.* **2004**, *25*, 31–43. [CrossRef]
95. Vouligny, É.; Domon, G.; Ruiz, J. An assessment of ordinary landscapes by an expert and by its residents: Landscape values in areas of intensive agricultural use. *Land Use Policy* **2009**, *26*, 890–900. [CrossRef]
96. Chhetri, P.; Arrowsmith, C. GIS-based modelling of recreational potential of nature-based tourist destinations. *Tour. Geogr.* **2008**, *10*, 233–257. [CrossRef]
97. Ruddell, E.; Gramann, J.; Rudis, V.; Westphal, J. The psychological utility of visual penetration in near-view forest scenic-beauty models. *Environ. Behav.* **1989**, *214*, 393–412. [CrossRef]
98. Roth, M.; Hildebrandt, S.; Röhner, S.; Tilk, C.; von Raumer, H.G.S.; Roser, F.; Borsdorff, M. Landscape as an area as perceived by people: Empirically-based nationwide modelling of scenic landscape quality in Germany. *J. Dig. Landsc. Archit.* **2018**, *3*, 129–137.
99. Thiele, J.; von Haaren, C.; Albert, C. Are river landscapes outstanding in providing cultural ecosystem services? An indicator-based exploration in Germany. *Ecol. Indic.* **2019**, *101*, 31–40. [CrossRef]
100. Cassatella, C.; Seardo, B.M. In search for multifunctionality: The contribution of scenic landscape assessment. In *Landscape Planning and Rural Development*; Cassatella, C., Ed.; Springer: Cham, Switzerland, 2014; pp. 41–60.
101. Ayad, Y.M. Remote sensing and GIS in modeling visual landscape change: A case study of the northwestern arid coast of Egypt. *Landsc. Urban Plan.* **2005**, *73*, 307–325. [CrossRef]
102. Opdam, P.; Luque, S.; Nassauer, J.; Verburg, P.H.; Wu, J. How can landscape ecology contribute to sustainability science? *Landsc. Ecol.* **2018**, *33*, 1–7. [CrossRef]
103. Doody, J.P. "Coastal squeeze": An historical perspective. *J. Coast. Conserv.* **2004**, *10*, 129–138. [CrossRef]
104. Powell, E.J.; Tyrrell, M.C.; Milliken, A.; Tirpak, J.M.; Staudinger, M.D. A review of coastal management approaches to support the integration of ecological and human community planning for climate change. *J. Coast. Conserv.* **2019**, *23*, 1–18. [CrossRef]
105. Dalton, T.; Thompson, R. Recreational boaters' perceptions of scenic value in Rhode Island coastal waters. *Ocean Coast. Manag.* **2013**, *71*, 99–107. [CrossRef]

 *water*

Article

Sensitivity of Storm-Induced Hazards in a Highly Curvilinear Coastline to Changing Storm Directions. The Tordera Delta Case (NW Mediterranean)

Marc Sanuy * and Jose A. Jiménez

Laboratori d'Enginyeria Marítima, Universitat Politècnica de Catalunya·BarcelonaTech, c/Jordi Girona 1-3, Campus Nord ed D1, 08034 Barcelona, Spain; jose.jimenez@upc.edu
* Correspondence: marc.sanuy@upc.edu; Tel.: +34-93-401-7392

Received: 26 March 2019; Accepted: 8 April 2019; Published: 10 April 2019

Abstract: Extreme coastal storms, especially when incident in areas with densely urbanized coastlines, are one of the most damaging forms of natural disasters. The main hazards originating from coastal storms are inundation and erosion, and their magnitude and extent needs to be accurately assessed for effective management of coastal risk. The use of state-of-art morphodynamic process-based models is becoming standard, with most being applied to straight coastlines with gentle slopes. In this study, the XBeach model is used to assess the coastal response of a curvilinear sensitive deltaic coast with coarse sediment and steep slopes (intermediate-reflective conditions). The tested hypothesis is that changes in wave direction may cause large variations in the magnitude of storm-induced hazards. The model is tested against field data available for the Sant Esteve Storm (December 2008), obtaining an overall BSS (Brier Skill Score) score on the emerged morphological response of 0.68. Later, the 2008 event is used as baseline scenario to create synthetic events covering the range from NE to S. The obtained results show that storm-induced hazards along a highly curvilinear coast are very sensitive to changes in wave direction. Therefore, even under climate scenarios of relatively steady storminess, a potential shift in wave direction may significantly change hazard conditions and thus, need to be accounted for in robust damage risk assessments.

Keywords: XBeach; inundation; erosion; BSS; Sant Esteve 2008; coarse sediment

---

1. Introduction

The impact of extreme storms on the coast is one of the costliest forms of natural disasters (Kron [1]; Bertin et al. [2]). In heavily urbanized coastal areas, such as the Mediterranean (in general) and the Catalan coast (in particular), where properties, infrastructures and businesses are located close to the shoreline, this kind of event usually results in the damage or destruction of exposed assets (Jiménez et al. [3]). These effects are the integrated consequences of two main storm-induced coastal hazards: inundation and erosion. In this context, an accurate assessment of the magnitude, location and extension of these hazards is becoming an essential part of the risk management process (e.g., Ciavola et al. [4,5]; Van Dongeren et al. [6], Jimenez et al. [7]; Plomaritis et al. [8], Harley et al. [9]) and, in this sense, the use of process-oriented models to forecast storm-induced morphodynamic changes under given scenarios is now standard (e.g., Roelvink et al., [10]; McCall, [11]; Van Dongeren et al. [12] and references therein, Dissanayake et al. [13]). Most of the studies on testing state-of-art morphodynamic process-based models have addressed cases characterized by straight coastlines and gentle slopes (i.e., conditions close to the comfort zone of the models) (e.g., McCall, [11]; Harter and Figlus [14]). However, applications to estimate costal hazards in hihgly curvilinear environments (e.g., deltaic cuspate coasts) have seldom been tested (e.g., Roelvink et al., [15]; Valchev et al., [16]; and Dissanayake et al. [13]). Furthermore, the effect of testing models based on surf-beat (i.e., the infragravity wave band) on steep slopes and coarse sediment has been recently undertaken mainly in 1D applications (e.g., Vousdoukas et al. [17]; Elsayed and Oumeraci [18]) but rarely so in fully 2DH (2-dimensional, depth-averaged) simulations.

Within this context, the magnitude of storm-induced hazards on a highly curvilinear coast by using XBeach is assessed in the present study. The relevance and main aim of this work is twofold: first, from a general standpoint, to test the use of Xbeach on a highly curvilinear coast characterized by coarse sediment reflective beaches, and second, from the local standpoint, to analyze the sensitivity of an already identified hotspot, the Tordera Delta (NW Mediterranean) (Jiménez et al. [7]), to assess storm impacts for different storm direction scenarios. Thus, the largest recorded storm in the area is used as base case scenario. It occurred in December 2008 and had the typical incoming direction of current climate conditions where eastern (E) incoming storms dominate (e.g., Mendoza et al. [19]). Existing storminess projections under climate change scenarios for the Western Mediterranean do not predict any increase in wave height (e.g., Lionello et al. [20]; Conte and Lionello [21]), but some projections identify potential changes in wave direction (Cases-Prat and Sierra [22,23]). Due to this and to the great sensitivity of cuspate coastlines to wave direction resulting from their curvature (e.g., Slott et al. [24]; Johnson et al. [25]), the study aims to assess the potential effects of changing wave direction on extreme storm-induced hazards for the Tordera Delta. The hypothesis to be tested is that changes in wave direction may cause large variations in the magnitude of storm-induced hazards. Other studies have included the sensitivity to incoming storm direction in their assessments, such as those by Mortlock et al. [26] in Australia, or de Winter and Ruessink [27] in Holland.

The article is arranged as follows: the second section introduces the study site and the data used, describes the Sant Esteve 2008 event, which is used as the base case storm-scenario, and presents the methodological part, i.e., the used morphodynamic model and the comparative assessment framework descriptions; the third section presents obtained results; and finally, the discussion and concluding remarks are presented in the fourth section.

2. Materials and Methods

2.1. Study Area

The Tordera Delta is located in the NW Mediterranean Sea approximately 60 km northwards of Barcelona (Figure 1). It is a coarse sand delta (i.e., sediment in the range between 0.8 and 1.6 mm) with an aerial surface of approximately 4.2 km^2 at the end of a small river basin of approximately 879 km^2 (Vila and Serra [28]). The Tordera river is dry during most part of the year, with long dry summers and episodic discharges after heavy rainfall (Martin-Vide and Llasat [29]). Coastal storms and heavy rainfall events are usually uncorrelated at the area and, in fact, during the simulated storm, significant river outflow was present after the storm peak, with a phase delay of 1 day (Sanchez-Vidal et al. [30]). It has a cuspate configuration with two well differentiated parts at each side of the river mouth (Figure 1). The northern part is fronted by the S'Abanell beach, which is oriented towards the E. It has a steep nearshore bathymetry without any relevant geomorphological features in shallow waters The northern hinterland presents a higher topography, except for the zone closest to the river mouth outlet. The southern part is fronted by the Malgrat de Mar beach and is oriented towards the S, which serves as natural protection of a lower hinterland. Shallow water bathymetry is characterized by the presence of a longshore bar running parallel to the coast from the river mouth to the SW, which encloses a plateau at 4 m water depth.

The delta coastline has been eroding during the last several decades as the net result of the littoral drift and the decrease of the Tordera river sediment output, with maximum measured retreats of approximately 120 m Jiménez et al. [31]. The combination of a progressively narrowing beach protecting a low-lying hinterland and this being mainly occupied by campsites makes this area a hotspot for storm-induced hazards (Jiménez et al. [7]) with different consequences depending on storm characteristics and beach morphology at the time of impact (see e.g., Sanuy et al. [32]). As Jiménez et al. [31] noted, under former accretive-stable deltaic conditions only extreme storms were able to exceed the capacity of protection provided by wide beaches, but under present medium/long-term erosion conditions, smaller storms have become able to exceed the dissipation capacity of the narrower beaches, increasing the frequency of storm-induced problems (e.g., Jiménez et al. [3]).

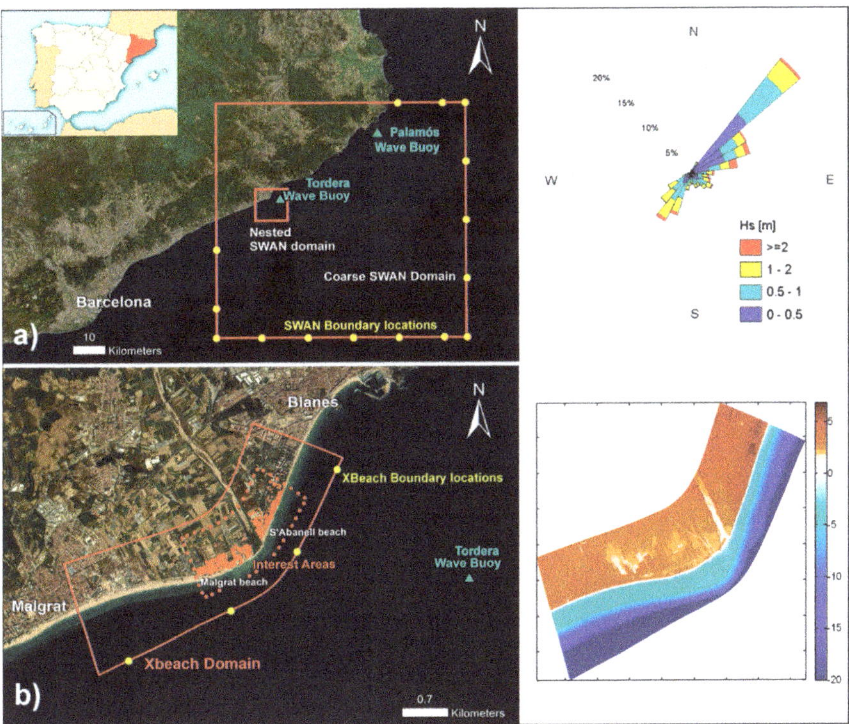

Figure 1. Tordera Delta study area, location and model set-up. (**a**) SWAN (Simulating Waves Nearshore) grids (red) with boundary wave-spectrum nodes (yellow), wave buoys (blue triangles), and significant wave height wave rose for the period 1948–2009 (Global Ocean Waves, GOW, Reguero et al. [33]) at the Tordera wave buoy location. (**b**) XBeach set-up, curvilinear grid domain (red) and interest areas (orange) and pre-storm topobathymetry.

Wave climate at the NW Mediterranean is characterized in the wave rose in Figure 1 (Global Ocean Waves, GOW dataset hindcast, Reguero et al. [33]). Storm events are defined as events with significant wave height (Hs) exceeding 2 m during a minimum of 6 h (Mendoza et al. [19]). The main incoming wave direction at the site is NE and E, with some events arriving from the S, especially during spring. Within this two main groups, some residual events, can also be found. Thus, in propagated conditions (nearshore area) storm conditions can be characterized with waves in the range NE-S being the two extremes of the range the most frequent situations. Nonetheless, some studies rise concerns at the Catalan coast about future climate-induced changes in storm direction, particularly a frequency transfer from the current main cluster (NE-E) towards the secondary one (S) (Casas-Prat et al. [22,23]). The area is micro-tidal, with storm surges having a limited role in storm-induced processes due to its relative weight when compared to the wave component. Notably, surges are uncorrelated with waves, are most frequently under 25 cm, with some extreme events showing maximum recorded surges around 50 cm (Mendoza and Jiménez [34]).

2.2. Data

The wave data used in this study to characterize the Sant Esteve storm and validate the models were measured by a directional wave buoy located off the Tordera Delta at approximately 70 m water depth (Figure 1) belonging to the XIOM (Xarxa d'Instruments Oceanogràfics i Meteorològics) network, which is no longer operative (Bolaños et al. [35]). The Tordera Buoy was a Datawell waverider, sampling during 20′ every hour, and thus providing sea states and statistics with an hourly time-step. It covers the period from 1984 to 2013.

Wave spectra from deepwaters and wind fields used to force the model chain were provided by Puertos del Estado, from the WAM (WAve Model, version 4, European Centre for Medium-Range Weather Forecasts, Reading, UK) and HIRLAM-AEMET (HIgh Resolution Local Area Modelling version 7, Agencia Estatal de METeorología, Madrid, Spain) models respectively. Both datasets have a temporal time-step of 1 h. Nearshore water levels were obtained from the same source, provided by the HAMSOM model (HAMburg Shelf Ocean Model, barotropic version, Center for Marine and Atmospheric Sciences. Institute of Oceanography, Hamburg, Germany) with the same temporal resolution.

Storm-induced topographic changes have been quantified by using LIDAR-derived topographies obtained before (16 October 2008) and after the storm impact (17 January 2009) by the Institut Cartogràfic i Geológic de Catalunya. The data were provided as high-resolution digital elevation models (DEMs) with a 1-m grid step, a maximum vertical precision of 2–3 cm and overall RSME of 6 cm (Ruiz et al. [36]).

The bathymetry of the study area has been obtained by combining an offshore grid with a spatial resolution of 0.28′ derived from the GEBCO (General Bathymetric Chart of Oceans) database [37], and a finer inner topography while nearshore bathymetry was built by combining the LIDAR-derived topography and multi-beam bathymetric data provided by the Ministry of Agriculture, Fish, Food, and Environment, covering the whole area from the +2 to the −50 m with a 5 × 5 m resolution. Multiple bathymetries were available from different years, including 2006 and 2010, and these information was merged to properly fit the 2008 LIDAR shoreline and link the emerged topography with the submerged bed.

2.3. The Sant Esteve 2008 Storm

The storm of reference used in this study was a V-class (extreme) event, according to the classification of storms in the NW Mediterranean of Mendoza et al. [19]. This storm, known as the Sant Esteve storm, occurred on 26 December 2008 in the northern part of the Catalan Sea. It was created by the presence of a low-pressure center located over the Balearic Sea, with a minimum pressure of 1012 hPa, along with a high pressure center over northern Europe (1047 hPa). This is one of the typical mechanisms of cyclogenesis in the Mediterranean (e.g., Trigo et al. [38]), and it is the most common situation originating extreme storms along the Catalan coast (Mendoza et al. [19]). Under these conditions, the action of very strong NE winds in the Gulf of Lyon (wind velocities up to 20 m s^{-1} were recorded at the coast) generated a wave field with a clear spatial pattern, with Hs values and power content decreasing from north to south along the Catalan coast (Jiménez [39]; Sánchez-Vidal et al. [30]). Thus, according to the data recorded by the Palamós buoy (see location in Figure 1), the storm lasted 73 h (Hs > 2 m) reaching a Hs of 7.5 m at the peak of the storm and a maximum wave height (Hmax) of 14.4 m. The associated return period of this storm was approximately 125 years according to the data on extreme climate obtained by Puertos del Estado [40] for this buoy and in light of data previous to the storm. The storm progressively lost strength as it moved south and, off the study site, the values recorded by the Tordera buoy showed a storm with a duration of 66 h, reaching a Hs at the peak of the storm of 4.65 m and Hmax of 8.0 m. Mean wave direction during the storm was E, which corresponds to the dominant direction during extreme storms in the area (Mendoza et al. [19]). Figure 2 shows the recorded values of wave parameters during the storm by the Tordera buoy (see location in Figure 1).

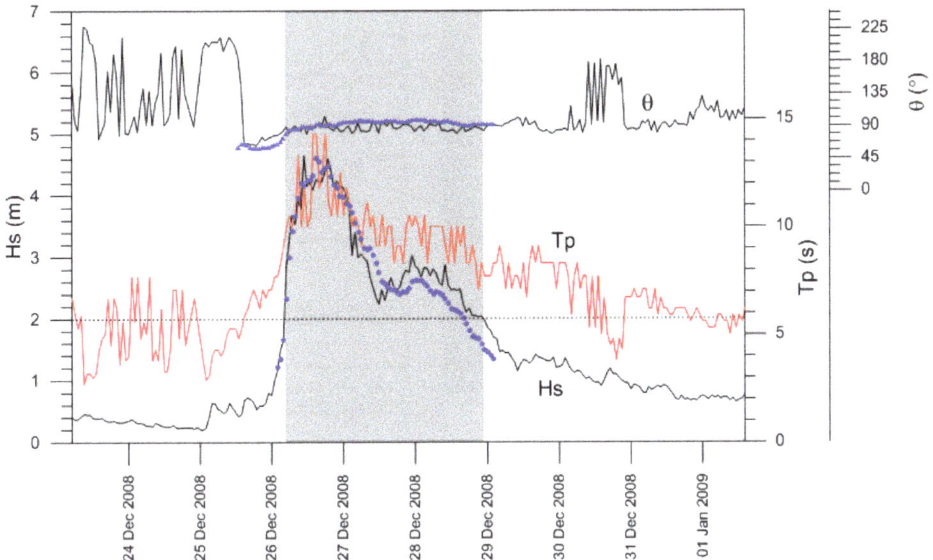

Figure 2. Wave records off the study site (Tordera buoy, see Figure 1) during the San Esteve storm. Blue dots indicate SWAN output at the same location.

The impact of the storm produced a significant coastal morphodynamic response in the form of erosion and overwash for most beaches along the northern part of the Catalan coast (Plana-Casado, [41]; Jiménez et al. [42]; Durán et al. [43]). Moreover, the magnitude of the storm was so considerable that many benthic ecosystems were also significantly affected (Sánchez-Vidal et al. [30]; Teixidó et al. [44]; Pagès et al. [45]).

Storm-induced topographic changes in the surroundings of the Tordera river mouth are shown in Figure 3. The observed response was different on both sides of the river, with the largest erosion taking place in the northern part, the s'Abanell beach. This beach is oriented to the East, nearly perpendicular to storm waves, and thus, the beach presented a generalized erosive behavior along its total extension (2.4 km). The volume of sediment eroded from the subaerial part of the beach was approximately 66,000 m^3, with a beach-averaged erosion rate of approximately 30 m^3/m and a maximum value of approximately 80 m^3/m at its northernmost part (section SB-1 in Figure 3). Storm-induced wave overtopping occurred along the entire beach, and in its southernmost part, close to the river mouth, overwash deposits up to 6 m^3/m were detected (section SB-3 in Figure 3). These volume changes resulted in a beach-averaged shoreline retreat of 11 m, with a maximum recession of approximately 30 m (Plana-Casado, [41]; Jiménez et al. [42]).

From the river mouth to the south, the coast experienced a different morphodynamic response. This section, the Malgrat de Mar beach, is oriented to the S, resulting in a large obliquity to E incoming waves during the storm. Just south of the river mouth a small post-storm accretion spot of approximately 7000 m^3 was detected. This seems to be related to the alongshore deposition of material eroded from the northern part. South of this area, the coastline shows a nearly generalized moderate erosion together with significant overwash deposits in the subaerial part of the beach (Figure 3). The spatially averaged erosion of the emerged beach was approximately 10 m^3/m (one third of that observed for the northern beach) whereas the averaged overwash accumulation was approximately 7.5 m^3/m (Plana-Casado, [41]; Jiménez et al. [42]).

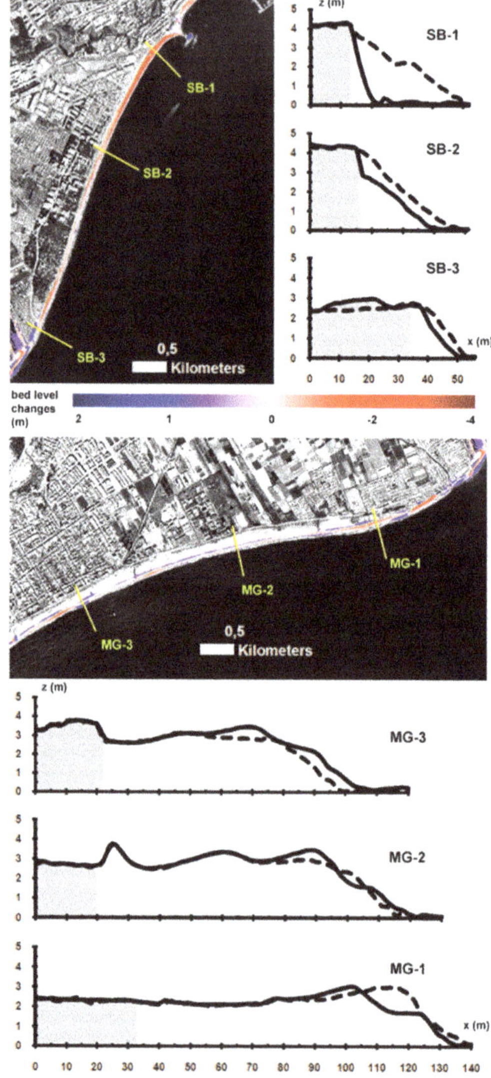

Figure 3. Morphologic changes after the St Esteve 2008 storm. Dashed line pre-storm and continuous line post-storm profiles. Grey shaded area represents de position of the promenade (north) or road (south).

2.4. Models

Simulations have been done by using a model setup composed of the SWAN model (Simulating Waves Nearshore, version Cycle III v41.01, Delft University of Technology, Deltares, The Netherlands) (Booij et al. [46]; Ris et al. [47], TU Delft [48]), which propagates waves to the coast, and of the XBeach model (eXtreme Beach behavior, Kingsday version, Deltares, Delft, The Netherlands) (Roelvink et al. [10]), which is used to assess two storm-induced coastal hazards: inundation and erosion.

SWAN (Simulating Waves Nearshore) is a third generation wave model which is based on the wave action balance equation. It simulates short-crested wind generated waves by incorporating wave–water interactions, wind growth and dissipation processes such as whitecapping, bottom friction, and depth-induced breaking. For more detailed insight into these mechanisms controlling energy and wave propagation processes the reader is referred to the SWAN manual and to SWAN scientific technical documents at http://swanmodel.sourceforge.net/. The implemented model version uses the Komen et al. [49] formulation to calculate whitecapping. This version permits counteracting the previously reported under-predictions of significant wave height and wave period in areas characterized by fetch-limited conditions under the influence of transient winds (see e.g., Bolaños [50]; Pallares et al. [51]), which are typical conditions for the NW Mediterranean coast.

The model has been implemented using a nested grid configuration (Figure 1). A coarse grid with a total extension of approximately 80 × 70 km is used to transfer offshore wave conditions to the study area. The bathymetry grid has a spatial resolution of 0.28′ whereas the wind field grid has a spatial resolution of 5′. This coarse setup is fed with wave spectra at 15 positions distributed along the offshore boundaries of the grid (Figure 1). The inner fine grid covers a domain of approximately 20 km × 26 km with a spatial resolution of 0.06′. This grid has been generated to properly reproduce the existing sharp changes in the bathymetry between intermediate-shallow waters due to the large steepness of the lower shoreface. This can permit a better simulation of wave propagation in the region close to the XBeach coastal grid (external boundary at 20 m water depth). A limitation of the SWAN model is its inability to simulate storm surges. To characterize storm surges in the study area we use water level predictions obtained with the HAMSOM model implemented by Puertos del Estado (Ratsimandresy, et al. [52]) at three locations close to the study site. The storm surge contribution to the total water level in the study area is of low magnitude and significantly smaller than wave-induced runup during storm conditions (e.g., Mendoza and Jiménez [34]). The validated SWAN model has been used to propagate waves from deep to shallow waters during the entire storm duration. Since a simultaneous time series of measured wave data was available within the modelled domain, it was possible to obtain transfer coefficients for wave conditions (wave height and direction) for any point in the grid with respect to buoy location. With this information, the recorded wave conditions at the buoy have been transferred to selected grid points to be used as input data for Xbeach modelling.

Once the forcing conditions during the storm have been propagated from deep water to nearshore, the remaining task is to propagate these conditions to the coast and to assess the magnitude of storm-induced hazards, i.e., erosion, overwash and overtopping, which is done with the XBeach model. XBeach is an open-source 2D depth averaged model which solves wave propagation, flow, sediment transport and bed level changes for varying wave and flow boundary conditions (Roelvink et al. [10]). It solves the time-dependent short wave action balance on the scale of wave groups, which allows for the reproduction of directionally spread infragravity motions (so called surf-beat) along with time-varying currents. The frequency domain is represented by a single representative peak frequency, assuming a narrow banded incident spectrum. Shallow water momentum and mass balance equations are solved to compute surface elevation and flow. Additionally, to solve the contribution of short waves to mass fluxes and return flows, XBeach uses the Generalized Lagrangian Mean formulation (Roelvink et al. [10]).

Sediment transport rates are calculated from the spatial variations in depth-averaged concentration, which are calculated from advection-diffusion equations with a source-sink term based on an equilibrium sediment concentration. The equilibrium concentration takes into account both the contribution of the suspended and bed loads by means of the Soulsby-Van Rijn formulation (Soulsby [53]) with a limitation of the maximum stirring velocity based on the Shields number at the start of the sheet flow (McCall et al. [11]). For deeper insight into the XBeach description, setup and equations, see Roelvink et al. [10].

The model has been implemented by using a curvilinear grid with variable cell size in both alongshore and cross-shore directions (see Figure 4). The extension of the mesh is approximately 1.5 km in the cross-shore direction, with cell size ranging from 5~6 m at the offshore boundary (20 m depth) to 0.7–0.8 m at the swash zone. In the alongshore direction the model has an extension of 4.5 km and the cell size ranges from 25 m at the lateral boundaries down to 2–3 m around the river mouth. The grid was obtained by means of the Delft3D-RGFGRID module, imposing consecutive maximum cell-size changes ~5–10% to ensure proper smoothing, while maintaining valid orthogonalisation. The final result has 669 alongshore by 568 cross-shore nodes. The model wave boundary conditions consist of wave characteristics specified at four different locations along the offshore boundary (see Figure 1), with a time-step of 1 h, which is the same resolution of the original data used to force SWAN. The use of four different locations aims to capture the difference in wave conditions on both sides of the river due to the different coastline orientation. Water level variations during storms to be simulated are directly introduced in the model from HAMSOM simulations, also with a time-step of 1 h. XBeach model computations (i.e., hydrodynamics and morphodynamcis) are performed with a temporal resolution of 1 s (dtbc = 1).

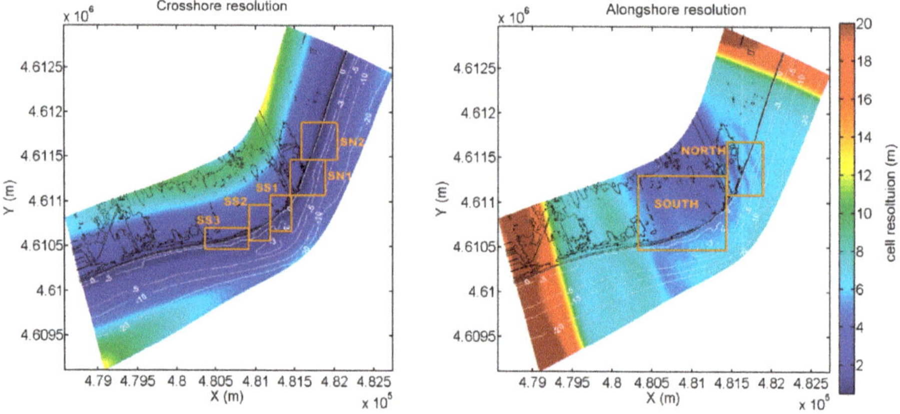

Figure 4. Distribution of cross-shore and alongshore grid resolution over the XBeach domain. Orange boxes show the location of post-processing subdomains, being SN and SS sectors (**left**) for the morpholodynamic analysis and NORTH, SOUTH sectors (**right**) for the inundation assessment.

The morphology of the study area, characterized by coarse sand and steep reflective beaches, makes XBeach modelling a demanding exercise in terms of predicting beach morphodynamic response during storms (see e.g., Vousdoukas et al. [17,54]). Notably, coarse sediment environments are characterized by a lower frequency of the avalanching processes, a greater importance of the bed load over the suspended transport, and higher importance of mechanisms such wave asymmetry, or water infiltrations and groundwater effects. All these are by default configured in XBeach to work for fine sand environments, and must be revised and modified for its application at the study site, modelled with a D50 of 1.3 mm and a D90 of 1.9 mm.

To properly reproduce morphodynamic changes, both the surf-beat and the non-hydrostatic modes of the XBeach model were initially tested. The surf-beat model was observed to under-predict overwash when using typical XBeach-grid resolution for straight and mild-slope coasts. This under-prediction is also caused by the lower contribution of infragravity waves to the total run-up in steep profiles (Wright and Short [55]). However, the surf-beat model accurately reproduced the alongshore patterns of erosion in the entire domain. In contrast, the non-hydrostatic model was observed to have better performance in reproducing wave-by-wave run-up but lower accuracy in reproducing the alongshore morphodynamic patterns of erosion and deposition, with a quite higher computational cost (since it requires denser grids). An additional difficulty was using XBeach with a curvilinear grid to properly reproduce alongshore processes driven by the change of coast orientation at both sites of the river mouth. The final adopted approach was to use the surf-beat model with a higher resolution than the typically used in straight and gentle-slope coasts, which considerably improved the overwash prediction. The average cross-shore resolution of 5.2 m (typically 20~25 m) at the offshore boundary, going down to 0.7 at the swash zone, ensured a better reproduction of wave propagation. This setting aimed to properly capture the abrupt changes in the bathymetry from the 20 m to the 5 m depth by using a larger number of cells. In addition, the average alongshore resolution around the river mouth, where the largest alongshore gradients are present, was increased up to 2.3 m (typically 5~10 m) (Figure 4).

2.5. Scenario Testing

Since the recorded Sant Esteve 2008 event had the characteristics of a V-class extreme storm in the area, with the largest recorded Hs and with a typical E direction, it was used as base case scenario (C0). From this, six additional scenarios were defined to test the effect of different incoming wave directions. The procedure consisted of simulating the exact same wave time series as the Sant Esteve 2008 storm (keeping the same Hs, Tp, and directional spreading) and only changing the mean wave direction. This approach permits maintaining the same storm wave intensity as the reference case at the offshore boundary but under a different direction. The tested conditions were two scenarios where wave direction was shifted 20° and 40° counter-clockwise to the North (C20− and C40−) and four scenarios shifting 20°, 40°, 60°, and 80° clockwise to the South (C20+, C40+, C60+, and C80+). Thus, seven different scenarios, including the baseline condition, were simulated and compared to assess the differences in storm-induced hazards under incoming directions ranging from ~60° N (C40−) to ~180° N (C80+). The scenarios have been chosen to cover the tipical range of incoming conditions at the −20 m depth with directional spans of 20°.

The magnitude of storm–induced hazards was quantified in different control sectors along the study area to capture main factors potentially affecting the beach response to different incoming directions. Thus, the morphodynamic response was analyzed in five sectors: two 250 m long sectors northward of the river mouth (SN1, SN1), and three sectors southward of the river mouth (Figure 4). These S sectors are as follows: (i) SS1, a 200-m-long stretch between the river mouth and an existing rigid structure at the shoreline; (ii) SS2, a 200-m-long sector, southward of the mentioned existing structure; and (iii) SS3, a 500-m-long stretch at the southernmost end. At each sector, three variables are used to characterize morphodynamic changes: erosion volume (m^3/m), overwash volume (m^3/m) and profile retreat (m). All of the variables are calculated within sectors from XBeach gridded output, from which sector-averaged values and standard deviations are derived. Erosion volumes are computed in the inner part of the beach, from the subaerial part down to the −2 m level, which roughly determines the water depth where main inner profile changes occur. Overwash volumes are computed as deposited sediment volumes in those parts of the subaerial beach where vertical growth is detected. The profile retreat is measured at three elevations at the beachface (1, 1.5, and 1.75 m above mean water level).

To characterize inundation, just two sectors were selected, one for the region at the north of the river mouth and another to the south (Figure 4). The selected variable to characterize inundation was the inundation surface (Ha) over different thresholds of inundation depth: 0.05, 0.25, 0.5, and 1 m.

3. Results

3.1. Base Case Scenario (C0). The Sant Esteve Storm

This base case scenario corresponds to the model validation using the recorded Sant Esteve 2008 Storm event. The SWAN model was validated with wave conditions recorded by the Tordera wave buoy and by comparing the measured and modelled wave conditions (Hs, θ), shown in Figure 2. As can be seen, the model reproduces well wave parameters during the storm, especially during the peak of the event, when simulated variables almost coincide with recorded ones. The obtained root mean squared error (RMSE) of the model during the entire duration of the storm is 0.53 m in Hs and 12.5 degrees in θ, with the largest contributions to the error taking place during the relaxation phase of the storm, whereas error values at the storm peak are significantly lower (Figure 2). Since the largest storm-induced morphodynamic changes occurred during the storm peak, we can accept that the use of the SWAN model to simulate wave propagation in the study area is acceptable for the purposes of this research.

To calibrate XBeach, the model's results were compared with LIDAR measurements of the emerged beach. The calibration of the model was performed by adopting a double approach: (i) by optimizing the Brier Skill Score (BSS), which quantifies model performance by comparing model output to the real post-storm LIDAR measurements of the emerged profile; (ii) by performing a qualitative assessment of the modelled features, such as alongshore and cross-shore patterns of bed level changes, magnitude, and location of the overwash deposits and validation of the inundation reach according to the available qualitative information on the event. The used BSS to characterize model predictive skill takes into account the measurement error (ΔZ_e) as in Harley and Ciavola [56], and thus, the BSS score is given by:

$$\text{BSS} = 1 - \left(\sum \left(|z_{mf} - z_{mod}| - \Delta Z_e \right)^2 / \left(\sum \left(z_{mf} - z_{mi} \right)^2 \right. \right. \tag{1}$$

where z_{mf} is the final LIDAR measured bed level, z_{mod} the final modelled bed level, and z_{mi} the initial bed level. Here, ΔZ_e is considered as the LIDAR measurement error (i.e., RSME of the overall LIDAR product), which is 0.06 m. According to Sutherland et al. [57], the classification of models' performance based on the BSS score can be considered to be very good for values over 0.4 and excellent for values over 0.5–0.6. Due to the morphology of the study area, which induces a differentiated morphodynamic response along the coast (Figures 5 and 6), three BSS were calculated: (i) a global BSS, which is calculated for the entire area; (ii) a local BSS at the north of the river mouth; and (iii) a local BSS southward of the river mouth. Since the BSS assessment can only be performed for the emerged profile (where pre- and post-storm topographic data exist), a qualitative assessment of the modelled submerged profile was also performed. To this end, we analyzed the final shape of the modelled submerged profile taking into account the expected typical morphodynamic response under storm conditions at the site.

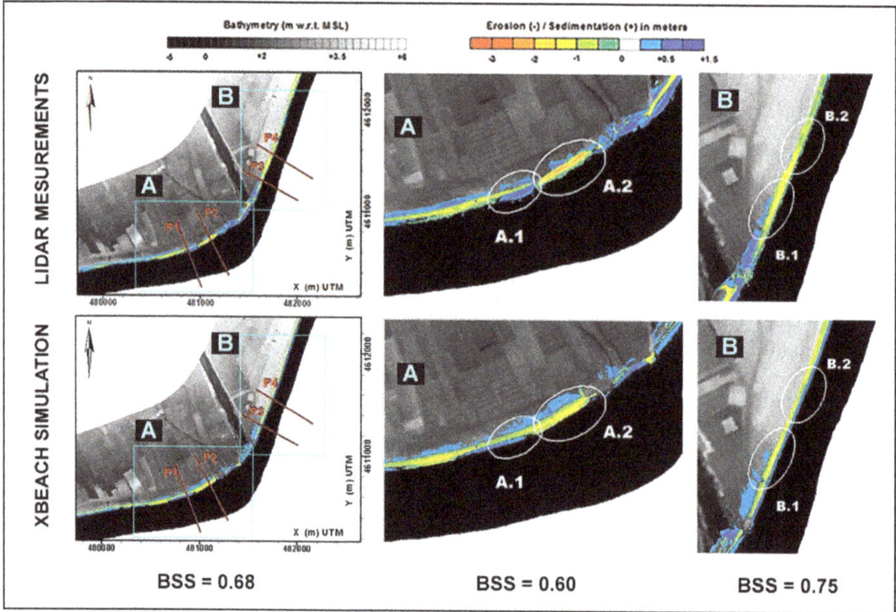

Figure 5. XBeach validation with LIDAR measurements of the emerged morphological changes. BSS: Brier Skill Score.

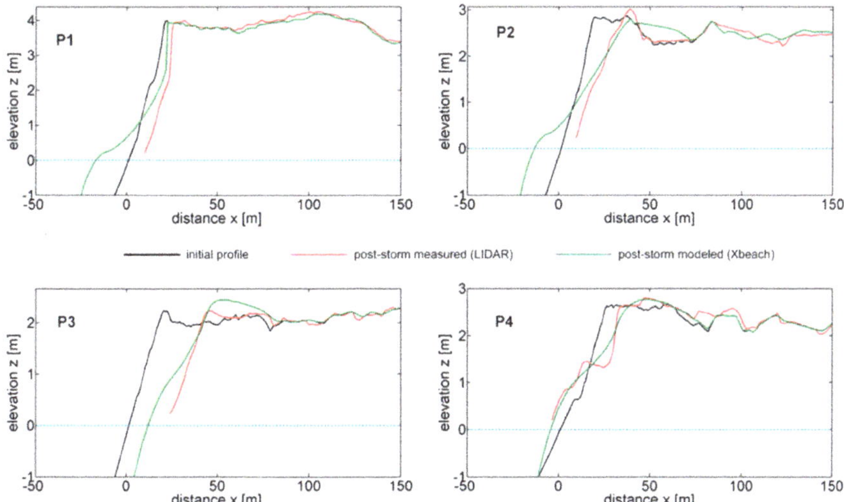

Figure 6. Comparison between XBeach simulated and measured beach profiles after the impact of the storm (see profiles location in Figure 5).

Although different combinations of model parameters resulted in BSS scores over 0.4 for the emerged profile, the qualitative assessment highlighted that some of them produced excessive deposition volumes in the submerged part. The final setup parameters, which resulted in an overall BSS of 0.68 (measured at the northern beach and first 600 m of the southern beach) and a meaningful qualitative simulation of the predicted submerged profile and alongshore bed level change patterns, are shown in Table 1 along with parameter description and tested ranges.

Table 1. Calibration parameters. Range of tested values and final parameter setup.

Parameter	Tested Values	Description	Final Set-up [1]
gamma	0.55–0.7	Breaker parameter in Baldock or Roelvink formulation (default = 0.55)	0.7
delta	0–0.5	Fraction of wave height to add to water depth in wave breaking formulations (default = 0)	0.5
facAs	0.2–0.7	Calibration factor time averaged flows due to wave asymmetry (default = 0.1)	0.6
facSk	0.2–0.7	Calibration factor time averaged flows due to wave skewness (default = 0.1)	0.6
wetslp	0.3–0.8	Critical avalanching slope under water (dz/dx and dz/dy) (default = 0.3)	0.7
gwflow	0 and 1	Turn on groundwater flow (default = 0)	1
sedcal	0.1–1	Sediment transport calibration coefficient per grain type (default = 1)	0.1

[1] based on best BSS score and qualitative assessment.

The implications of the selected values are as follows. First, the increase in wave-attack on the coast improves the behavior of the model in terms of reproducing observed overwash deposits (gamma and delta). Second, the contribution of avalanching to bed level changes was limited, taking into account the steepness of the site (wetslp). Third, the non-linearity effect on sediment transport for steep profiles (facAs and facSk) may be taken into account as reported in Elsayed and Oumeraci [18]. Fourth, sediment particle mobilization may be limited (sedcal). With respect to this, other authors have already reported an excess of modelled sediment suspension because the shear stress values required to initiate particle motion are higher than those predicted by using the Shields curve, as noted in Elsayed and Oumeraci [18] or McCall [11]. Fifth and finally, the groundwater module was turned on (gwflow) because the role of infiltration is more significant in coarse sediment environments, such as the Tordera Delta, where grains are close to gravel-size. The value of the permeability factor has been estimated using the Kozeny–Carman formula as described in Carrier [58]. For the grain size in the study area, the value of the permeability factor was estimated to be 0.0058 m/s. The results of the storm-induced morphological changes simulated with the final adjusted set of model coefficients are shown in Figures 5 and 6. As it can be seen, the measured changes in beach elevation at both sites of the river mouth are well reproduced by the model, with a BSS of 0.68 when the entire area is considered. The model also properly reproduces the differentiated response at both sites of the river mouth, mimicking the effects of change in coastal orientation with respect to the storm wave direction and the differences in coastal morphology.

Figure 7 shows the modelled depth-averaged wave-induced circulation during the peak of the storm, where a different circulation pattern is observed at both sides of the river mouth. Mean XBeach output (average conditions at 1 h output time-step) is used to average the current velocity and sediment transport during 4 h around the maximum peak of the event. At the northern part, where the beach is fully exposed to storm waves (i.e., a relatively small obliquity during the storm) and without any submerged morphological features, wave-induced circulation shows a typical quasi-uniform longshore current structure along the beach until the river mouth at the southern end. At the southern part, the coast is partially sheltered from storm waves (i.e., a large wave obliquity during the storm) and there is an alongshore bar running parallel to the shoreline with varying crest levels. This bar delimits a shallow shelf of approximately 4 m water depth to the shoreline. In this area, the induced circulation pattern during the peak of the storm shows a longshore current directed towards the south close to the shoreline and a local inversion of the current towards the N over the shallow shelf (Figure 7).

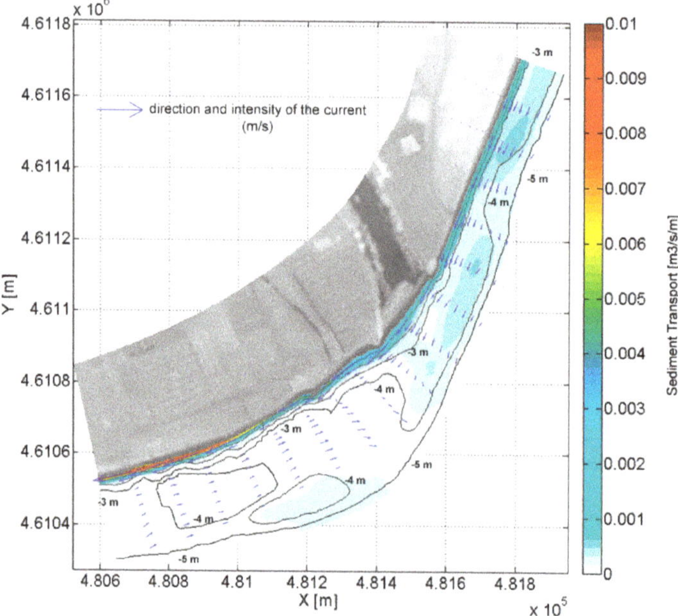

Figure 7. Simulated depth-averaged currents and sediment transport. Mean XBeach output at each 1 h time-step is averaged during the 4 h of maximum storm intensity.

With respect to morphodynamic changes (Figures 5 and 6), the model predicts, for the northern part, a nearly continuous erosion along the beach without significant alongshore gradients in sediment transport, and with significant sediment mobility down to about −5 m depth. The model properly reproduced the observed increase in erosion magnitude from B2 (P4) to B1 (P3), as well as most of the overwash deposits, which increase towards the south as the height of the berm decreases. In summary, the model reproduced well the observed variability in beach erosion and overwash along this northern site with a local BSS of 0.75. Southwards of the river mouth (Malgrat Beach), induced sediment transport and beach erosion present significant alongshore variations, with the largest erosion taking place just southwards of an existing structure located at the northern part of the sector, A2 (P2), which should act as a local boundary condition. This local effect can be seen in Figure 7 where a gradient in the longshore sediment transport, starting at the rigid structure north of A2, is detected. Longshore transport rates are larger than in the north, and mainly concentrated down to −3 m water depth. The model reproduced observed beach topographic changes with large erosion and overwash deposits due to a relatively low beach berm (A2, P2). At the southernmost part, erosion is only taking place at the upper level of the profile, whereas part of the material is deposited around zero level (A1, P1). The obtained local BSS for this sector was 0.60.

3.2. The Effects of Wave Direction on Storm-Induced Hazards

As was previously mentioned, once the morphodynamic model was calibrated, it was used to analyze the sensitivity of the area to changes in wave direction during storm impacts. Simulated morphodynamic changes and inundation for cases C20− (~80° N) to C40+ (~140° N) at northern and southern parts of the study area are shown in Figure 8, whereas the variation of integrated control variables for each sector can be seen in Figures 9 and 10 for inundation and morphodynamic changes respectively.

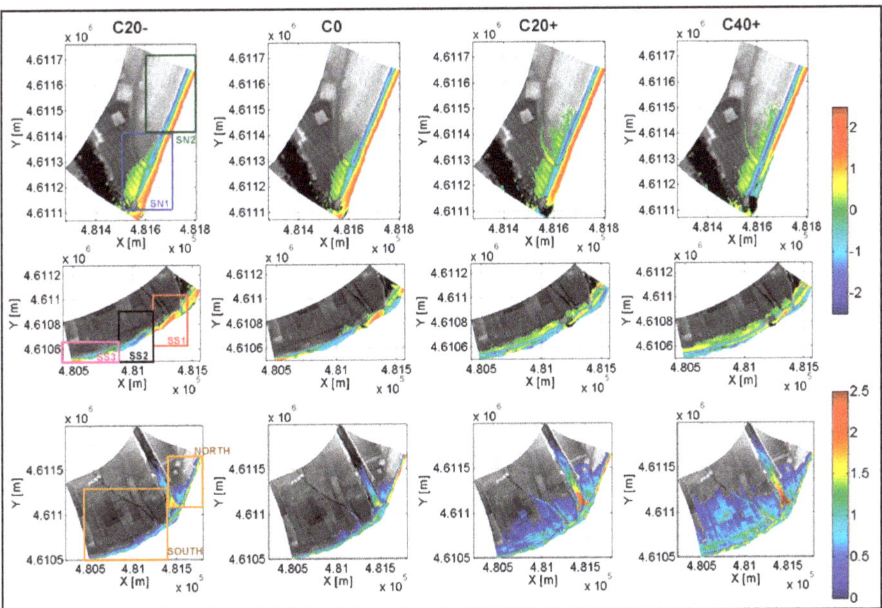

Figure 8. XBeach-simulated inundation depth (m) (**bottom**) and bed level changes (m) northwards (**top**) and southwards (**middle**) of the Tordera River.

With respect to inundation, the beach northward of the river mouth (Figure 9a) experiences an increase of the predicted inundation surface from C40− to C20+, when the surface reaches its maximum extension (5.47 Ha with 1 Ha over 0.5 m depth). Wave direction in this scenario corresponds to a nearly normal wave attack on the local coastline orientation. As incoming wave direction continues shifting towards the south, the predicted inundation surface progressively decreases, reaching values for the C80+ case similar to those observed for the C40− case. At the southern coast, the obtained pattern is significantly different (Figure 9b). Thus, for wave direction scenarios dominated by NE components (C40−, C20−) no significant inundation is observed, and this is consistent with a large sheltering from highly oblique wave incidence. For wave directions shifting from the base scenario to the south (C0 to C80+), the predicted inundated surface increases, reaching a maximum value for scenarios C60+ and C80+ of approximately 74 Ha (for an inundation depth above 0.05 m) and 33 Ha (for an inundation depth of 0.5 m), the C80+ case. Taking into account the total inundation of the study area, we can state that the magnitude of the inundation hazard significantly increases as wave direction shifts to the south. The southern coast is the most affected in terms of the magnitude of expected changes due to the local low-lying topography.

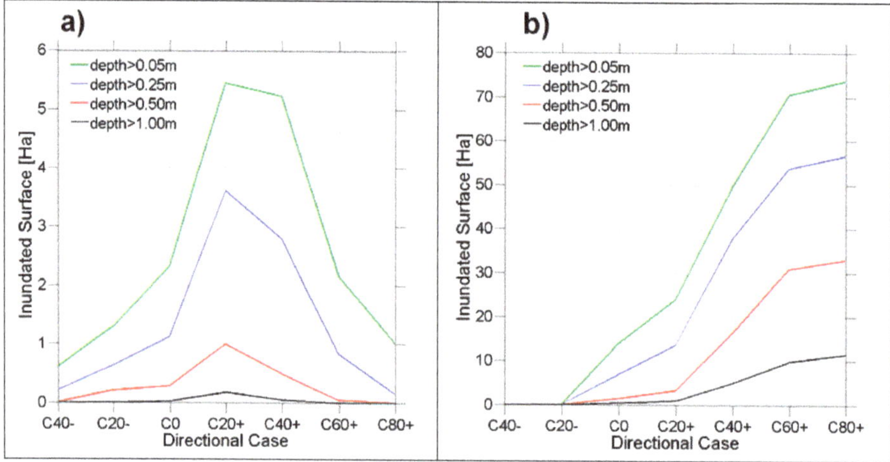

Figure 9. Variation of simulated inundated surface for different inundation depths as a function of the simulated storm direction for the northern (**a**) and southern (**b**) control areas (see Figure 4 or Figure 8 for location).

Regarding morphological changes, the northern beach (S'Abanell) shows a relatively low sensitivity to wave direction for scenarios C40− and C20+ when erosion volumes and profile retreats are considered (Figure 10). For this range of wave directions, the local wave-induced circulation is characterized by a southward directed longshore current along the beach (see Figures 7 and 11) which turns north when the wave direction is C20+ (Figure 11). As the incoming wave direction turns southwards, a stronger north-directed alongshore current is induced with velocity increasing along the beach (Figure 11) which results in an increasing erosion and profile retreat. Figure 11 also shows hatching areas in the velocity field which originate in those areas where wave incidence is orthogonal to the nearshore bathymetry, characterized by a heterogeneous bar in front of the coast. This beach erosion increase is particularly observed at the southernmost end of the section (SN1), just northwards of the river mouth, where an existing revetment acts as a boundary condition (barrier) for northwards directed transport. This overall modelled behavior is consistent with local field observations, where the northern part of the beach (out of the domain) experiences a significant sediment deposition under the impact of southern storms, bringing sediment from the south.

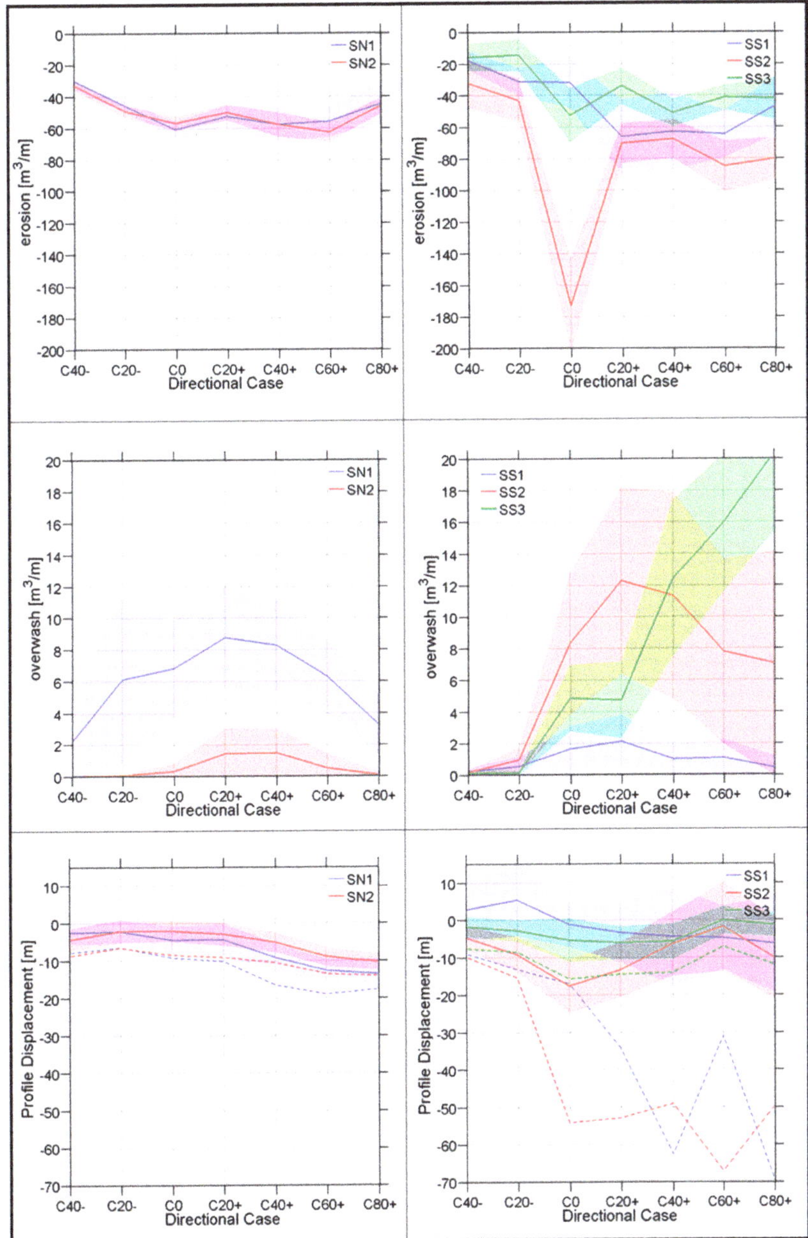

Figure 10. Variation of simulated morphodynamic parameters (see text for description) at selected control areas the N (**left**) and S (**right**) coasts for tested storm directions (see Figure 4 or Figure 8 for location). Continuous lines denote variable mean, shaded areas represent standard deviation and dashed lines indicate maximum profile retreat at each sector.

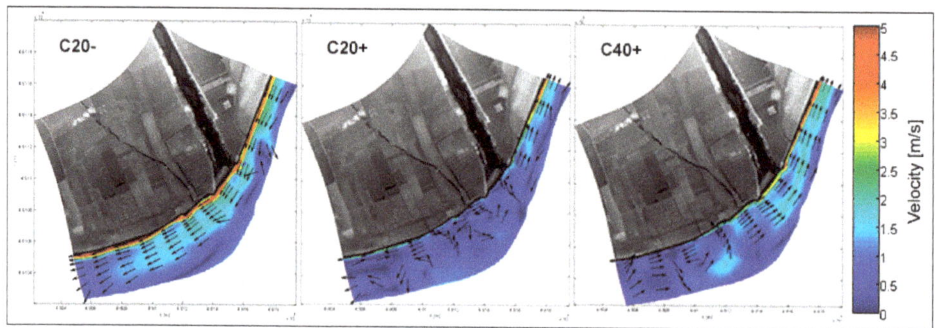

Figure 11. XBeach simulated depth-averaged currents. Mean XBeach output at each 1 h time-step is averaged during the 4 h of maximum storm intensity.

Overwash deposits along the northern section present a variation pattern with wave direction consistent with modelled inundation. The largest overwash verifies in the southern end of this sector (SN1) due to its lower beach berm. Thus, maximum overwash deposition verifies at SN1 under C20+ and C40+, which were the scenarios producing the largest inundations. As wave obliquity increases (scenarios C60+ and C80+), overwash significantly decreases, which is also in agreement with the observed inundation decrease under these conditions.

Southwards of the river mouth (Malgrat de Mar), the analysed sectors show a differentiated response due to the local effect of the existing building in the coastline which acts as a boundary condition for sediment transport. The northernmost end, SS1, just southward of the river mouth, experiences accretion (i.e., positive profile displacement) under C40− and C20− storms, which would correspond to the deposition of sediments brought by the southward directed longshore transport from the northern beach. As wave direction shifts to the south (from C20+ to C80+) this area becomes slightly erosive. This morphodynamic behavior is consistent with the observed switch in alongshore current direction for scenarios C20+ and C40+ (Figure 11). The central sector SS2, located just southwards of the mentioned obstacle, shows a different morphological response with storm direction to that observed in SS1. Thus, the largest erosion rate verifies under C0 due to combined effect of the orientation of the coast, a gap in the submerged bar and the presence of the hard structure at the northern end of the sector. This can be seen in Figure 7, where a significantly large gradient in sediment transport rates under C0 is observed. As wave direction shifts towards south, erosion rates significantly decrease, reaching a nearly null integrated value under C40+ to C80+ scenarios. The largest erosion under scenarios C20− to C0 in this sector, in comparison with the rest of the southern coast, seems to reflect the effect of the revetment on enhancing local downcoast erosion, similar to the well-known flanking effects in seawalls (e.g., Kraus and McDougal [59]). The southernmost sector, SS3, is far from the local effect of the revetment, so much so that the aforementioned downcoast erosion enhancement is not detected and, under highly oblique storms, erosion rates are significantly lower. Total volumetric changes in this sector present a relative low sensitivity to changes in wave direction for southern wave scenarios (C20+ to C80+).

Regarding overwash, the southernmost section SS3 presents a wave-direction influence similar to the influence observed for inundation, with overwash rates growing as wave direction turns to the South. The sectors closer to the river mouth SS2 and SS1 present a similar response with a maximum for C20+ and C40+ scenarios, when the wave directions are close to local orthogonality (Figure 8), and they are significantly larger for SS2 (~12 m^3/m) than SS1 (~2 m^3/m).

4. Discussion and Conclusions

In this study, the potential effects of changing wave direction for the storm-induced hazards on a highly curvilinear coarse sandy coastline have been assessed. This sensitivity test has been selected because although storminess projections under climate change scenarios for the Western Mediterranean do not predict any increase in wave height (e.g., Lionello et al. [20]; Conte and Lionello [21]), some existing projections identify potential changes in wave direction (Casas-Prat and Sierra [22,23]). These changes in wave direction may have significant implications for coastal sediment transport and coastal stability, as has been confirmed for the interannual changes influenced by El Niño (e.g., Barnard et al. [60]). Moreover, regarding cuspate coastlines such as the study area, their greater sensitivity, due to their curvature, results in even more significant implications (e.g., Slott et al. [24]; Johnson et al. [25]).

The tested hypothesis is that changes in wave direction may cause large variations in the magnitude of storm-induced hazards. This effect has also been addressed in other studies such as those by Mortlock et al. [26] and de Winter and Ruessink [27], which specifically analyzed the effects of changes in wave direction on the storm-induced hazards in the SE Australian and Holland coasts respectively. To this end and to isolate the influence of wave direction, we used a recorded long-return period storm as a base case scenario and we built test scenarios just by changing wave direction while maintaining the other wave parameters as recorded during the base storm (wave height and period).

In any case, tested conditions have not been designed to be used as climate change induced projections, as that may require the proper forecasting of regional wave conditions under given climate scenarios (e.g., Casas Prat and Sierra [23]). These have to be considered from the perspective of coastal risk management, in which a set of possible conditions are analyzed to characterize coastal vulnerability and resilience to inform risk management under uncertainty (see e.g., Hinkel et al. [61] for an application of this perspective to analyze sea level rise). In the study area, the current storm wave conditions depend on direction, with largest wave height and power being associated with NE-E waves, whereas S storms are less frequent and present a smaller associated power (e.g., Sánchez-Arcilla et al. [62]; Mendoza et al. [19]). To assess the potential variability on storm-induced hazards, tested scenarios were built by just changing wave direction while the remaining recorded parameters (representative of a worst case scenario, according to recorded conditions) were maintained.

This analysis has been performed by using the SWAN and XBeach models to simulate storm-induced hazards. Both models were calibrated by using data recorded during the impact of an extreme storm recorded in December 2008, which is used as the base case scenario. Although it is desirable to use more than one event to properly calibrate/validate the models (e.g., Ranashinge [63]), data availability during storm conditions was restricted to this event. However, on the positive side, it has to be considered that this storm was the largest event recorded in the area and representative of extreme storms with a very long return period (Mendoza et al. [19]) under current climate conditions. The SWAN model was very successful in simulating wave conditions during the development phase of the storm up to the pass of the peak of the storm, with the larger differences between measured and simulated waves being detected during the relaxation phase of the storm, when most of the induced changes had already occurred. The default parametrization of the XBeach model had to be adapted for application at the side to represent the effects the coarse-sand environment. Sediment transport was limited by using the sedcal parameter, avalanching was limited by increasing the critical slope, wave asymmetry was increased as suggested in literature for steep slopes (Elsayed and Oumeraci [18]) and groundwater effects were included. Gamma and delta wave breaking parameters were also tuned (Table 1). Calibrated parameters setup for XBeach in the study area led to a BSS score of 0.68 in spite of the out-of-comfort tested conditions (i.e., highly curvilinear coast, steep beach, coarse sediment). Although the predictive skill was very good for the northern and southern beaches, the model performance was better in the northern domain (BSS = 0.75) than in the southern one (BSS = 0.60), since this last area presented a significantly larger obliquity to wave direction during the storm, and a more complex bathymetry.

The obtained results show a very high sensitivity of storm-induced processes, i.e., inundation and erosion, to changes in storm wave direction. With respect to inundation, expected changes in hazard magnitude are very significant, especially in the southern part of the study area, since its morphology is characterized by a lower berm, and its low-lying unprotected hinterland makes this area sensitive to storm flooding (Jiménez et al. [7]). Thus, as storm waves turn from the base case (C0) to the south, the inundated surface along this southern beach dramatically increases due to its direct exposure to the south. On the contrary, a potential shift of wave direction to the N will have a positive impact on inundation in this area, since it will be more sheltered from wave action. At the northern beach, the largest increase in inundation hazard verifies under C20+ and C40+ scenarios when waves face nearly orthogonally to the coastline, although due to local morphology, the affected surface is much lower than in the southern beach. The hinterland of the study area is mostly occupied by agriculture land and, in the outer fringe just behind the shoreline, by campsites. In this sense, to transfer the potential change in hazard magnitude to changes in damage risk, it should be important to consider not only the change in direction but also its seasonality. Thus, risk may vary dramatically between the summer season (when the campsite facilities are used by visitors) and the rest of the year when only installations will be affected (e.g., Merz et al. [64]). An analysis of the risk associated with storm-induced inundation for different storm conditions can be seen in Sanuy et al. [32].

Similarly, storm-induced morphodynamic changes are more sensitive to directional changes on the southern beach, where the magnitude of the changes is larger. The beaches at the south of the river mouth present a larger spatial variability than those in the north due to the presence of a local boundary condition in form of a revetment at the shoreline. This revetment, which modifies local longshore transport, significantly enhances downcoast erosion under storm conditions. This induces a southwards directed longshore sediment transport while simultaneously promoting the accumulation of upcoast sediment. This contrasting behavior is particularly observed in the base case scenario which seems to represent the optimum conditions for longshore sediment transport in the area, thus inducing the largest changes in the surroundings of the structure.

In a particular case, under C40− and C20− scenarios, when wave direction turns north, the beach sector just south of the river experiences an important sediment accumulation due to the apparently efficient transfer of sediment from the northern beach across the river mouth and the partial barrier effect of the existing revetment.

The magnitude of the erosional response along the two control sectors in the northern beach is similar, although a higher variability is detected in the area closest to the river mouth. In general, there is a slight increase in erosion rates as wave direction turns south. This variation should be indicative of the role of longshore sediment fluxes during storm conditions. Thus, as the controlled northern area is just besides the river mouth, where there is another structure acting as a boundary condition, the increase in longshore sediment transport as waves turn S (scenarios from C20+ to C80+) will increase sediment losses, which will be transported further to the north. This behavior is currently observed in the northernmost part of this beach (out of the control zone in Figure 10) which experiences sediment accumulation under the impact of southern storms.

As expected, changes in the magnitude of overwash deposits follow observed changes in inundation, i.e., they increase as wave direction turns to the south, with maximum values around C20+ and C40+. The exception to this is the predicted changes in the southernmost sector, which present the largest overwash for C80+ conditions. The spatial variability in the northern beach is significantly lower than in the south, with small variations in magnitude across the tested range. Moreover, and reflecting the observed differences in inundation, the magnitude of overwash deposits is much higher in the southern sector.

Finally, and as a concluding remark, this analysis has shown that storm-induced hazards along a highly curvilinear coast are extremely sensitive to changes in wave direction. This means that even under a climate scenario of relatively steady storminess (wave power and frequency), a potential shift in wave direction may significantly change hazard conditions and, in consequence, need to be accounted for in robust damage risk assessments. To this end, an analysis such as the one presented here also permits an assessment of how coastal geomorphology modulates induced changes. In the study area, the low-lying nature of the southern beach and its orientation with respect to the current dominant storm direction make this area much more sensitive to directional changes. This is especially relevant from the coastal management standpoint because this area has been already identified as a hotspot for storm impacts under current conditions. The use of detailed process-based models has permitted the identification and quantification of the drastic increase in sensitivity when anthropogenic perturbations are present along the coast. These perturbations act as boundary conditions modifying local hydrodynamics and associated transport. For the case study analyzed here, the obtained results clearly identify the hazardous potential of the existing revetment in the southern beach, which has also been identified under current conditions, suggesting that its removal will soften the estimated morphodynamic response.

Author Contributions: Conceptualization, J.A.J. and M.S.; methodology, J.A.J. and M.S.; software, M.S.; validation, M.S.; formal analysis, M.S.; investigation, M.S. and J.A.J.; data curation, M.S.; writing—original draft preparation, M.S. and J.A.J.; writing—review and editing, M.S. and J.A.J.; supervision, J.A.J.; project administration, J.A.J.; funding acquisition, J.A.J.

Funding: This study was conducted in the framework of the RISC-KIT (Grant No 603458) and the M-CostAdapt (CTM2017-83655-C2-1-R) research projects, funded by the EU and the Spanish Ministry of Economy and Competitiveness (MINECO/AEI/FEDER, UE) respectively. The first author was supported by a PhD grant from the Spanish Ministry of Education, Culture and Sport.

Acknowledgments: Lidar data were obtained thanks a collaboration agreement between the Institut Cartogràfic i Geològic de Catalunya (ICGC) and the LIM/UPC, with the aim of analysing the usefulness of LiDAR data for monitoring coastal processes. The authors also express their gratitude to the Generalitat de Catalunya and Puertos del Estado for supplying wave data and to the Ministry of Agriculture, Fish, Food and Environment for the bathymetric data used in this study.

Conflicts of Interest: The authors declare no conflict of interest.

References

1. Kron, W. Coasts: The high-risk areas of the world. *Nat. Hazards* **2013**, *66*, 1363–1382. [CrossRef]
2. Bertin, X.; Li, K.; Roland, A.; Zhang, Y.J.; Breilh, J.F.; Chaumillon, E. A modeling-based analysis of the flooding associated with Xynthia, central Bay of Biscay. *Coast. Eng.* **2014**, *94*, 80–89. [CrossRef]
3. Jiménez, J.A.; Sancho-García, A.; Bosom, E.; Valdemoro, H.I.; Guillén, J. Storm-induced damages along the Catalan coast (NW Mediterranean) during the period 1958–2008. *Geomorphology* **2012**, *143–144*, 24–33. [CrossRef]
4. Ciavola, P.; Ferreira, O.; Haerens, P.; Van Koningsveld, M.; Armaroli, C.; Lequeux, Q. Storm impacts along European coastlines. Part 1: The joint effort of the MICORE and ConHaz Projects. *Environ. Sci. Policy* **2011**, *14*, 912–923. [CrossRef]
5. Ciavola, P.; Ferreira, O.; Haerens, P.; Van Koningsveld, M.; Armaroli, C. Storm impacts along European coastlines. Part 2: Lessons learned from the MICORE project. *Environ. Sci. Policy* **2011**, *14*, 924–933. [CrossRef]
6. van Dongeren, A.; Ciavola, P.; Martinez, G.; Viavattene, C.; Bogaard, T.; Ferreira, O.; Higgins, R.; McCall, R. Introduction to RISC-KIT: Resilience-increasing strategies for coasts. *Coast. Eng.* **2018**, *134*, 2–9. [CrossRef]
7. Jiménez, J.A.; Sanuy, M.; Ballesteros, C.; Valdemoro, H.I. The Tordera Delta, a hotspot to storm impacts in the coast northwards of Barcelona (NW Mediterranean). *Coast. Eng.* **2018**, *134*, 148–158. [CrossRef]
8. Plomaritis, T.A.; Costas, S.; Ferreira, Ó. Use of a Bayesian Network for coastal hazards, impact and disaster risk reduction assessment at a coastal barrier (Ria Formosa, Portugal). *Coast. Eng.* **2018**, *134*, 134–147. [CrossRef]
9. Harley, M.D.; Turner, I.L.; Middleton, J.H.; Kinsela, M.A.; Hanslow, D.; Splinter, K.D.; Mumford, P. Observations of Beach Recovery in SE Australia Following the June 2016 East Coast Low. In Proceedings of the Australasian Coasts & Ports 2017: Working with Nature, Australia, Cairns, 21–23 June 2017; p. 559.
10. Roelvink, D.; Reniers, A.; van Dongeren, A.; van Thiel de Vries, J.; McCall, R.; Lescinski, J. Modelling storm impacts on beaches, dunes and barrier islands. *Coast. Eng.* **2009**, *56*, 1133–1152. [CrossRef]
11. McCall, R.T.; Van Thiel de Vries, J.S.M.; Plant, N.G.; Van Dongeren, A.R.; Roelvink, J.A.; Thompson, D.M.; Reniers, A.J.H.M. Two-dimensional time dependent hurricane overwash and erosion modeling at Santa Rosa Island. *Coast. Eng.* **2010**, *57*, 668–683. [CrossRef]

12. Van Dongeren, A.; Roelvink, D.; McCall, R.; Neferhoff, K.; van Rooijen, A. Modeling the morphological impacats of coastal storms. In *Coastal Storms*; Ciavola, P., Coco, G., Eds.; John Wiley & Sons Ltd.: Hoboken, NJ, USA, 2017; pp. 195–216.

13. Dissanayake, P.; Brown, J.; Karunarathna, H. Modelling storm-induced beach/dune evolution: Sefton coast, Liverpool Bay, UK. *Mar. Geol.* **2014**, *357*, 225–242. [CrossRef]

14. Harter, C.; Figlus, J. Numerical modeling of the morphodynamic response of a low-lying barrier island beach and foredune system inundated during Hurricane Ike using XBeach and CSHORE. *Coast. Eng.* **2017**, *120*, 64–74. [CrossRef]

15. Roelvink, D.; Stelling, G.; Hoonhout, B.; Risandi, J.; Jacobs, W.; Merli, D. Development and Field Validation of a 2Dh Curvilinear Storm Impact Model. *Coast. Eng.* **2012**, *1*, 120. [CrossRef]

16. Valchev, N.; Eftimova, P.; Andreeva, N. Implementation and validation of a multi-domain coastal hazard forecasting system in an open bay. *Coast. Eng.* **2018**, *134*, 212–228. [CrossRef]

17. Vousdoukas, M.I.; Ferreira, Ó.; Almeida, L.P.; Pacheco, A. Toward reliable storm-hazard forecasts: XBeach calibration and its potential application in an operational early-warning system. *Ocean Dyn.* **2012**, *62*, 1001–1015. [CrossRef]

18. Elsayed, S.M.; Oumeraci, H. Effect of beach slope and grain-stabilization on coastal sediment transport: An attempt to overcome the erosion overestimation by XBeach. *Coast. Eng.* **2017**, *121*, 179–196. [CrossRef]

19. Mendoza, E.T.; Jimenez, J.A.; Mateo, J. A coastal storms intensity scale for the Catalan sea (NW Mediterranean). *Nat. Hazards Earth Syst. Sci.* **2011**, *11*, 2453–2462. [CrossRef]

20. Lionello, P.; Boldrin, U.; Giorgi, F. Future changes in cyclone climatology over Europe as inferred from a regional climate simulation. *Clim. Dyn.* **2008**, *30*, 657–671. [CrossRef]

21. Conte, D.; Lionello, P. Characteristics of large positive and negative surges in the Mediterranean Sea and their attenuation in future climate scenarios. *Glob. Planet. Chang.* **2013**, *111*, 159–173. [CrossRef]

22. Casas-Prat, M.; Sierra, J.P. Trend analysis of wave direction and associated impacts on the Catalan coast. *Clim. Chang.* **2012**, *115*, 667–691. [CrossRef]

23. Casas-Prat, M.; Sierra, J.P. Projected future wave climate in the NW Mediterranean Sea. *J. Geophys. Res. Ocean.* **2013**, *118*, 3548–3568. [CrossRef]

24. Slott, J.M.; Murray, A.B.; Ashton, A.D.; Crowley, T.J. Coastline responses to changing storm patterns. *Geophys. Res. Lett.* **2006**, *33*, 1–6. [CrossRef]

25. Johnson, J.M.; Moore, L.J.; Ells, K.; Murray, A.B.; Adams, P.N.; MacKenzie, R.A., III; Jaeger, J.M.; MacKenzie, R.A.; Jaeger, J.M. Recent shifts in coastline change and shoreline stabilization linked to storm climate change. *Earth Surf. Process. Landf.* **2015**, *40*, 569–585. [CrossRef]

26. Mortlock, T.R.; Goodwin, I.D.; McAneney, J.K.; Roche, K. The June 2016 Australian East Coast Low: Importance of wave direction for coastal erosion assessment. *Water* **2017**, *9*, 121. [CrossRef]

27. de Winter, R.C.; Ruessink, B.G. Sensitivity analysis of climate change impacts on dune erosion: Case study for the Dutch Holland coast. *Clim. Chang.* **2017**, *141*, 685–701. [CrossRef]

28. Vila, I.; Serra, J. Tordera River Delta system build up (NE Iberian Peninsula): Sedimentary sequences and offshore correlation. *Sci. Mar.* **2015**, *79*, 305–317. [CrossRef]

29. Martín-Vide, J.; Llasat, M.C. Las Precipitaciones Torrenciales en Cataluña. *Serie Geográfica* **2000**, *9*, 17–26.

30. Sanchez-Vidal, A.; Canals, M.; Calafat, A.M.; Lastras, G.; Pedrosa-Pàmies, R.; Menéndez, M.; Medina, R.; Company, J.B.; Hereu, B.; Romero, J.; et al. Impacts on the deep-sea ecosystem by a severe coastal storm. *PLoS ONE* **2012**, *7*, e30395. [CrossRef]

31. Jiménez, J.A.; Gracia, V.; Valdemoro, H.I.; Mendoza, E.T.; Sánchez-Arcilla, A. Managing erosion-induced problems in NW Mediterranean urban beaches. *Ocean Coast. Manag.* **2011**, *54*, 907–918. [CrossRef]

32. Sanuy, M.; Duo, E.; Jäger, W.S.; Ciavola, P.; Jiménez, J.A. Linking source with consequences of coastal storm impacts for climate change and risk reduction scenarios for Mediterranean sandy beaches. *Nat. Hazards Earth Syst. Sci.* **2018**, *18*, 1825–1847. [CrossRef]

33. Reguero, B.G.; Menéndez, M.; Méndez, F.J.; Mínguez, R.; Losada, I.J. A Global Ocean wave (GOW) calibrated reanalysis from 1948 onwards. *Coast. Eng.* **2012**, *65*, 38–55. [CrossRef]

34. Mendoza, E.T.; Jiménez, J.A. Clasificación de tormentas costeras para el litoral catalán (Mediterráneo NO). *Ecnología y Ciencias del Agua* **2008**, *23*, 21–32.

35. Bolaños, R.; Jorda, G.; Cateura, J.; Lopez, J.; Puigdefabregas, J.; Gomez, J.; Espino, M. The XIOM: 20 years of a regional coastal observatory in the Spanish Catalan coast. *J. Mar. Syst.* **2009**, *77*, 237–260. [CrossRef]

36. Ruiz, A.; Kornus, W.; Talaya, J. Coastal applications of Lidar in Catalonia. In Proceedings of the 6th European Congress on Regional Geoscientific Cartography and Information Systems, Munich, Germany, 9–12 June 2009.

37. General Bathymetric Chart of the Oceans (GEBCO). 2014. Available online: https://www.gebco.net (accessed on 10 February 2019).

38. Trigo, I.F.; Bigg, G.R.; Davies, T.D. Climatology of Cyclogenesis Mechanisms in the Mediterranean. *Mon. Weather Rev.* **2002**, *130*, 549–569. [CrossRef]

39. Jiménez, J.A. Characterising Sant Esteve's storm (26th December 2008) along the Catalan coast (NW Mediterranean). In *Assessment of the Ecological Impact of the Extreme Storm of Sant Esteve's Day (26 December 2008) on the Littoral Ecosystems of the North Mediterranean Spanish Coasts*; Mateo, M.A., Garcia Rubies, T., Eds.; Final Report (PIEC 200430E599); Centro de Estudios Avanzados de Blanes, Consejo Superior de Investigaciones Científicas: Blanes, Spain, 2012; pp. 31–44.

40. Puertos del Estado, Madrid, Extremos Máximos de Oleaje (Altura Significante). Boya de Palamós. 2006. Available online: http://www.puertos.es/es-es/oceanografia/Paginas/portus.aspx (accessed on 8 February 2019).

41. Plana-Casado, A. Storm-Induced Changes in the Catalan Coast Using Lidar: The St. Esteve Storm (26/12/2008) Case. Master's Thesis, Faculty of Civil Engineering, Universitat Politècnica de Catalunya, Barcelona, Spain, 2013. Available online: http://hdl.handle.net/2099.1/23343 (accessed on 5 December 2016).

42. Jiménez, J.A.; Plana, A.; Sanuy, M.; Ruiz, A. Morphodynamic impact of an extreme storm on a cuspate deltaic shoreline. In Proceedings of the 34th International Costal Engineering Conference (2014 ASCE), Seoul, Korea, 15–20 June 2014.

43. Durán, R.; Guillén, J.; Ruiz, A.; Jiménez, J.A.; Sagristà, E. Morphological changes, beach inundation and overwash caused by an extreme storm on a low-lying embayed beach bounded by a dune system (NW Mediterranean). *Geomorphology* **2016**, *274*, 129–142. [CrossRef]

44. Teixidó, N.; Casas, E.; Cebrián, E.; Linares, C.; Garrabou, J. Impacts on Coralligenous Outcrop Biodiversity of a Dramatic Coastal Storm. *PLoS ONE* **2013**, *8*, e53742. [CrossRef] [PubMed]

45. Pagès, J.F.; Gera, A.; Romero, J.; Farina, S.; Garcia-Rubies, A.; Hereu, B.; Alcoverro, T. The Mediterranean Benthic Herbivores Show Diverse Responses to Extreme Storm Disturbances. *PLoS ONE* **2013**, *8*, e62719. [CrossRef] [PubMed]

46. Booij, N.; Ris, R.C.; Holthuijsen, L.H. A third-generation wave model for coastal regions. I-Model description and validation. *J. Geophys. Res.* **1999**, *104*, 7649–7666. [CrossRef]

47. Ris, R.C.; Holthuijsen, L.H.; Booij, N. A third-generation wave model for coastal regions: Verification. *J. Geophys. Res.* **1999**, *104*, 7667–7681. [CrossRef]

48. TU Delft, SWAN Simulating Waves Nearshore. 2016. Available online: http://www.swan.tudelft.nl/ (accessed on 13 April 2016).

49. Komen, G.J.; Hasselmann, K.; Hasselmann, K. On the Existence of a Fully Developed Wind-Sea Spectrum. *J. Phys. Oceanogr.* **1984**, *14*, 1271–1285. [CrossRef]

50. Bolaños, R. Tormentas de Oleaje en el Mediterráneo: Física y Predicción. Ph.D. Thesis, Universitat Politecnica de Catalunya, Barcelona, Spain, 2004.

51. Pallares, E.; Sánchez-Arcilla, A.; Espino, M. Wave energy balance in wave models (SWAN) for semi-enclosed domains-Application to the Catalan coast. *Cont. Shelf Res.* **2014**, *87*, 41–53. [CrossRef]

52. Ratsimandresy, A.W.; Sotillo, M.G.; Carretero Albiach, J.C.; Álvarez Fanjul, E.; Hajji, H. A 44-year high-resolution ocean and atmospheric hindcast for the Mediterranean Basin developed within the HIPOCAS Project. *Coast. Eng.* **2008**, *55*, 827–842. [CrossRef]

53. Soulsby, R.L. *Dynamics of Marine Sands*; Thomas Telford: London, UK, 1997.

54. Vousdoukas, M.I.; Almeida, L.P.; Ferreira, Ó. Modelling storm-induced beach morphological change in a meso-tidal, reflective beach using XBeach. *J. Coast. Res.* **2011**, *64*, 1916–1920.

55. Wright, L.D.; Short, A.D. Morphodynamic variability of surf zones and beaches: A synthesis. *Mar. Geol.* **1984**, *56*, 93–118. [CrossRef]

56. Harley, M.D.; Ciavola, P. Managing local coastal inundation risk using real-time forecasts and artificial dune placements. *Coast. Eng.* **2013**, *77*, 77–90. [CrossRef]

57. Sutherland, J.; Peet, A.H.; Soulsby, R.L. Evaluating the performance of morphological models. *Coast. Eng.* **2004**, *51*, 917–939. [CrossRef]

58. Carrier, W.D. Goodbye, Hazen; Hello, Kozeny-Carman. *J. Geotech. Geoenviron. Eng.* **2003**, *129*, 1054–1056. [CrossRef]
59. Kraus, N.C.; McDougal, W.G. The Effects of Seawalls on the Beach: Part I, an Updated Literature Review. *J. Coast. Res.* **1996**, *12*, 691–701.
60. Barnard, P.L.; Allan, J.; Hansen, J.E.; Kaminsky, G.M.; Ruggiero, P.; Doria, A. The impact of the 2009-10 El Niño Modoki on U.S. West Coast beaches. *Geophys. Res. Lett.* **2011**, *38*. [CrossRef]
61. Hinkel, J.; Jaeger, C.; Nicholls, R.J.; Lowe, J.; Renn, O.; Peijun, S. Sea-level rise scenarios and coastal risk management. *Nat. Clim. Chang.* **2015**, *5*, 188. [CrossRef]
62. Sánchez-Arcilla, A.; González-Marco, D.; Bolaños, R. A review of wave climate and prediction along the Spanish Mediterranean coast. *Nat. Hazards Earth Syst. Sci.* **2008**, *8*, 1217–1228. [CrossRef]
63. Ranasinghe, R. Assessing climate change impacts on open sandy coasts: A review. *Earth-Sci. Rev.* **2016**, *160*, 320–332. [CrossRef]
64. Merz, B.; Thieken, A.; Gocht, M. Flood Risk Mapping at the Local Scale: Concepts and Challenges. In *Flood Risk Management in Europe. Advances in Natural and Technological Hazards Research*; Begum, S., Stive, M.J.F., Hall, J.W., Eds.; Springer: Dordrecht, The Netherlands, 2007; Volume 25. [CrossRef]

Article

Impacts of Sea Level Rise and River Discharge on the Hydrodynamics Characteristics of Jakarta Bay (Indonesia)

Martin Yahya Surya [1], Zhiguo He [1,2], Yuezhang Xia [1,2,*] and Li Li [1,2]

[1] Ocean College, Zhejiang University, Zhoushan 316021, China
[2] State Key Laboratory of Satellite Ocean Environment Dynamics (Second Institute of Oceanography, MNR), Hangzhou 310058, China
* Correspondence: yzxia@zju.edu.cn; Tel.: +86-13588373750

Received: 8 May 2019; Accepted: 3 July 2019; Published: 5 July 2019

Abstract: Jakarta city has been vulnerable to sea level rise and flooding for many years. A Giant Seawall (GSW) was proposed in Jakarta Bay to protect the city. The impacts of sea level rise and river discharge on the tidal dynamics in Jakarta Bay and flooding areas in Jakarta city were investigated using the finite-volume coastal ocean model (FVCOM). Model results showed that the bay is diurnally dominated by the K1 tidal component. The diurnal tides propagate westward, while the semidiurnal tides propagate eastward in the bay. The rise of sea level increases the diurnal tidal component and the inundation areas due to the increased tidal forcing: when considering a sea level rise of 0.6 m, the K1 amplitude increases by ~1% (0.25 cm) near the coastline and the current magnitude increases by 16.6% (0.05 m/s). The inundation area increases with the sea level rise in the low land elevation areas occurring near the coastlines: the inundation area increased by 29.68 km^2 (7.1%) with a sea level rise of 0.6 m. The increase of river discharge amplified the diurnal tidal component as well as the inundation areas at the river mouth due to increased fluvial forcing: if 10 times the mean river discharge occurs, the K1 amplitude increases by ~1% (0.25 cm) and the current magnitude increases by 100% (0.4 m/s), and the inundation areas increase by 26.61 km^2 (6.2%). The K1 tidal phase remains almost unchanged under both the sea level rise and river discharge conditions. The combined increase of sea level rise and the river discharge amplifies the inundation areas and the tidal currents due to increased tidal and fluvial forcing. The construction of GSW would decrease the tidal prism and dissipation effects of the bay, thus slightly increasing the K1 amplitude of the tidal level: by less than 1% (0.2 cm). There would be no significant change of phase lag for the K1 component. Although this study is site specific, the findings could be applied more widely to any open-type bays.

Keywords: Jakarta Bay; sea level rise; river discharge; flooding; tides; currents

1. Introduction

Jakarta Bay is located on the northwest coast of Java Island, extending from 5.8° S to 6.2° S and 106.6° E to 107.1° E [1]. It is a shallow bay with an average depth of about 18 m and an area of 662 km^2 [2]. The bay is bordered by Pasir Cape to the west and Karawang Cape to the east, with a total coastline length of about 72 km.

Jakarta Bay is essential for the development of Jakarta city, the capital city of Indonesia. Jakarta city is the economic, cultural and political center of Indonesia. There are 10 million people in Jakarta city and more than 28 million people in the satellite cities of Jakarta [3]. The large population and the rapid development of the city have brought increasing urban development issues. The demographic change influenced by the urbanization causes large-scale rural land conversion and new town development in Jakarta city. This massive urbanization has led to some environmental issues. One of the environmental problems is related to flooding and sea level rise. Flood, in terms of river flooding and coastal flooding, is the most common natural disaster affecting populations in Asia [4]. For example, extreme rainfall events will increase in frequency and severity within the twenty-first century and lead to extreme river discharges [5].

Jakarta city experiences annual flooding disasters due to heavy rains, high river discharges, and high tides. An extreme flood occurred in 2007, when north Jakarta was hit by unusually high tides and heavy rains. The total inundation area in the extreme flood covered 400 km^2. Approximately 70% of Jakarta was flooded with water depth of up to 4 m. This extreme flood cost was around 890 million USD in direct losses and more than 620 million USD in indirect losses [6]. Nearly 600,000 people were displaced and 79 people died. Many roads were closed, including the highway to the international airport, and electricity and telephone lines were cut. Some researchers believed that land subsidence contributed to the heavy flood in 2007. The land subsidence due to groundwater extraction was the main cause for lowering the land topography. At that time, the rate of land subsidence was about 1–10 cm per year [7].

As well as the flooding issue, the problem of rising sea levels has become a major concern in Jakarta Bay. During the 20th century, the mass loss of glaciers and ice caps contributed to sea level rise [5]. These losses were estimated to be 0.5 ± 0.18 mm in sea level equivalent (SLE) per year between 1961 and 2003. Global warming of ocean waters produces water molecule expansions that are responsible for sea level rise. The average contribution of the thermal expansion to sea level rise is 0.42 ± 0.12 mm.

To overcome the flooding and sea level rise, a Giant Seawall (GSW) at Jakarta Bay has been proposed to protect the city from flood events increasing due to subsidence, exceptional high tides, and sea level rise [8]. The GSW project will include sea wall construction, land reclamation, toll roads, and a port extension in Jakarta Bay. The GSW will offer a long-term protection and flooding solution, and at the same time facilitate social-economic development. The GSW project may also have large impacts on tidal level, flow pattern, current and flooding area in Jakarta Bay. There have been many studies around Jakarta Bay, but relatively few focused on river discharge and subsequent sea level rise effect. Most of the applied models of Jakarta Bay have taken into account only tides and winds as the predominant forces to study circulation and hydrodynamics.

Ningsih's model for Java Sea showed that monsoon wind plays important roles in the general circulation of this area [9]. The flow pattern in Jakarta Bay is mainly driven by monsoon winds with current moving easterly during northwest monsoons (NWM), and reversed during southeast monsoons (SEM) [10–12]. In general, the southeast monsoons (SEM) are stronger than northwest monsoons (NWM) [13]. During flood conditions, the westward current induces a clockwise circulation through Jakarta Bay, and an eastward counter-clockwise circulation at ebb conditions [1,14].

Land subsidence in Jakarta is the predominant driver exacerbating coastal floods, followed by the influence of sea level rise [4]. The potential flood area in the 2025–2050 period is estimated to be 3.4 times greater than during the 2000–2025 period [15]. The sea level rise effects on tidal hydrodynamics along the northern Gulf of Mexico have been observed and modelled [16]. Tidal amplitudes within the bays increased by as much as 67% (10 cm) in some areas. Most of the bays would experience faster tidal propagation due to changes in harmonic constituent phases in future scenarios. The tidal inundation increases along the NGOM study area, especially in low-lying marsh areas and barrier islands with low dune elevations.

In recent years, estuaries and coastal system have been modelled successfully using the finite-volume coastal ocean model (FVCOM). Fukuoka Bay has been modelled using the FVCOM and Princeton Ocean Model (POM) [17]. A comparison of these two models showed that FVCOM provides a better result because FVCOM can describe the coastline geometry better. Benoa Bay, which has shallow water depth, has also been modelled with FVCOM [18]. FVCOM can successfully produce the residual current and vertical salinity profile, which is significantly affected by fresh water discharge. In Saemangeum in Korea, FVCOM has been used to successfully explain the mechanism of decreasing tidal current and tidal amplitude after the construction of a sea dike [19]. Thus, FVCOM is a powerful tool for modelling the hydrodynamics in Jakarta Bay. The unstructured grid of FVCOM would be suitable for the irregular coastline and bathymetry of Jakarta Bay.

In this paper, an FVCOM model for Jakarta Bay was used to construct tidal dynamics and the flooding occurrence in Jakarta city. The impacts of the river discharges and sea level rise on tidal dynamics and flooding areas were numerically examined, and the effects of the GSW on tides and flooding were investigated.

2. Materials and Methods

The influence of river discharge and sea level rise on the hydrodynamics of the bay was investigated through numerical modelling. The hydrodynamic model (Figure 1) in this paper was based on the model previously established by Rusdiansyah et al. [1], with the model domain being expanded to cover the entire area as described in the Master Plan National Capital Integrated Coastal Development [8]. The impacts of river discharges and sea level rises were analysed using numerical experiments in this study.

2.1. Hydrodynamics of Jakarta Bay

Jakarta Bay is an open type bay (Figure 1) with a water depth less than 60 m. Surface currents in Jakarta Bay are mainly in an east to west direction during spring tides, with peak current speed occurring at the bay mouth and near the river mouths (Figure 2a,b) [1]. Inside the bay, the current speed is mainly below 0.2 m/s. The ebbing currents are slightly larger than the flood currents due to the river discharge. The current speeds near the bottom level reach 0.1 m/s in Jakarta Bay during spring flood tides (Figure 2c,d). During spring ebbing tides, the bottom currents are small (within 0.07 m/s) in almost the entire bay.

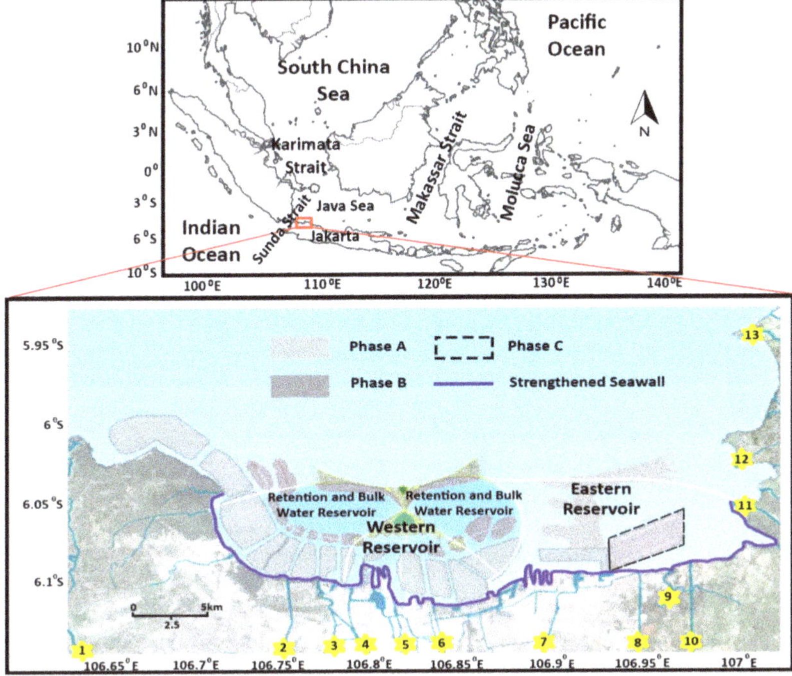

Figure 1. Map of the location of Jakarta Bay, and the Giant Seawall (GSW) as described in the NCICD Master Plan (Source: Master Plan NCICD). Numbers 1 to 13 correspond to rivers.

To overcome the flooding issues in Jakarta, Indonesia's government proposed the Giant Seawall (GSW) construction through the Master plan National Capital Integrated Coastal Development [8]. The GSW construction is considered through three phases: phase A, B, and C. In phase A the existing sea wall and river embankments are strengthened (the solid blue line in Figure 1), including land reclamation (the light grey color along the coast in Figure 1). In phase B, the outer sea wall is constructed and land is reclaimed (the dark grey color in Figure 1). This phase B is mainly determined by the required storage capacity of giant water reservoir between the coastline and seawall. There are two reservoirs in the GSW to adjust the water discharges from rivers. Phase C covers the long-term development in the east of Jakarta Bay (the dashed lines).

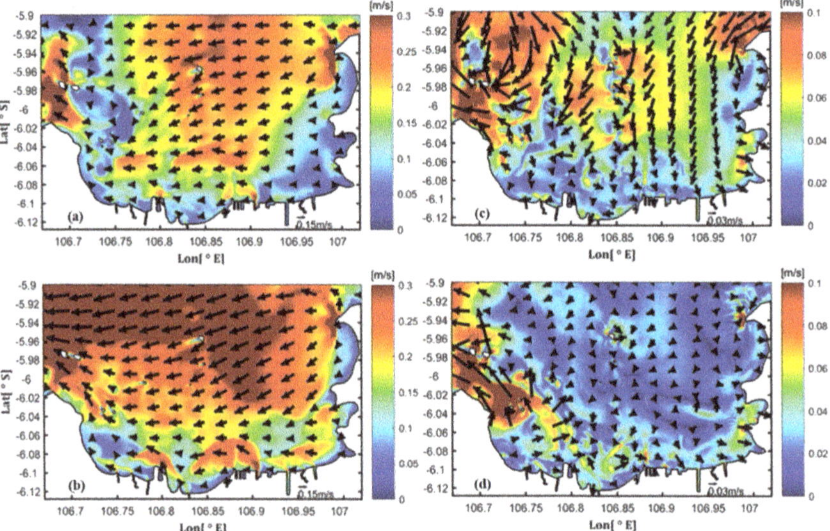

Figure 2. Currents (m/s) during spring tides in Jakarta Bay: (**a**) flooding and ebbing (**b**) ebb currents near the water surface; (**c,d**) show flooding and ebbing currents near the bottom level.

2.2. Tides in Jakarta Bay

The tidal level can be decomposed into different tidal components using harmonic analysis with the least squares method in MATLAB [20]. There are eight astronomical tidal components, named M2, S2, N2, K2, K1, O1, P1, and Q1 tides. When tides propagate into shallow water areas, the shallow water components are generated. For example, the MF, MM, M4, MS4, MN4 tides are all shallow water tides.

The Global Sea Level Observing System (GLOSS) by the Intergovernmental Oceanographic Commission (IOC) provides the sea surface level data for Kolinamil Station, Jakarta. The measurements at Kolinamil Station are from 2 June to 30 June 2015. The maximum tidal range at Kolinamil Station is 0.87 m. The maximum tidal range at Jakarta Bay is about 1 m with the highest level at around 0.6 m and the lowest level at around −0.4 m during spring tides relative to the mean sea level.

Defant (1961) [21] classified tidal type based on the amplitudes of tidal harmonic components called Formzahl number. The Formzahl number (F) is obtained from:

$$F = \frac{Amp_{K1} + Amp_{O1}}{Amp_{M2} + Amp_{S2}} \tag{1}$$

where Amp_{K1}, Amp_{O1}, Amp_{M2} and Amp_{S2} indicate the amplitude of the corresponding tidal component. The classification of the Formzahl number is as follow: $F < 0.25$ is semidiurnal type, $0.25 < F < 3$ is mixed type, and $F > 3$ is diurnal type.

According to the Formzahl number (*F*) by Defant (1961) [21], the tidal pattern in Jakarta bay is diurnal indicated by the Formzahl number of 3.95 observed at Kolinamil Station. Hence, the tides in Jakarta Bay are diurnal.

2.3. Description of the Flow Model

The finite-volume coastal ocean model (FVCOM) was used to simulate the hydrodynamics of Jakarta Bay. The finite volume approach and unstructured meshes with 3-D primitive equations make FVCOM ideally suited for coastal areas [22]. The unstructured grid of FVCOM provides topography flexibility that is adequate for the irregular morphology of Jakarta Bay. A σ-coordinate transformation system in a vertical direction is used to get a clear representation of the irregular bottom topography. The σ-coordinate transformation is defined as:

$$\sigma = \frac{z - \zeta}{D} \tag{2}$$

where total water column depth is $D = H + \zeta$, while H is bottom depth (relative to $z = 0$) and ζ is the height of the free surface (relative to $z = 0$).

Under the σ-coordinate system, the continuity and momentum equations can be written as Equations (2)–(5), and salinity, temperature, and density as Equations (6)–(8):

$$\frac{\partial D}{\partial t} + \frac{\partial uD}{\partial x} + \frac{\partial vD}{\partial y} + \frac{\partial \omega}{\partial \sigma} = 0 \tag{3}$$

$$\frac{\partial uD}{\partial t} + \frac{\partial u^2 D}{\partial x} + \frac{\partial uvD}{\partial y} + \frac{\partial u\omega}{\partial \sigma} - fvD = -gD\frac{\partial \zeta}{\partial x} - \frac{gD}{\rho_0}\left[\frac{\partial}{\partial x}\left[D\int_\sigma^0 \rho' d\sigma'\right] + \sigma\rho'\frac{\partial D}{\partial x}\right] + \frac{1}{D}\frac{\partial}{\partial \sigma}\left(K_m\frac{\partial u}{\partial \sigma}\right) + DF_u \tag{4}$$

$$\frac{\partial vD}{\partial t} + \frac{\partial uvD}{\partial x} + \frac{\partial v^2 D}{\partial y} + \frac{\partial v\omega}{\partial \sigma} + fuD = -gD\frac{\partial \zeta}{\partial y} - \frac{gD}{\rho_0}\left[\frac{\partial}{\partial y}\left[D\int_\sigma^0 \rho' d\sigma'\right] + \sigma\rho'\frac{\partial D}{\partial y}\right] + \frac{1}{D}\frac{\partial}{\partial \sigma}\left(K_m\frac{\partial v}{\partial \sigma}\right) + DF_v \tag{5}$$

$$\frac{\partial wD}{\partial t} + \frac{\partial uwD}{\partial x} + \frac{\partial vwD}{\partial y} + \frac{\partial w\omega}{\partial \sigma} = -gD\frac{\partial \zeta}{\partial x} + \frac{gD}{\rho_0}\left[\sigma\rho'\frac{\partial D}{\partial \sigma}\right] + \frac{1}{D}\frac{\partial}{\partial \sigma}\left(K_m\frac{\partial w}{\partial \sigma}\right) + DF_w \tag{6}$$

$$\frac{\partial SD}{\partial t} + \frac{\partial SuD}{\partial x} + \frac{\partial SvD}{\partial y} + \frac{\partial S\omega}{\partial \sigma} = \frac{1}{D}\frac{\partial}{\partial \sigma}\left(K_h\frac{\partial S}{\partial \sigma}\right) + DF_s \tag{7}$$

$$\frac{\partial TD}{\partial t} + \frac{\partial TuD}{\partial x} + \frac{\partial TvD}{\partial y} + \frac{\partial T\omega}{\partial \sigma} = \frac{1}{D}\frac{\partial}{\partial \sigma}\left(K_h\frac{\partial T}{\partial \sigma}\right) + D\hat{H} + DF_T \tag{8}$$

$$\rho = \rho(S, T) \tag{9}$$

where x, y, and σ are the respective east, north, and vertical axes in the σ-coordinate; u, v, and ω are velocity components; S and T are salinity and temperature; ρ is total density; ρ' is the reference density of ρ_0; f is the Coriolis parameter; g is gravitational acceleration; K_m is thermal vertical eddy viscosity and K_h is thermal vertical eddy diffusion; F_u, F_v, F_w, F_S, F_T are horizontal momentum, salt diffusion, and thermal terms.

The finite volume discrete method is applied in the model. The Mellor and Yamada level 2.5 (MY-2.5) turbulent closure scheme [23] was used for vertical mixing and the Smagorinsky scheme for horizontal mixing [24]. More details regarding the model can be found in Chen et al. (2006) [22].

2.4. Flow Model Setup

The computational domain covers Jakarta Bay, extending from 5.8° S to 6.15° S and 106.6° E to 107.1° E (Figure 3). This domain is covered by two grids, one with no GSW (Figure 3a) and one with GSW (Figure 3b). The model grid for Jakarta Bay with no GSW (Figure 3a) consists of 63,932 nodes and 127,273 elements, while the grid for Jakarta Bay with a GSW (Figure 3b) consists of 62,880 nodes and 122,986 elements. The computational grid has a high resolution varying from ~2.3 km at the open ocean boundary to ~25 m near the coastlines. Land areas of ~4.5 km from the nearest coastline have resolutions varying from ~25 m to ~300 m. Ten uniform vertical layers in a sigma coordinate system are set in the model according to sensitivity analysis.

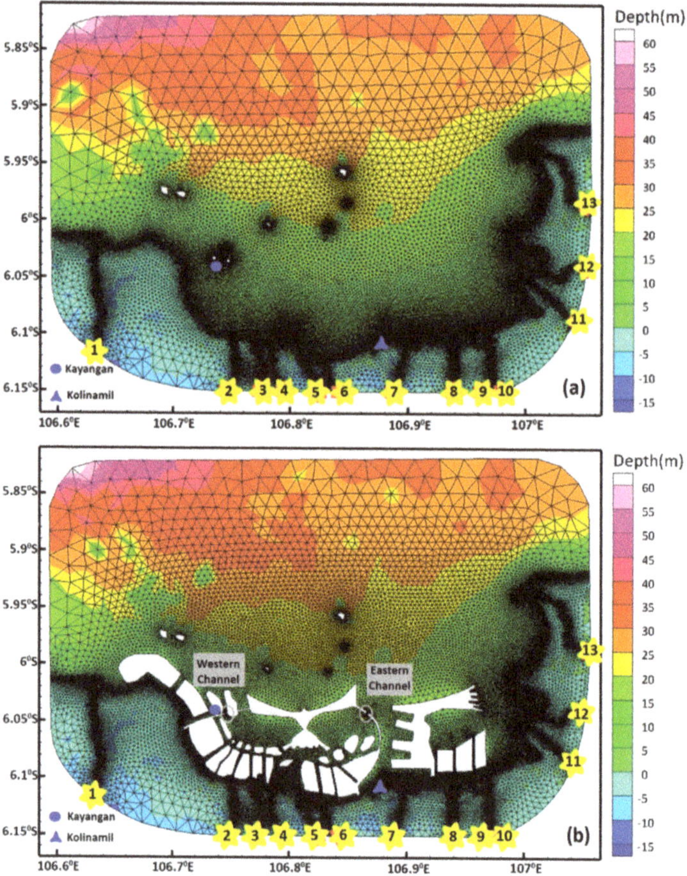

Figure 3. Model Grid for Jakarta Bay (**a**) reference scenario 1 and (**b**) Jakarta Bay after the Giant Seawall (GSW) construction (scenario 3). The numbers indicate rivers. The blue dot and triangle indicate field stations.

Coastline data were extracted from the Navy Chart Indonesia and Google Earth (2014). Bathymetrical data were obtained from survey data provided by the Ministry of Public Works and combined with nautical charts of Jakarta Bay. Jakarta Bay is a shallow water bay with maximum depth around 60 m. Land topography was obtained from the Consortium for Spatial Information (CGIAR-CSI) (http://srtm.csi.cgiar.org). The SRTM (Shuttle Radar Topographic Mission) digital elevation data were used for this land topography with a high spatial resolution of 90 m. The CGIAR-CSI SRTM data product applied a hole-filling algorithm to provide continuous elevation values [25].

The observation data of tidal elevation and currents from two field stations (Figure 3) were used to validate the numerical model. The blue dot represents Kayangan Station, which provided tidal elevation and current data, while the blue triangle represents Kolinamil Station, which provided tidal elevation data.

The numerical simulation is driven by ocean tides, winds, and river discharges. The open boundary condition is used for ocean tides as the main forcing, while the closed boundary condition is used for river discharges as the main forcing. The wind data were used as the atmospheric forcing at this research. Other types of external forcing such as precipitation, evaporation, heat flux, and groundwater input were not included in the simulation. Initial conditions for tidal levels and currents were set to zero.

Oceanic tides play an important role when considering the flooding effect. The tide in the Java Sea is a diurnal type with maximum tidal range around 1 m during spring tides. Tidal forcing data were obtained from the TPXO (http://volkov.oce.orst.edu/tides/tpxo8_atlas.html) [26]. Thirteen tidal constituents included eight basic components (M2, S2, N2, K2, K1, O1, P1, and Q1) and five shallow water components (MF, MM, M4, MS4, MN4).

Jakarta Bay itself is sheltered by many islands and so wind plays a very limited role [27]. However, a monsoon wind still impacts the dynamic circulation in Jakarta Bay. Wind data were derived from the European Center for Medium-Range Weather Forecast (ECMWF). ECMWF can be accessed at ERA-Interim reanalysis (http://apps.ecmwf.int). The spatial and temporal resolutions of the wind data were $0.125° \times 0.125°$ and 6 h, respectively.

Thirteen rivers discharge into Jakarta Bay. The total mean river discharge from these 13 rivers was around 205 m^3/s throughout the year 2012 [28]. Detailed information of the annual distribution of river discharge can be seen in Table 1.

Table 1. River discharge in Jakarta Bay based on Wulp et al. (2016) [28]. For rivers location see Figure 3.

No	Name	Discharge (m^3/s)	No	Name	Discharge (m^3/s)
1	Cisadane	36.3	8	Cakung	0.2
2	Cengkareng	2.4	9	Blencong	0.1
3	Banjir Kanal Barat	0.5	10	Banjir Kanal Timur	0.2
4	Muara Angke	0.1	11	CBL	16.6
5	Muara Baru	0.1	12	Keramat	0.8
6	Ciliwung	8.2	13	Citarum	136.5
7	Sunter	2.7			

Sea surface salinity in the western Java Sea varied from 30.6–32.6 PSU [29], while sea surface temperatures in northern Java ranged from 29–31 °C [30]. In this state, the initial salinity and temperature were set as constant values of 31.5 PSU and 31 °C, respectively. In this paper, water density was assumed to be homogeneous. The authors assumed that the Manning coefficient for bottom roughness varied from 0.004 at the open ocean to 0.008 near the islands and coastal region. The model ran for 17 days starting from 3 to 19 June 2015. Courant Friedrich Levy (CFL) was used as the numerical stability condition to calculate the external and internal step. The external time step numerical stability is defined as:

$$\Delta t_E \leq \frac{\Delta L}{\sqrt{gD}} \tag{10}$$

where Δt_E is the time step of the external model and ΔL (computational length scale) is the shortest edge from individual triangular grid element, and D is the local depth. The internal step is defined as:

$$\Delta t_I \leq \frac{\Delta L}{C_I} \tag{11}$$

where Δt_I is the time step of the internal model, and C_I is the maximum phase speed of the internal gravity waves. Since C_I is usually smaller than $C_E = \sqrt{gD}$, Δt_I could be larger than Δt_E. The external time step was 0.15 s and the internal time step was 5 s based on the numerical stability condition.

2.5. Scenarios

The prevention of future coastal flooding in Jakarta is vital because of the phenomenon of future sea level rises and the frequent occurrence of high river discharge conditions. The Intergovernmental Panel on Climate Change (IPCC) reported that the temperature in Indonesia will increase from 1.3 °C to 4.6 °C. As the temperature increases, sea surface levels will also increase. The rate of sea level rise accelerated between the mid-19th and the mid-20th centuries based on tidal gauge and geological data [5]. The TOPEX/POSEIDON satellite is joint venture between NASA (National Aeronautics and Space Administration) and CNES (National Centre for Space Studies). The satellite observed that the rate of sea level rises along the coast of Indonesia was approximately 7 mm/year between 1992 and 2015 and that value was used in this paper. There are three scenarios for the sea level; the first is the scenario 1 (reference model), the second is with a sea level increase of 25 cm (scenario 2a), and the third is with a sea level increase of 60 cm (scenario 2b).

On the other hand, high river discharge occurs due to high precipitation and intraseasonal effects such as La Niña. High precipitation usually occurs under the northwest monsoon (NWM) conditions from December to February [13]. There are two high river discharge scenarios, 1.5 times the average discharge (scenario 1a) and 10 times the average discharge (scenario 1b). Scenario 1a is linked to the northwest monsoon (NWM) conditions, while scenario 1b is assumed for extreme river discharge. These scenarios for river discharge are combined with the scenarios for sea level rise to give a view of the areas affected by coastal flooding.

The impact of the GSW on coastal flooding was also included in this model. The effectiveness of the GSW in preventing the city from coastal flooding was examined using river discharges (scenarios 3a and 3b) and the sea level rises (scenarios 4a and 4b), presented in Table 2.

Table 2. Descriptions of different scenarios for Jakarta Bay. Scenario 1 is the reference scenario before the construction of the GSW and covers the 3–19 June 2015 period.

Scenarios	Descriptions
1	Jakarta Bay without GSW (reference scenario)
1a	With 1.5 times mean river discharge based on scenario 1
1b	With 10 times mean river discharge based on scenario 1
2a	Sea level +25 cm based on scenario 1
2b	Sea level +60 cm based on scenario 1
3	Add the GSW based on scenario 1
3a	With 1.5 times mean river discharge based on scenario 3
3b	With 10 times mean river discharge based on scenario 3
4a	Sea level +25 cm based on scenario 3
4b	Sea level +60 cm based on scenario 3

The numerical scenarios are listed in Table 2, showing Jakarta Bay (scenario 1) as the reference scenario before the construction of the GSW. This scenario covers a period from 3 to 19 June 2015.

3. Model Validation

The reference model was validated by comparing with the observation data. Data from two field stations, Kayangan Station (tidal and current) and Kolinamil Station (tidal), were used to validate the model. Data from the Kolinamil Station were obtained from http://www.ioc-sealevelmonitoring.org. Tidal and current data at Kayangan Station were collected by the Ministry of Public Works. Tidal data from the Kayangan Station were collected from 2 to 15 June 2015 and current data from 12 to 14 June 2015. During the June period, Jakarta experienced a dry season because of the southeast monsoons (SEM) conditions. Wind direction moved westward with maximum wind speeds at the simulation period around 4 m/s.

Ma et al. (2011) [31] suggested a statistical method to assess model skill for verification. This skill indicates the corresponding degree of deviation from prediction and observation data. A skill value of 1 means perfect agreement between model and observation.

$$Skill = 1 - \frac{\sum_{i=1}^{N}\left|X_{mod} - X_{obs}\right|^2}{\sum_{i=1}^{N}\left(\left|X_{mod} - \overline{X_{obs}}\right| + \left|X_{obs} - \overline{X_{obs}}\right|\right)^2} \tag{12}$$

Tidal verification data at Kolinamil Station started from 6 June 2015 to 19 June 2015 (Figure 4a), while the data at Kayangan Station were from 6 to 15 June 2015 (Figure 4b). The skill parameters for the tidal level calibration at Kolinamil Station and Kayangan Station were 0.97 and 0.96, respectively. The tidal data at Kolinamil Station and Kayangan Station were reproduced by the model with root mean square error (RMSE) of 0.08 m and 0.09 m, respectively. These two tidal verifications show good agreement between the model and observation data.

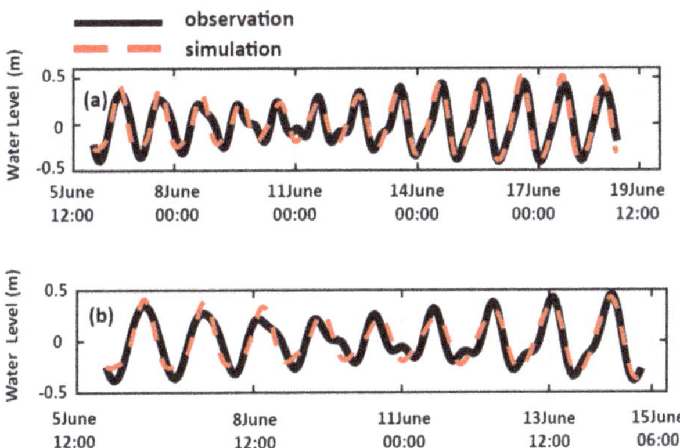

Figure 4. Comparisons of tidal level between the model results and the field data at (**a**) Kolinamil Station and (**b**) Kayangan Station.

Current verification was conducted at Kayangan Station, where the data were collected from 12 to 14 June 2015 (Figure 5). The skill scores between the modelled and the observed current velocity and direction were 0.96 and 0.88, respectively. The velocity and direction at Kayangan Station had RMSE of 0.015 m/s and 50°, respectively. These skill values indicate that the model has good agreement with the observation data.

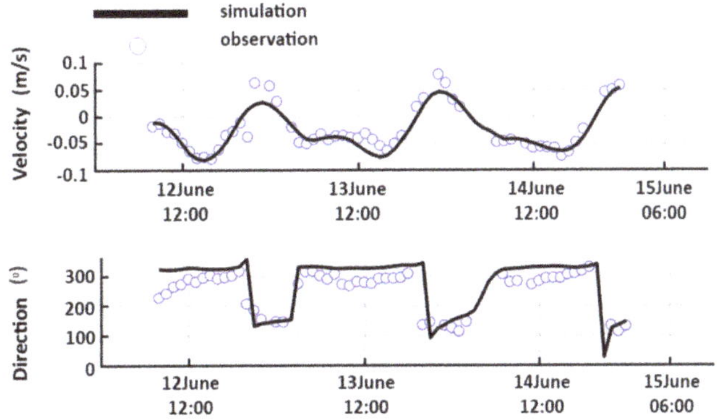

Figure 5. Comparisons of current speeds and directions between the modelled and observed data at Kayangan Station (12–14 June 2015).

4. Results

4.1. Tidal Dynamics in Jakarta Bay

Sea surface level data in the reference model were analyzed using harmonic analysis [20] to examine the main components of tides in Jakarta Bay. The four main astronomical tidal constituents (K1, O1, M2, and S2) explain about 80% of the total variances of water level in the bay. The diurnal components (K1 and O1) account for more than 70% of the tides, with tidal magnitudes of approximately 0.25 m and 0.13 m, respectively. The semidiurnal tides just account for a small part of the tides in the bay, with amplitudes of M2 and S2 tides of ~0.04 m and ~0.07 m, respectively (Figure 6).

Model results showed that diurnal tides and semidiurnal tides propagate differently. Diurnal tides mainly propagate westward, while semidiurnal tides propagate eastward. The different propagation pattern of diurnal and semidiurnal components is caused by the topography of Jakarta Bay, as the bay is located at the Java Sea and near to the Sunda Strait. Ray et al. (2005) [32] reported that the semidiurnal response is clearly dominated by the large tides from the Indian Ocean, while, in contrast, the diurnal tides come from the Pacific Ocean. These diurnal tides propagate westward in the Java Sea. The diurnal tides from the Java Sea propagate westward and meet waves coming from the South China Sea and the Karimata Straits [13,32,33]. The semidiurnal tides penetrate north and merge with the diurnal tides at western Kalimantan. In Jakarta Bay, the diurnal components are mainly driven by tides that come from the Pacific Ocean. The semidiurnal components are driven by tides that come from Indian Ocean. The shallow bathymetry and complicated coastlines generate the distribution pattern of the tides in the bay.

K1 and O1 tides share a similar pattern of propagating westward, from the Pacific Ocean through the Makassar Strait and the Molucca Sea. The semidiurnal components show another behavior. The eastward propagation of an M2 tide comes from the Indian Ocean through the Sunda Strait, while an S2 tide shows a more complicated pattern. An S2 tide comes from the Sunda Strait then propagates northward, and from the Karimata Strait then propagates southward. The complex patterns of tides in the bay are affected by the standing waves which occur in the west part of Java, and the amphidormic point occurring at the Sunda Strait.

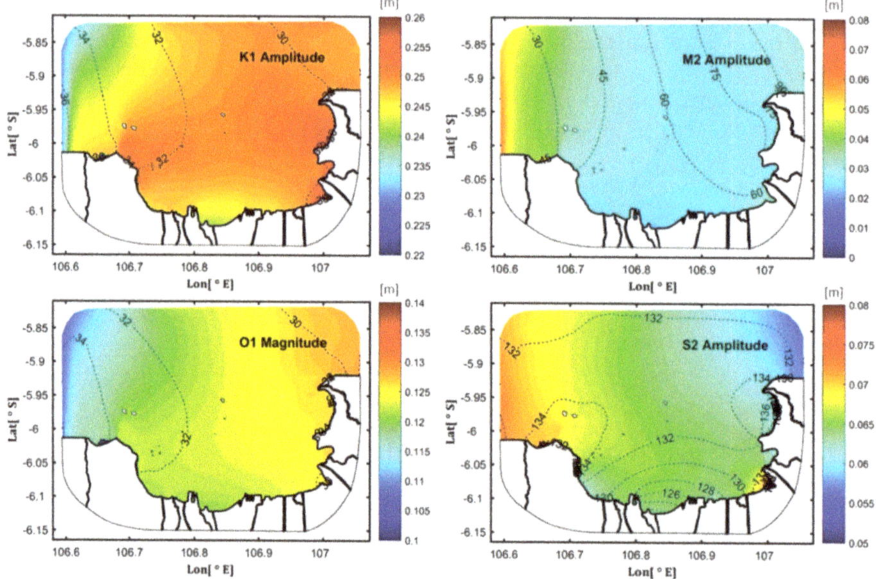

Figure 6. Co-tidal charts of the main tidal constituents (K1, O1, M2, and S2).

In the reference model (scenario 1), the total inundation areas of Jakarta Bay amount to 32.36 km^2, which is 7.7% of the total land area in the study domain (420.13 km^2). These inundation areas relate to the land topography around the coastlines of the bay. Some parts of the topography are already under mean sea water level. Most of the inundated parts occur at the eastern part (near the river number 13) and central part of Jakarta (between the river number 6 and 7). In the eastern part, the Citarum River (river number 13) has the highest river discharge of all 13 rivers in Jakarta Bay. The topography of eastern Jakarta also has the same elevation as the coastal areas at around 0 m. In the central part of Jakarta Bay, the elevations of most of the inundated areas are below mean sea water level. The inundated areas are caused by sea water overtopping, as the rivers only have small amounts of river discharge.

Jakarta Bay, without GSW, had the highest water elevation of approximately 0.6 m at high spring tides. In scenario 1a, there is almost no change of water elevation in the entire bay. In scenario 1b, the water elevation along the coastline increased by less than 2.08% (1.25 cm) at the western part of the bay (from river numbers 1 to 6), and 0.9% (0.5 cm) at the center part of bay (from river numbers 6 to 11), and 3.33% (2 cm) at the eastern part of the bay (from the Keramat River to the Citarum River). In scenario 2a, there is a slight change of water elevation increase, 0.33% (0.2 cm) at the center and eastern parts of bay (from river numbers 6 to 13). In scenario 2b, water elevation increases by 4.17% (2.5 cm) at the center part of the bay (near river numbers 2 to 7) and reaches a maximum, 6.67% (4 cm), at the eastern part of the bay (near river number 13).

4.2. Impacts of River Discharge on Hydrodynamics in Jakarta Bay

The impacts of the river discharge in Jakarta Bay were tested in scenarios 1a and 1b. Wulp et all (2016) [28] reported that in wet seasons mean river discharge in Jakarta increased 1.5 times than the averaged river discharge. If scenario 1a were applied, the total inundation areas of Jakarta Bay would be 33.88 km^2 (8.06% of the total land areas in the study domain), increasing by 1.55 km^2 compared with the reference scenario 1. The inundation areas slightly increase near the eastern part of Jakarta where the Citarum River is located.

Under extreme conditions, when the river discharge is 10 times the normal one (scenario 1b), flooded areas in Jakarta Bay are 58.52 km^2 (13.93% of the total land area), an increase of 26.61 km^2 with respect to scenario 1 (Figure 7). The inundation areas mainly occur in the western and eastern part of Jakarta. A major cause of this extreme condition is the Cisadane and Citarum rivers (river numbers 1 and 13, respectively), which have larger river discharges than the others. The central part of Jakarta is relatively unaffected in this scenario due to the small amount of river discharge nearby. As the river discharge increases, the inundation areas become larger, especially in the eastern part of Jakarta.

To further examine the impacts of river discharge on tidal amplitude, the dominant astronomical tidal component K1 was chosen to represent the changes in the bay. In scenario 1a, the K1 tidal amplitude increases only near the river mouths (river numbers 11 to 13) by 0.8% (0.2 cm). In scenario 1b, the K1 amplitude increases by 1% (0.25 cm) in the western and eastern parts (river mouth numbers 1, 11, and 13) and 0.5% (0.125 cm) in the center part (river number 6) of the bay. There is no significant change for the phase lags in K1 in scenario 1a. In scenario 1b, the phases for K1 advanced, meaning that high tides would occur earlier than in the reference scenario 1.

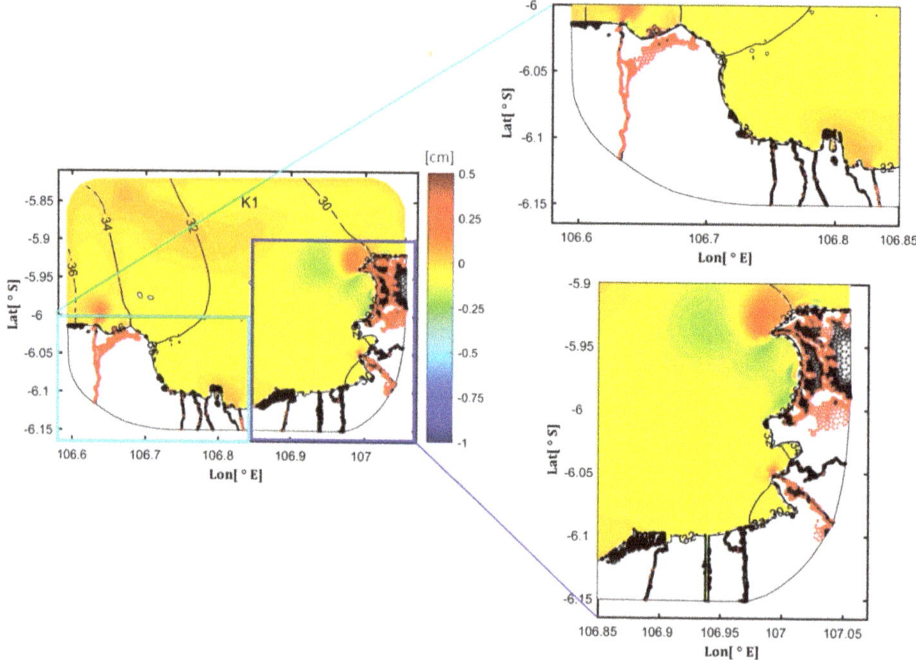

Figure 7. Changes in K1 tidal amplitudes (cm) and inundation areas in Jakarta Bay between scenario 1b and scenario 1. Black dots indicate the inundation area in scenario 1, and red dots indicate the increased inundation area in scenario 1b. Color contours indicate the changes in K1 amplitude (cm).

Surface current directions in Jakarta Bay are largely westward during flood spring tides. The surface current pattern changes slightly due to the variations of river discharge (Figure 8). River discharge strengthens the current magnitude at the eastern part of Jakarta Bay and weakens the current magnitude at the central and western parts of Jakarta Bay. In scenario 1a, current velocities decrease by 33.33% (0.1 m/s) at the center of the bay. In scenario 1b, current velocities increase by around 50% (0.2 m/s) at the center of bay and by around 100% (0.4 m/s) at river mouths (numbers 1, 11, and 13).

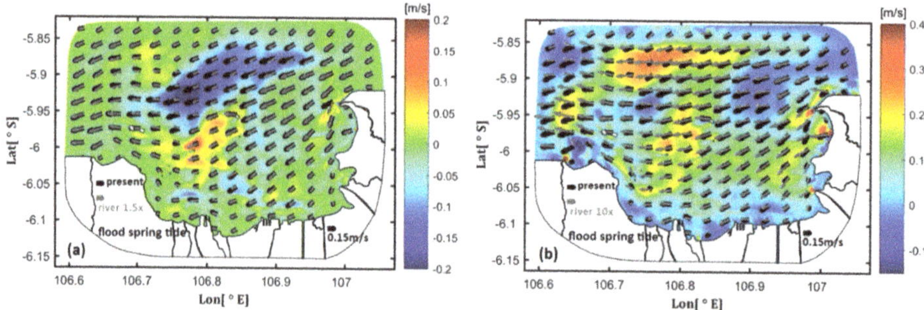

Figure 8. Differences in surface current velocity (m/s) and direction under flood condition spring tide (**a**) scenario 1a and (**b**) scenario 1b. Black vectors are for scenario 1, and grey vectors are for scenario 2a/2b. Color contours indicate the change in magnitude (m/s) of the surface current velocity.

4.3. Impacts of Sea Level Rise on Hydrodynamics

The impacts of sea level rise on the hydrodynamics of Jakarta Bay were tested in scenarios 2a and 2b (Figure 9). In scenario 2a (0.25 m sea level rise), the total inundation areas were 43.37 km^2 (10.32% of the total land area), increasing by 11.01 km^2 from the reference scenario 1. In scenario 2b (0.60 m sea level rise), total inundation areas were 62.04 km^2 (14.77% of the total land area), with an increase of 29.68 km^2 from the reference scenario 1. The flood areas of Jakarta Bay occurred near the coastal region from the western to the eastern part.

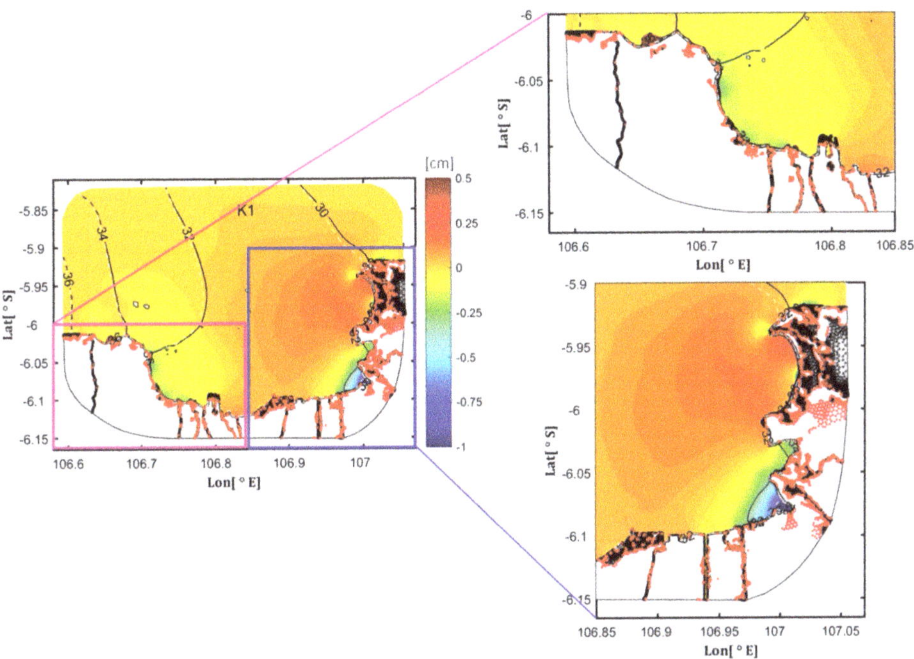

Figure 9. Changes in K1 tidal amplitudes (cm) and inundation areas in Jakarta Bay between scenario 2b and scenario 1. Black dots indicate the inundation area in scenario 1, and red dots indicate the increased inundation area in scenario 2b. Color contours indicate the changes in K1 amplitude (cm).

When the sea level rise is 0.25 m (scenario 2a), the K1 tidal amplitude increases by around 1% (0.25 cm) at the eastern part of the bay (from river numbers 11 to 13) and decreases by about 2% (0.5 cm) between river numbers 10 and 11. There is a small phase lag change for the K1 component, compared with that in the scenario 1.

When the sea level rise is 0.60 m (scenario 2b), the increase of the K1 amplitude is about 1% (0.25 cm) near the coastlines at the eastern part of the bay, which is almost the same as that in scenario 2a. On the other hand, the K1 amplitude decreases between river numbers 10 and 11 by around 3% (0.75 cm), which is larger than that in scenario 2a. The K1 tidal phase remains almost unchanged. The K1 tidal amplitude increases with the rate of sea level rise.

The pattern of surface currents remains largely the same as the sea level rises (Figure 10). A greater rate of sea level rise strengthens the current magnitude in the entire bay, except in the middle part of the bay. Under a sea level rise of 0.25 m (scenario 2a), current velocities increase by 8.3% (0.025 m/s) in the entire bay and decrease by 13.33% (0.04 m/s) in the middle of the bay. Under conditions of sea level rise of 0.6 m (scenario 2b), current velocities would increase by 16.6% (0.05 m/s) near the coastal region and decrease by 24.9% (0.075 m/s) in the middle of the bay.

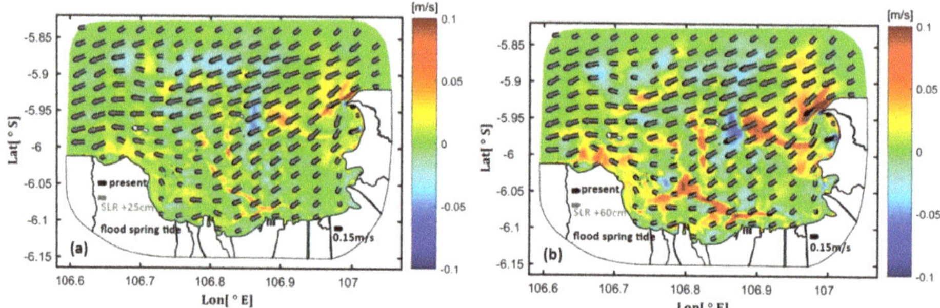

Figure 10. Differences in surface current velocity (m/s) and direction under flood condition spring tide (**a**) scenario 2a and (**b**) scenario 2b. Black vectors are for scenario 1, and grey vectors are for scenario 2a/2b. Color contours indicate the change in magnitude (m/s) of the surface current velocity.

5. Discussion

This research improves the understanding of sea level rise and river discharge effect on the Jakarta Bay hydrodynamics. As the Jakarta Bay tidal range is no more than 2 m, it can be classified as a microtidal region. Moreover, it is likely that the other microtidal regions with similar characteristics to Jakarta Bay will experience the same conditions under the effects of sea level rise. These findings can be used to assess future hydrodynamic conditions due to sea level rise effects. The projected change of tidal hydrodynamics and inundation areas in this study will be helpful to adopt sound management strategies and adaptation plans.

The impacts of the GSW on the Jakarta Bay flooding area are shown in Table 3. To overcome flooding disasters, the construction of the GSW has been proposed for Jakarta Bay. The construction of the GSW will slightly increase the maximum tidal range by 0.01 m [1] because of the reduced tidal prism by 20% and increased tidal choking in the bay.

Table 3. Flooding areas of Jakarta Bay due to sea level rise and river discharge.

Scenarios	Flooding Area (without GSW, km²)	Scenarios	Flooding Area (with GSW, km²)
1	32.36	3	33.47
1a	33.88	3a	36.7
1b	58.52	3b	420.13
2a	43.37	4a	44.05
2b	62.04	4b	63.56

In scenario 3, the total inundation areas of Jakarta Bay are 33.47 km² (7.97% of the total land area), increasing by 1.11 km² from scenario 1. There would be a slight change in flooding area after the GSW had been constructed.

After the construction of the GSW, a slight change would occur to tidal amplitudes during high spring tides near the coastline, compared with scenario 1 (Figure 11). Tidal amplitudes would increase by around 1 cm (2% from scenario 1) at the eastern reservoir. At the western reservoir, tidal amplitudes near the coastal area would increase by around 3 cm (5.88% from scenario 1) and 1 cm (2% from scenario 1) near the channel on the left side of the western reservoir. There would no change in tidal amplitude on the right side of the western reservoir. Tidal amplitudes in the eastern part of the bay (from river numbers 12 to 13) will decrease by around 1 cm (2% from scenario 1).

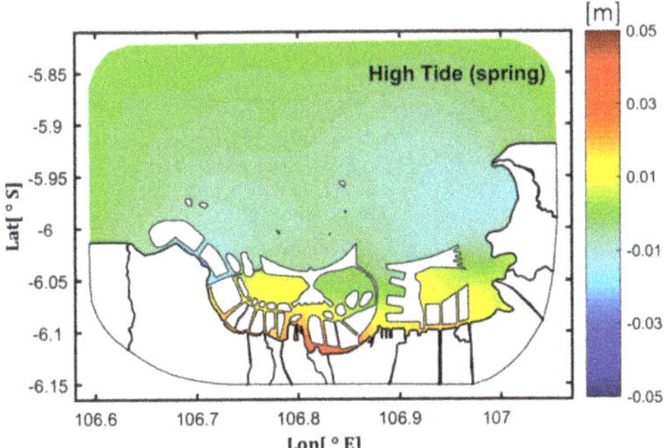

Figure 11. Changes in water elevation (m) between scenario 1 and after the Giant Seawall (GSW) construction (scenario 3).

5.1. Effects of the Giant Seawall on Sea Level Rise

The effects of the GSW on sea level rise were tested in scenarios 4a and 4b (Figure 12a). The inundation area in scenario 4a would be 44.05 km^2 (10.49% of the total land area in the study domain), increasing by 10.58 km^2 from scenario 3 (the reference model with GSW). In scenario 4b, the total inundation area would be 63.56 km^2 (15.13% of the total land area in the study domain) increasing by 30.09 km^2 from scenario 3. The inundation areas in scenarios 4a and 4b would increase by less than 1% compared with scenarios 2a and 2b. The construction of the Giant Seawall in Jakarta Bay would need further design improvement to prevent a flooding hazard due to sea level rise. Furthermore, the implementation of the GSW design should also consider strengthening and increasing the current seawall height on the shore to prevent sea water overtopping the wall.

Compared with scenario 3, when the sea level rises by 0.25 m in scenario 4a, the K1 tidal amplitude slightly increases by around 0.2% (0.05 cm) in the eastern reservoir and on the left side of the western reservoir, and decreases by 0.65% (0.1 cm) near the Cisadane River (river number 1). There are small phase lag changes for the K1 component near the reclamation island areas outside the western reservoir.

When the sea level rises by 0.60 m in scenario 4b, the K1 tidal amplitude increases by around 0.4% (0.1 cm) in the eastern reservoir (from river numbers 9 to 11) and by 1.22% (0.3 cm) near the coastlines inside the western reservoir. The K1 tidal amplitude also decreases by 0.8% (0.2 cm) on the right side of the western reservoir, the eastern reservoir (from river numbers 7 to 8), and the mouth of the Cisadane, Keramat, and Citarum rivers (river numbers 1, 12, and 13). There is no significant change of phase lag for the K1 component.

5.2. Effects of the Giant Seawall on River Discharge

The impacts of a Giant Seawall on river discharge were tested in scenarios 3a and 3b (Figure 12b). When river discharge is increased 1.5 times (scenario 3a), inundation areas are 36.7 km^2 (8.74% of the total land area in study domain) increasing by 2.82 km^2 from scenario 3. The inundation areas in scenario 3a would increase by 0.6% compared with scenarios 1a. The K1 amplitude remains almost the same in the entire bay. There is a small increase of around 0.42% (0.1 cm) at river mouth numbers 11 and 13. There are no significant changes in phase lag for the K1 component in scenario 3a compared to scenario 3.

If river discharge is 10 times higher (scenario 3b), all the areas are flooded due to a blocking of the tidal inflow. Further design improvement is needed for the construction of the GSW to prevent a flooding hazard due to high river discharge, especially under extreme conditions (25-year return period). The designed dike height should be increased along the rivers to prevent overland flow from high river discharge. Protection of the inundation area after construction of the GSW under sea level rise and high river discharge can be considered in mitigation strategies. Overall, this research will help to improve coastal management in Jakarta Bay.

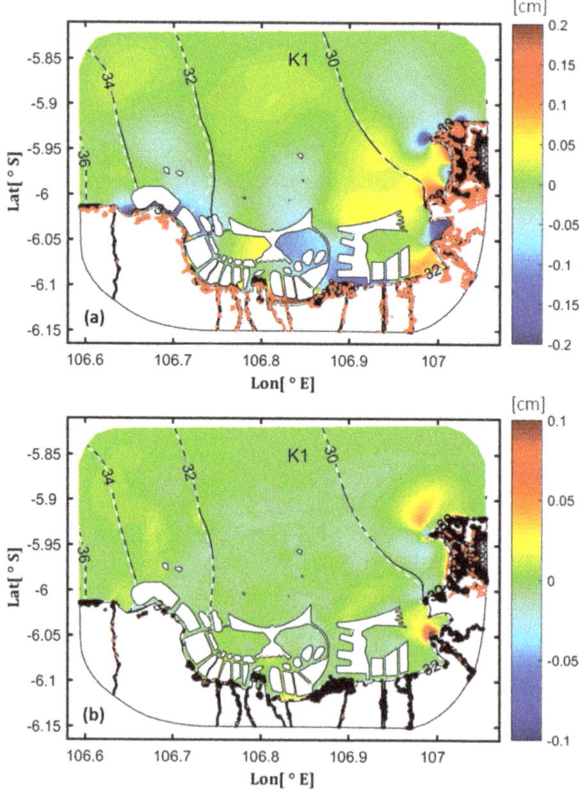

Figure 12. Changes in tidal amplitudes (cm) and inundation areas (red dots) in Jakarta Bay: (**a**) K1 tidal component between scenario 4b and scenario 1 and (**b**) K1 tidal component between scenario 3a and scenario 1. Color contours indicate the changes in amplitude (cm), and the black/white dashed lines indicate the phase change.

6. Concluding Remarks

The impacts of sea level rise and river discharge on the hydrodynamics of Jakarta Bay have been investigated using FVCOM. Sea surface level and current data at two field stations were used to validate the model. The model reproduced tidal elevation and current very well, with statistical skill parameters of 0.96, 0.96 and 0.88 for elevation, current velocity and direction, respectively. Numerical experiments were designed to examine the contribution of sea level rise and river discharge to the changes of hydrodynamics in Jakarta Bay.

The distribution of the amplitudes and phases of diurnal and semidiurnal tides in Jakarta Bay are quite different to other bays as the bay is located in the Java Sea and near the Sunda Strait. The diurnal tidal wave moves westward, while the semidiurnal tidal wave moves eastward. Diurnal tides come from the Pacific Ocean through the Makassar Strait and Mollusca Sea and propagate westward into the Java Sea. In contrast, semidiurnal tides come from the Indian Ocean through the Sunda Strait and propagate northward and from the Karimata Strait, propagating southward. The shallow bathymetry and complicated coastlines contribute to generating the distribution pattern of the tides in the bay.

Future sea level rise and river discharge variations will affect the hydrodynamics and coastal flooding areas in Jakarta Bay. The increase of river discharge amplifies the tidal components as well as the inundation areas near the river mouth due to increased fluvial forcing: the K1 amplitude increases by ~1% (0.25 cm), the current magnitude increases by 100% (0.4 m/s), and the inundation area increases by 26.61 km^2 under the 10 times of mean river discharge. The rise of sea level increases the diurnal tidal component due to the increased tidal forcing: when sea level rise is 0.6m, the K1 amplitude increases by ~1% (0.25 cm) near the coastline and current magnitude increases by 16.6 % (0.05 m/s). The increasing sea level rise and river discharge would amplify the inundation areas and the tidal currents due to the increased tidal forcing and the river discharge amounts.

The GSW construction would slightly increase the water level by around 1 cm at the eastern reservoir due to the reduced tidal prism, while in the western reservoir, the water level would increase by up to 3 cm. If sea level rise occurs together with flooding, the K1 amplitude would slightly increase by less than 1% (0.2 cm), due to the reduction of the dissipation effects. The design of the GSW will need further improvement to prevent flooding hazards in the event of sea level rise and the effect of high river discharge. Strengthening the current seawall on the shore and increasing the dike height along the rivers would help to prevent a flooding hazard in Jakarta Bay. The outcome of this research improves our understanding of the sea level rise and river discharge effects in open type bays.

Author Contributions: Conceptualization, Y.X. and L.L.; Methodology, M.Y.S.; Software, M.Y.S.; Validation, M.Y.S., Y.X. and L.L.; Formal Analysis, M.Y.S., L.L.; Investigation, M.Y.S., Z.H. and L.L.; Resources, M.Y.S., L.L.; Data Curation, L.L.; Writing-Original Draft Preparation, M.Y.S.; Writing-Review & Editing, L.L. and Y.X.; Visualization, M.Y.S.; Supervision, Y.X. and Z.H.; Project Administration, Z.H.; Funding Acquisition, Z.H. and L.L.

Funding: This study was supported by the National Key Research and Development Program of China (2017YFC1405101) and was partially supported by the Natural Science Foundation of Zhejiang Province (LR16E090001) and the Bureau of Science and Technology of Zhoushan (2018C81036). Martin was supported by a scholarship from Zhejiang University for a Master's Degree program at the Ocean College of Zhejiang University, China.

Acknowledgments: We thank the Ministry of Public Works, Indonesia, for providing the tide, current, and bathymetry data.

Conflicts of Interest: The authors declare no conflict of interest.

References

1. Rusdiansyah, A.; Tang, Y.; He, Z.; Li, L.; Ye, Y.; Surya, M.Y. The impacts of the large-scale hydraulic structures on tidal dynamics in open-type bay: Numerical study in Jakarta Bay. *Ocean Dyn.* **2018**, *68*, 1141–1154. [CrossRef]

2. Sagala, S.; Lassa, J.; Yasaditama, H.; Hudalah, D. The Evolution of Risk and Vulnerability in Greater Jakarta: Contesting Government Policy. *Inst. Resour. Governance Social Chang.* **2015**, *2*, 1–18.

3. Central Bureau of Statistics. *Preliminary Results of Population Census*; Central Bureau of Statistics: Denpasar, Indonesia, 2010.

4. Budiyono, Y.; Aerts, J.; Brinkman, J.; Marfai, M.A.; Ward, P. Flood Risk Assessment for Delta Mega-cities: A Case Study of Jakarta. *Nat. Hazards* **2015**, *75*, 389–413. [CrossRef]

5. IPCC. *Summary for Policymakers Climate Change the Physical Science Basic Contribution of Working Group 1 to the Fourth Assessment Report of the Intergovernmental Panel on Climate Change*; Cambridge University Press: Cambridge, UK, 2017.

6. Bappenas. *Laporan Perkiraan Kerusakan dan Kerugian Pasca Bencana Banjir Awal Februari 2007 Di wilayah Jabodetabek (Jakarta, Bogor, Depok, Tangerang, dan Bekasi)*; Kementian Negara Perencanaan Pembangunan Nasional/BAPPENAS: Jakarta, Indonesia, 2007.

7. Abidin, H.Z.; Andreas, H.; Djaja, R.; Darmawan, D.; Gamal, M. Land subsidence and characteristics of Jakarta between 1997 and 2005, as estimated using GPS surveys. *GPS Solut.* **2008**, *12*, 23–32. [CrossRef]

8. Ministry of Economic Affairs. *Master Plan National Capital Integrated Coastal Development (NCICD)*; Ministry of Economic Affairs: Jakarta, Indonesia, 2014.

9. Ningsih, N.S. Three-Dimensional Model for Coastal Ocean Circulation and Sea Floor Topography Changes: Application to the Java Sea. Ph.D. Thesis, Kyoto University, Kyoto, Japan, 2000.

10. Koropitan, A.F.; Ikeda, M.; Damar, A.; Yamanaka, Y. Influences of physical processes on the ecosystem of Jakarta Bay: A coupled physical–ecosystem model experiment. *ICES J. Mar. Sci.* **2009**, *66*, 336–348. [CrossRef]

11. Sachoemar, S.I.; Yanagi, T. Seasonal variation of water characteristics in northern coastal area of Java. *Societe Franco-Japonaise d'Oceanographie* **2001**, *39*, 77–85.

12. Setiawan, A.; Putri, M.R. Study of current circulation in Jakarta Bay. In Proceedings of the 3rd International Symposium on Advanced and Aerospace Science and Technology in Indonesia, Jakarta, Indonesia, 31 August–3 September 1998.

13. Wyrtki, K. *Physical Oceanography of the Southeast Asian Waters*; Naga Report 2; Scripps Institution of Oceanography: La Jolla, CA, USA, 1961.

14. Koropitan, A.F.; Ikeda, M. Three-dimensional modeling of tidal circulation and mixing over the Java Sea. *J. Oceanogr.* **2008**, *64*, 61–80. [CrossRef]

15. Takagi, H.; Esteban, M.; Mikami, T.; Fujii, D. Projection of coastal floods in 2050 Jakarta. *Urban Clim.* **2006**, *17*, 135–145. [CrossRef]

16. Passeri, D.L.; Hagem, S.C.; Plant, N.G.; Bilskie, M.V.; Medeiros, S.C.; Alizad, K. Tidal hydrodynamics under future sea level rise and coastal morphology in the Northern Gulf of Mexico. *Earth's Future* **2016**, *4*, 159–176. [CrossRef]

17. Aoki, K.; Isobe, A. Application of Finite Volume Coastal Ocean Model to hindcasting the wind-induced sea-level variation in Fukuoka Bay. *J. Oceanogr.* **2007**, *63*, 333–339. [CrossRef]

18. Gede, H.I.; Koji, A. Numerical study on tidal currents and seawater exchange in the Benoa Bay, Bali, Indonesia. *Acta Oceanol. Sin.* **2004**, *33*, 90–100.

19. Park, Y.G.; Kim, H.Y.; Hwang, J.H.; Kim, T.; Park, S. Dynamics of dike effects on tidal circulation around Saemangeum, Korea. *Ocean Coast. Manag. J.* **2014**, *102*, 572–582. [CrossRef]

20. Pawlowicz, R.; Beardsley, B.; Lentz, S. Classical tidal harmonic analysis includirng error estimates in MATLAB using T_TIDE. *Comput. Geosci.* **2002**, *28*, 929–937. [CrossRef]

21. Defant, A. *Physical Oceanography*; Pergamon Press: New York, NY, USA, 1961.

22. Chen, C.; Beardsley, R.C.; Cowles, G. *An Unstructured Grid, Finite Volume Coastal Ocean Model: FVCOM User Manual*, 2nd ed.; Technical Report 06-0602; University of Massachusetts-Dartmouth: New Bedford, MA, USA, 2006.

23. Mellor, G.L.; Yamada, T. Development of turbulence closure model for geophysical fluid problems. Review of geophysics and space physics. *Rev. Geophys.* **1982**, *20*, 851–875. [CrossRef]

24. Smagorinsky, J. General circulation experiments with the primitive equations. *Mon. Weather Rev.* **1963**, *91*, 99–164. [CrossRef]

25. Jarvis, A.; Reuter, H.I.; Nelson, A.; Guevara, E. *Holefilled SRTM for the Globe Version 4. CGIAR-CSI SRTM 90 m Database*; International Center for Tropical Agriculture: Cali, Colombia, 2008.

26. Egber, G.D.; Erofeeva, S.Y. Efficient inverse modelling of barotropic ocean tides. *J. Atmos. Ocean. Technol.* **2002**, *19*, 183–204. [CrossRef]

27. Radjawane, I.M.; Riandini, F. Numerical Simulation of Cohesive Sediment Transport in Jakarta Bay. *Int. J. Remote Sens. Earth Sci.* **2009**, *6*, 65–76. [CrossRef]

28. Wulp, S.A.; Damar, A.; Ladwig, N.; Hesse, K.-J. Numerical simulation of river discharges, nutrient flux and nutrient dispersal in Jakarta Bay, Indonesia. *Mar. Pollut. Bull.* **2016**, *110*, 675–685. [CrossRef]

29. Soeriaatmadja, R.E. Surface salinities in the Straits of Malacca. *Mar. Res Indones.* **1956**, *2*, 27–55. [CrossRef]

30. Trisakti, B.; Sulma, S.; Budhiman, S. Study of sea surface temperature (SST) using Landsat-7 ETM (In comparison with sea surface temperature of NOAA-12 AVHRR. In Proceedings of the Thirteenth Workshop of OMISAR (WOM-13) on Validation and Application of Satellite Data for Marine Resources Conservation, Bali, Indonesia, 5–9 October 2004.

31. Ma, G.; Shi, F.; Liu, S.; Qi, D. Hydrodynamic modelling of Changjiang Estuary: Model skill assessment and large-scale structure impacts. *Appl. Ocean Res.* **2011**, 69–78. [CrossRef]

32. Ray, R.D.; Egbert, G.D.; Erofeeva, S.Y. A Brief Overview of Tides in the Indonesian Seas. *Oceanography* **2005**, *18*, 4. [CrossRef]

33. Gordon, A.L. Oceanography of the Indonesian Seas and their throughflow. *Oceanography* **2005**, *18*, 14–27. [CrossRef]

Article

Spatial Variability of Beach Impact from Post-Tropical Cyclone Katia (2011) on Northern Ireland's North Coast

Giorgio Anfuso [1,*], Carlos Loureiro [2,3], Mohammed Taaouati [4], Thomas Smyth [5] and Derek Jackson [6]

1   Department of Earth Sciences, Faculty of Marine and Environmental Sciences, University of Cádiz, Polígono del Río San Pedro s/n, 11510 Puerto Real, Spain
2   Biological and Environmental Sciences, Faculty of Natural Sciences, University of Stirling, Stirling, Scotland FK9 4LA, UK; carlos.loureiro@stir.ac.uk
3   Geological Sciences, School of Agricultural, Earth and Environmental Sciences, University of KwaZulu-Natal, Durban 4001, South Africa
4   Department of Exact Sciences, National School of Architecture, Tetouan 93000, Morocco; mtaaouati@gmail.com
5   Department of Biological and Geographical Sciences, School of Applied Sciences, University of Huddersfield, Huddersfield HD1 3DH, UK; T.AG.Smyth@hud.ac.uk
6   School of Geography & Environmental Sciences, Ulster University, Coleraine, Co. Londonderry, Northern Ireland BT52 1SA, UK; d.jackson@ulster.ac.uk
*   Correspondence: giorgio.anfuso@uca.es; Tel.: +34-956016167

Received: 17 March 2020; Accepted: 9 May 2020; Published: 13 May 2020

Abstract: In northern Europe, beach erosion, coastal flooding and associated damages to engineering structures are linked to mid-latitude storms that form through cyclogenesis and post-tropical cyclones, when a tropical cyclone moves north from its tropical origin. The present work analyses the hydrodynamic forcing and morphological changes observed at three beaches in the north coast of Northern Ireland (Magilligan, Portrush West's southern and northern sectors, and Whiterocks), prior to, during, and immediately after post-tropical cyclone Katia. Katia was the second major hurricane of the active 2011 Atlantic hurricane season and impacted the British Isles on the 12–13 September 2011. During the Katia event, offshore wave buoys recorded values in excess of 5 m at the peak of the storm on the 13 September, but nearshore significant wave height ranged from 1 to 3 m, reflecting relevant wave energy dissipation across an extensive and shallow continental shelf. This was especially so at Magilligan, where widespread refraction and attenuation led to reduced shore-normal energy fluxes and very minor morphological changes. Morphological changes were restricted to upper beach erosion and flattening of the foreshore. Longshore transport was evident at Portrush West, with the northern sector experiencing erosion while the southern sector accreted, inducing a short-term rotational response in this embayment. In Whiterocks, berm erosion contributed to a general beach flattening and this resulted in an overall accretion due to sediment influx from the updrift western areas. Taking into account that the post-tropical cyclone Katia produced £100 m ($157 million, 2011 USD) in damage in the United Kingdom alone, the results of the present study represent a contribution to the general database of post-tropical storm response on Northern European coastlines, informing coastal response prediction and damage mitigation.

Keywords: wave energy; Hurricane Katia; longshore transport; dissipative

1. Preamble: Katia Cyclone Description

Hurricane Katia's formation was instigated by a wide low-pressure system on the 28th of August 2011, offshore of the western coast of Africa (Figure 1). The low pressure system moved westward and, on 29 August, acquired sufficient convective intensity to be designated as a tropical depression when it was located about 695 km southwest of the south westernmost Cape Verde Islands (https://www.nhc.noaa.gov, accessed on 18 April 2020). The depression moved to the west-northwest for the next 24 h and gradually strengthened, becoming a tropical storm on 30 August about 787 km southwest of the Cape Verde Islands. The cyclone maintained a west-northwest trajectory at around 27.8 km h^{-1} for the next two days and steadily strengthened to reach hurricane intensity on the Saffir-Simpson Hurricane Wind Scale by 1 September when it was located about 2176 km east of the Leeward Islands. After achieving hurricane status, Katia turned northwest and continued to strengthen and reached hurricane category 4 status on 5 September with wind peak intensity of 220 km h^{-1} and a central low pressure of 942 mb when the hurricane was located about 870 km south of Bermuda. Such extreme conditions lasted one day only and the hurricane then slowed down and gradually turned north-east on 9 September (Figure 1). After this, the wind field expanded and weakened. When Katia was located about 650 km northwest of Bermuda, it turned toward the east-northeast and increased in speed, to approximately 92 km h^{-1}, arriving over the cold sea-surface temperatures (22 °C) of the North Atlantic Ocean. The combination of cold water and strong vertical wind shear favoured the quick transition from a hurricane status into a powerful post-tropical low-pressure system by 1200 UTC 10 September when it was located about 463 km south-southeast of Cape Race, Newfoundland. On 11 September, Katia cyclone, a large and powerful post-tropical storm, turned north-east towards the northern British Isles with an average velocity of 85 km h^{-1}. The post-tropical cyclone reached the northern coast of Scotland on 12 September and produced sustained gale-force winds across most of the British Isles and hurricane-force wind gusts in Scotland, Northern Ireland, and northern England with average wind speed from 101 to 188 km h^{-1} with peak values of 212 km h^{-1} recorded in North Wales. On the 13 September the cyclone continued north-eastward and dissipated over the North Sea. In Europe, the post-tropical cyclone Katia impacted numerous locations, downing trees, bringing down power lines, and leaving thousands without electricity. In the United Kingdom the storm was responsible for two deaths and caused approximately £100 m ($157 million 2011 USD) in damage [1].

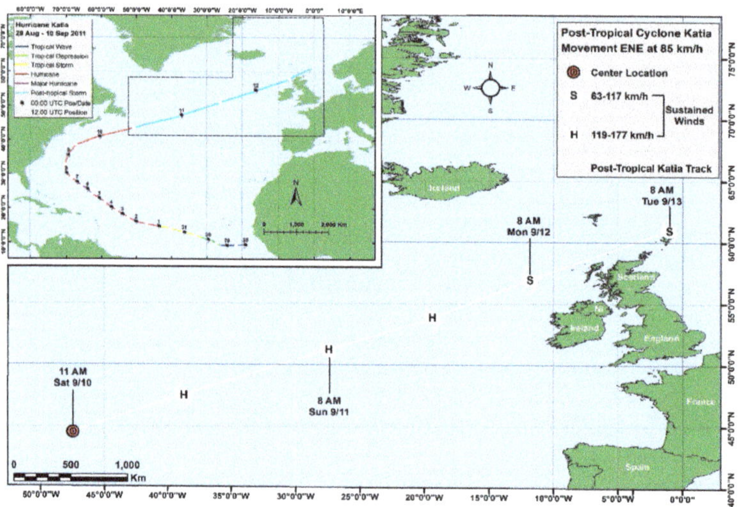

Figure 1. Track of Katia cyclone obtained from NOAA [1].

2. Introduction

Coastal development continues to increase and some 50% of the world's coastline is currently under pressure from excessive development [2,3], mainly in the form of tourism, one of the world's largest industries [4,5]. In Europe, the rapid expansion of urban artificial surfaces in coastal zones during the 1990–2000 period [6], has occurred in the Mediterranean and South Atlantic areas, namely Portugal (34% increase) and Spain (18%), followed by France, Italy and Greece. Ireland, a more peripheral holiday beach destination, has also had significant development of urban artificial surfaces in the coastal zones. In Northern Ireland, in 2018, 2.8 million visitors and over 2 million Northern Ireland residents took an overnight trip to the region, spending an unprecedented £968 million, £42 million more compared with 2017 [7].

Activities related with tourism can be significantly affected by the impacts of storms and hurricanes, producing damages to recreational and protective structures, with associated reduction of beach width and general aesthetics [8]. Over the past century, several storms and hurricanes have caused vast economic loses along with large numbers of deaths along the world's coastlines [9]. According to Dolan and Davis [10], the most powerful storms that have struck the Atlantic coast of USA, occurred on the 7–9 March 1962 (the "Ash Wednesday" storm), 7–11 March 1989, and on the 28 October–1 November 1991 (the "All Hallows' Eve").

Morton and Sallenger [11] investigated hurricanes and tropical cyclones that impacted the Gulf of Mexico and the Atlantic coast of USA, evaluating damages and washover penetration linked to hurricanes Carla (1961), Camille (1969), Frederic (1979), Alicia (1983) and Hugo (1989). According to Sallenger [12], the greatest coastal changes are recorded when the beach system completely submerges in an inundation regime: it can take place locally on a barrier island, incising a new inlet, as happened during Hurricane Isabel (2003) in North Carolina, Charley and Ivan in Florida (2004), and Katrina (2005) in Alabama. Additionally, inundation can submerge tens of kilometres of coast as occurred on the Bolivar Peninsula (TX, USA) during hurricane Ike (2008) and in Louisiana during Rita and Katrina events in 2005.

In northern Europe, damages to coastal structures, beach erosion and flooding inundation are often associated with mid-latitude (or extratropical) storms that form through cyclogenesis in the mid-latitude westerly wind belt and, secondly, post-tropical cyclones that form when a tropical cyclone moves north from its tropical origin [13].

In terms of the effects of mid-latitude storms, Bonazzi et al. [14] reconstructed spatial maps of peak gust footprints for 135 of the most important damaging events in 15 European countries in the past four decades, i.e., 1972–2010. The most important storms were 87J, Daria, Vivian, Anatol, Lothar, Martin, Erwin/Gudrun, Kyrill, Emma, Klaus and Xynthia. They observed 64% of events used in their analysis occurred during North Atlantic Oscillation positive phase (NAO+) months and their inter-annual variability, described by the NAO, modulated the main orientation of the storm tracks and the frequency of storm events.

On the Atlantic edge of Europe, Anfuso et al. [15] characterized, using the Storm Power Index [10], the distribution of storms in the Gulf of Cadiz during the 1958–2001 period and, highlighted particularly stormy years, e.g., years characterized by more storms and extended storm durations. Rangel-Buitrago and Anfuso [16] and Anfuso et al. [15] also observed the most powerful stormy years in Cadiz occurred every 5–6 years (e.g., in 1995–1996, 2002–2003, 2009–2010) with a 7–8-year periodicity recorded by Ferreira et al. [17] and Almeida et al. [18] in Faro (Southern Portugal).

The energetic conditions recorded in the Cadiz Gulf area during the 1995–1996 period also corresponded with similar weather conditions observed over the same period in Wales (UK) by Phillips [19] and Phillips and Crisp [20]. Dodet et al. [21] highlighted a highly unusual sequence of extratropical storms over the 2013–2014 winter period along Europe's Northeast Atlantic region, incorporating wave analysis over the period 2002–2017 for the northwest of Ireland, the Bay of Biscay and west of Portugal and beach erosion/recovery in five beaches in SW England, Brittany and the Bay of Biscay (France) that were surveyed on a monthly basis for more than 10 years. That winter recorded the most energetic conditions along the Atlantic coast of Europe since at least 1948 [22] resulting in most of western Europe' coastlines being severely impacted [23–25] albeit with the exception of beaches in NW Northern Ireland (this field site).

Santos et al. [26] examined waves around UK that exceeded the 1 in 1-year return level analysed from 18 different buoy records for the period from 2002 to mid-2016, de-clustered into 92 distinct storm events. The majority of events were observed between November and March, with large inter-annual differences in the number of events per season associated with the West Europe Pressure Anomaly. The 2013/2014 storm season represented an outlier in terms of the number of wave events, their temporal clustering and return levels.

The strength and position of extratropical cyclones is influenced largely by the pattern of atmospheric circulations over the North Atlantic Basin, which, in turn, are reflected in the signal and strength of the North Atlantic Oscillation Index [27]. However, Atlantic tropical cyclones that move northward from the tropics and undergo extratropical transition may also cause high-impact weather events in Western Europe [28]. Tropical cyclones generated in the Atlantic basin drift westward at tropical latitudes within the easterly Trade winds, and migrate northward affecting the east coast of the US. Usually once every 1–2 years, these cyclones move eastward undergoing extratropical transition and reach western Europe as post-tropical storms often with hurricane force-winds [29]. In Northwest Ireland, the earliest reported high-magnitude event was the 'Night of the Big Wind', which was probably the tail-end of a hurricane reported in January 1839. Cooper and Orford [30] described the occurrence and impacts of post-tropical cyclones on the British Isles using historical and contemporary information. Examining the period between 1922 and 1998, they identified nine major tropical cyclones that traversed the Atlantic and impacted the British Isles. MacClenahan et al. [31] identified Hurricane Debbie (September 1961) as the largest storm that impacted Ireland during the second half of the 20th Century. Recently, Guisado-Pintado and Jackson [32,33] described the effects of the post-tropical Storm Ophelia (2018) and Storm Hector (2019) in Ireland's NW Donegal coast.

The post-tropical cyclone Katia impacted the British Isles during the 12–13 September 2011 causing £100 m ($157 million, 2011 USD) in damage. The present work analyses what, if any, morphological changes occurred during the Katia cyclone in three beaches in the north coast of Northern Ireland, taking into account that storm-induced waves persisted until the 15 September, a couple of days after the cyclone dissipation. The importance of the present study lies in the necessity of understanding and predicting morphological changes associated with the impacts of hurricanes and intense post-tropical storms that, while infrequent, can sometimes have significant impacts on exposed coastal areas of the British Isles and cause relevant economic losses [30]. The behaviour of these coastal systems can be greatly affected in the future due to observed and modelled changes in frequency and intensity of extreme storms, and particularly the poleward migration of the maximum intensity of tropical cyclones as a result of global climate changes [34]. There is also concern that a possible change in hurricane tracks could lead to such destructive events impacting more frequently Southern European coasts, resulting in potentially more dramatic responses [35,36]. The results of the present study contributes to our understanding of beach and coastal response to post-tropical storm events along the coast of Northern Ireland and adds information to the general database of storm response on coastlines of this nature, informing damage mitigation and coastal response prediction.

3. Study Area

This paper examines the morphological change in three sandy beach sites, Magilligan Strand, Portrush (West Strand) and Whiterocks (eastern section of Curran Strand) (Figure 1, Table 1). These beaches are located on a high wave energy, microtidal, 20 km section of Northern Ireland's northern coastline [37,38]. Magilligan Strand, the most westerly beach studied, is part of a 10 km long, dissipative beach that stretches from Magilligan Point in the west to Downhill in the east. The area of beach monitored at Magilligan has been accreting since 1980 [39] and is backed by large dunes, approximately 10 m in height, which are densely vegetated with *Ammophila arenaria*.

Table 1. Main attributes of selected sites.

Site Name Location	Sand Grain Size (mm)	Beach Slope (tan β)	Tidal Range (m)	Beach Type	Local Geomorphology
Magilligan	0.17	0.0375	1.6	Intermediate to dissipative	Extensive dune systems, tidal inlet, sand ridge plain
Portrush northern	0.186	0.0320	1.5	Dissipative	Modified dunes, human modification (sea wall) along coastline
Whiterocks	0.197	0.0352	1.5	Intermediate	Extensive dune system behind beach

West Strand (Portrush) is an 850 m long, concave shaped beach bounded by basalt headlands to the northeast and southwest. Two sectors have been investigated within this pocket beach, one at the northern and one at the southern part and respectively noted as 'Portrush southern' and 'Portrush northern' sectors. The beach has undergone significant development beginning in 1825 when a jettied harbour was constructed against the north eastern headland [40]. In the 1960s, a promenade and car park were constructed on the dune complex behind the beach and a recurved seawall constructed on the back beach replaced the natural foredune. This development resulted in significant lowering of the beach surface elevation [41].

Whiterocks beach is located at the easternmost extremity of Curran Strand and is the most easterly study site investigated. Curran Strand is a 3 km long beach constrained by a basalt headland to the west and chalk cliffs to the east. The beach is convex in shape due to the sheltering effect of the Skerries islands located approximately 1.5 km offshore, however wave refraction around the islands produces high energy waves at the eastern extremity of the beach. The section of beach monitored at Whiterocks is backed by chalk boulders and a single steep vegetated foredune ranging from 6–25 m in height behind which a golf course has been constructed.

High-resolution multibeam bathymetric data for this coastline, collected in the framework of the Joint Irish Bathymetric Survey completed in September 2008, demonstrates an irregular and dynamic configuration of the continental shelf and shoreface of the North Coast of Northern Ireland, with tidal banks and sand waves indicating complex flow patterns and active sediment transport pathways (Figure 2). The substratum of the shelf and shoreface of this coastal area is predominantly composed of fine to medium sand sediments, with most exposed bedrock and stony outcrops close to the shore [42]. A wide and relatively flat shoreface extends for over 6 km with depths of less than 15 m offshore Magilligan beach, flanked by the Tuns Bank, a large ebb-delta associated with the Foyle River. The shoreface of Portrush beach is much narrower and steep, with a relatively linear configuration and reaching depths in excess of 18 m approximately 1.2 km seaward of the beach. The offshore shelf and shoreface at Whiterocks presents a complex configuration, influenced by the presence of the Skerries islands and their influence on wave, tidal and sediment transport fluxes. The most exposed section of Whiterocks shoreface is relatively similar to Portrush beach, reaching depths in excess of 20 m approximately 2 km seaward of the beach.

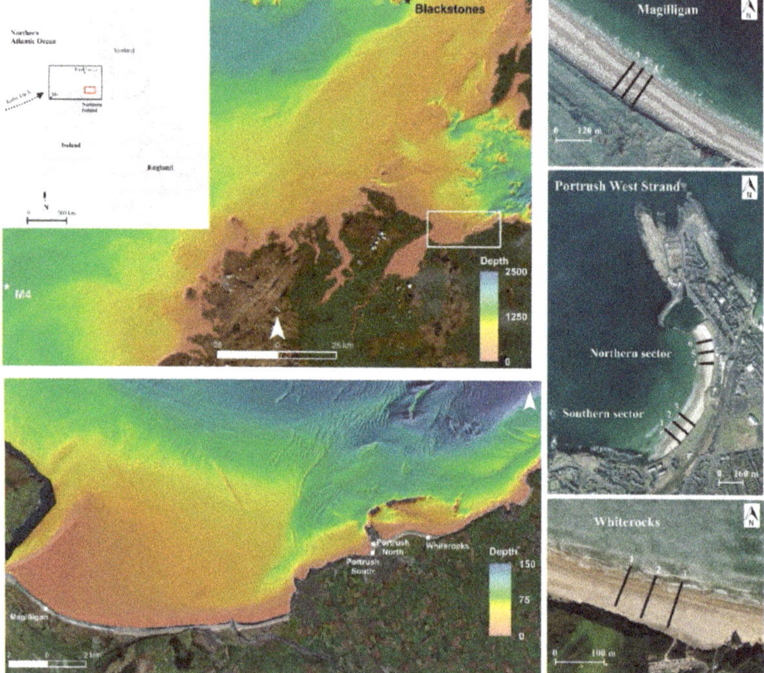

Figure 2. Location map with used grid for wave propagation and offshore buoys.

4. Methodology

A beach monitoring program was undertaken to investigate the impacts of storm Katia by surveying beach morphology before, during and after the storm. Surveys were conducted at Magilligan from 7 to 14 September 2011 while surveys at Portrush West Strand (at two sectors) and at Whiterocks beach took place between the 12 and 15 September 2011. Beach morphological changes were determined through topographic surveys extending from the back beach to low water level using a differential GPS (Trimble 4400) with 1–3 cm accuracy. Survey data were used to quantify the impact of Katia on beach morphology and volumetric changes at each sector. Cross-shore profiles were also extracted and their vertical morphological variability analysed to identify the main active zones [43,44].

Morphological and volumetric variations were compared with nearshore wave forcing to assess the process-response relationships in the monitored beaches, using the nearshore wave power, wave steepness and alongshore wave energy flux. These parameters have been extensively used for exploring morphological changes in wave-dominated beaches and found to be relevant indicators to understand beach erosion and accretion [45]. Here, we computed wave forcing indicators based on shallow water wave parameters obtained from high-resolution nearshore wave modelling using SWAN [46,47]. SWAN was implemented using a nested modelling scheme and forced in the western and northern boundaries with observed wave parameters measured at the M4 offshore buoy (55° N, 10° W), maintained by the Irish Marine Institute, and the Blackstones buoy (56°03′ N, 7°03′ W) operated by CEFAS (Figure 2). Waves were initially propagated over a large-scale computational grid with a resolution of 250 m and using the 2018 EMODnet bathymetry dataset [48] that extended from the buoy locations all the way into the north coast region (Figures 2 and 3a), in order to obtain the boundary conditions for a finer resolution run focused on the study area (Figures 2 and 3b). The model was run with an hourly timestep from 00:00 on the 7 September 2011 to 23:00 on the 16 September 2011. The nested nearshore wave runs were performed using a 5 m high-resolution computational domain, implemented with a detailed bathymetric grid based on JIBS multibeam dataset for the North Coast (Figure 2). The nearshore runs were performed for the exact same time period indicated above, using wave spectra obtained from the large-scale run. The nested runs considered variable water levels obtained from the hourly records of Portrush Tide Gauge. SWAN was implemented in third generation, 2D stationary mode, using a JONSWAP spectral shape to represent the wave field, directional discretization in regular classes of 5° and frequency discretization in 33 logarithmic distributed classes between 1 and 0.03 Hz. Following Loureiro et al. [49] and Matias et al. [50], SWAN runs used default parameters for wave growth, whitecapping dissipation, depth-induced breaking according to the β-kd model for surf-breaking [51], triad and quadruplet wave-wave interactions. Outputs from SWAN provided wave conditions for the nearshore area in each survey site, extracted for a single point in front of the beach in 4 to 5 m water depth.

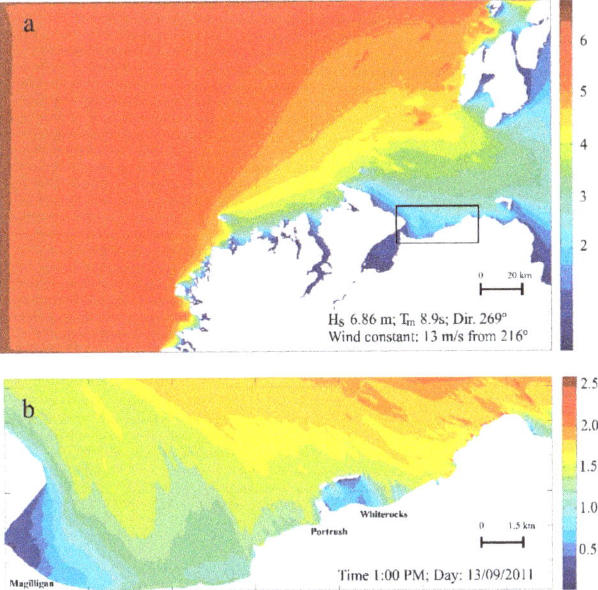

Figure 3. Modeled wave heights (m) for peak offshore storm conditions (**a**) and nearshore wave heights (m) (**b**) during above offshore wave conditions.

Based on SWAN outputs, wave forcing indicators for analyzing process-response relationships were computed assuming the shallow water approximations for linear wave theory following Komar [52]. Wave power (P_s) provides an indication of the rate at which energy is transferred by moving waves and is widely recognized as an important parameter for exploring wave-induced morphological change in sandy beaches (e.g., [45,52]). Wave power was computed according to:

$$P_s = EC_g \tag{1}$$

where E is wave energy, computed according to:

$$E = (1/8)\,\rho\,g\,H_s^2 \tag{2}$$

where ρ is water density (1025 kg/m^3), g gravitational acceleration (9.81 m/s) and H_s is nearshore wave height. Wave group velocity (C_g), was also obtained using the shallow water approximation:

$$C_g = \sqrt{g\,h} \tag{3}$$

where h is the water depth. Wave steepness (L_s) was determined according to:

$$L_s = T \sqrt{g\,h} \tag{4}$$

where T is the nearshore wave period.

Recognizing the importance of wave direction in combination with wave energy in driving longshore sediment transport and alongshore variable morphological changes during energetic conditions, particularly along indented or embayed coastlines (e.g., [45,53]), the alongshore component of the wave energy flux was also computed according to Komar's [52] approximation, given by:

$$P_l = P_s \sin \alpha_b \cos \alpha_b \tag{5}$$

where α_b is the wave breaking angle, determined according to the nearshore wave direction and beach orientation.

Wave steepness, obtained from the ratio of wave height (H_s) with wave length (L_s) was also computed for analyzing process-response relationships, considering the established association among high steepness waves (H/L > 0.02), offshore sediment transport and beach erosion in contrast to low steepness waves (H/L < 0.02), that are associated with onshore sediment transport and beach accretion [54].

5. Results

5.1. Wave Energy Spatial and Temporal Distribution

Wave conditions differed significantly between the offshore location where the wave records were obtained in the western coasts of Ireland and Scotland, and the nearshore areas adjacent to the monitored sites on the north coast of Northern Ireland (Figure 3).

Storm waves generated by the Katia post-tropical cyclone lasted until the 15 September, two days after the cyclone dissipated. Significant wave heights ranged from 1 to 3 m in the nearshore region, while offshore the wave buoys recorded values in excess of 5 m at the peak of the storm on 13 September 2011. Significant wave attenuation across the wide and irregular shelf and shoreface of Northern Ireland is evident from the exposed open ocean locations of the buoys to the relatively sheltered north coast area (Figures 2 and 3a,b). Considering in higher detail the variability within the north coast high resolution grid (Figure 3b), it is observed that wave heights also change significantly between the western, more protected area, towards the eastern more exposed one. Water levels recorded during the storm at Portrush's tidal gauge and modelled wave characteristics in each investigated site are presented in Figure 4. Maximum water levels were recorded on the 13 September 2011 (Figure 4a), with a storm surge effect ranging between 0.2 and 0.5 m induced by the low atmospheric pressure during the passage of the storm on the 13 September. Nearshore wave heights, even during the most energetic period of the storm, recorded between the 13 and 14 September and shown in Figures 3 and 4b, are relatively low, ranging from around 0.8 m in Magilligan to around 1.6 m in Portrush.

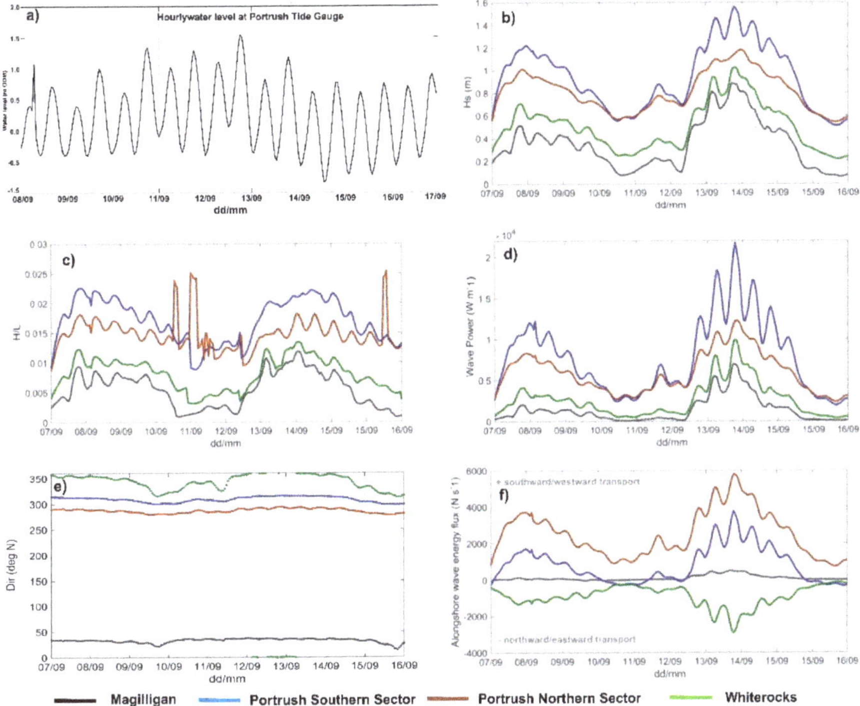

Figure 4. Water levels (**a**) at the Portrush tide gauge were reported as values above Ordnance Datum Malin (equivalent to Mean Sea Level). Wave height (**b**), wave steepness (**c**), wave power (**d**), wave direction (**e**) and alongshore wave energy flux (**f**) at different investigated sites.

This pattern is confirmed by the comparison of the time series for the different sites of interest; wave heights during storm Katia were not considerably high, having undergone significant attenuation and dissipation as they propagated through the shallow and irregular shelf of the north coast of Ireland. This is particularly noticeable for Magilligan, fronted by a wide and shallow shoreface, and Whiterocks, partially protected by the Skerries islands, while the more exposed Northern and Southern sectors of Portrush experienced more energetic conditions. Wave steepness (Figure 4c) displays significant spatial variability, while temporal changes largely reflected the variation in wave height during the surveyed period (Figure 4b). Lowest values were recorded in the most sheltered location, Magilligan, where the steepness ranged from ca. 0.002, during less energetic conditions (i.e., 11–12 September) to ca. 0.007 during more energetic conditions (i.e., 8–10 and 13–15 September periods). Steeper waves were observed at Portrush southern sector (Figure 4c), with values of ca. 0.015 and 0.025 for less and more energetic conditions, respectively. Such an increase in wave steepness, on the order of 0.05, from lower to moderate steepness conditions, was also observed in the other study sites (Figure 4c).

Spatial and temporal variability in wave power also reflected the changes in nearshore wave height, with distinct differences recorded among the different surveyed locations. The lowest values were observed at Magilligan whilst the highest were found at Portrush southern sector (Figure 4d). The dependence of nearshore wave power on water depth is particularly evident in Portrush southern sector, with clear temporal variation associated with the tide-induced changes in water level during the most energetic wave conditions. The influence of wave direction on morphological changes experienced in the four beaches during the storm, explored through the angle of approach and alongshore wave energy flux, indicate that relevant southward fluxes were experienced in the more exposed Portrush Northern and Southern sectors, while at Whiterocks alongshore fluxes were easterly directed. In Magilligan, waves arrived fully refracted and shore normal, inducing negligible alongshore wave energy fluxes (Figure 4e,f).

5.2. Morphological and Volumetric Beach Changes

5.2.1. Magilligan

During the 7–10 September 2011, the beach showed limited elevation changes (on the order of 5–10 cm), which were largely uniform along both the cross-shore, i.e., along the dry beach and the foreshore and the longshore direction (Figure 5a). Volumetric variation indicates some general beach erosion (Table 2). The survey carried out on the 14th, i.e., after the most energetic waves impacted the beach (Figure 4b,d), showed longshore uniform erosion in the dry beach, with vertical erosion of ca. 10 cm, and an equivalent accretion at the central part of the beach according to a beach pivoting mechanism. Volumetric changes reflected a general, very small, accretion (Figure 5b, Table 2), with an alongshore uniform pattern consistent with the shore normal waves that impacted this beach during the storm (Figure 4e,f).

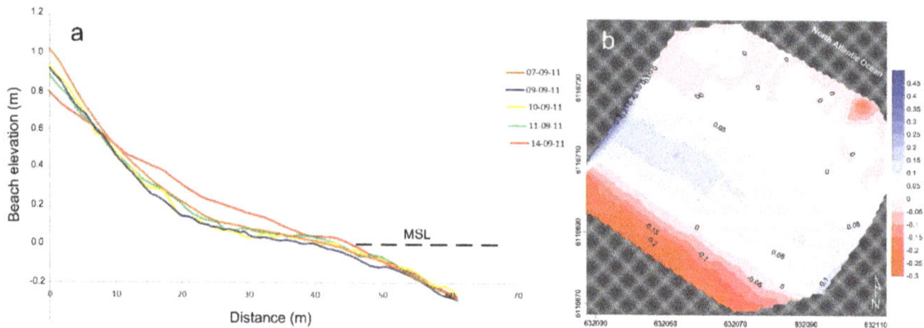

Figure 5. (**a**) Profile 2 evolution and (**b**) 3D morphological changes at Magilligan.

Table 2. Volumetric variation (m^3) between surveys and at the end of the monitoring program.

Location/Date	7 to 9 September	9 to 10	10 to 11	11 to 14	Whole Period 7th to 14th
Magilligan	−184.3	+82.9	+20.7	+50.9	+64.4
Location/date	**12 to 13 September**	**13 to 14**	**14 to 15**	**-**	**Whole Period 12th to 15th**
Portrush southern	−1.1	+512.3	−226.9	-	+123.9
Portrush northern	−432.6	+253.5	+9.5	-	−233.3
Whiterocks	+316.5	+158.6	+109.2	-	+623.2

5.2.2. Portrush, Southern Sector

During the initial phases of the storm, the beach presented very small morphological and volumetric changes (Figure 6 and Table 2) with a shore normal directed energy flux (Figure 4f). A uniform accretion along the cross-shore profile of ca. 10 cm was recorded on the 14 September (Figure 6a) and corresponded with a volumetric increase of 512.3 m^3 (Table 2). This was linked to the sediment supply from the Portrush northern sector due to northerly approaching waves that induced a southward directed flux (Figure 4f). During the last stages of the storm, approximately 5 cm of vertical erosion was observed in different parts of the profile, especially in the central and lower parts (Figure 6a and Table 2) probably due to the reduction of sediment inputs availability from the northern sector. Overall, from the 12 to 15 September, the beach presented a vertical accretion of 10–15 cm especially in the central-upper part (Figure 6b) associated with a volumetric accretion of 123.9 m^3, likely due to sediment supplied from the northern sector that recorded erosion (Figure 7).

The uppermost part of the beach showed different behavior, as its dynamic is strongly affected by a backing concave concrete seawall.

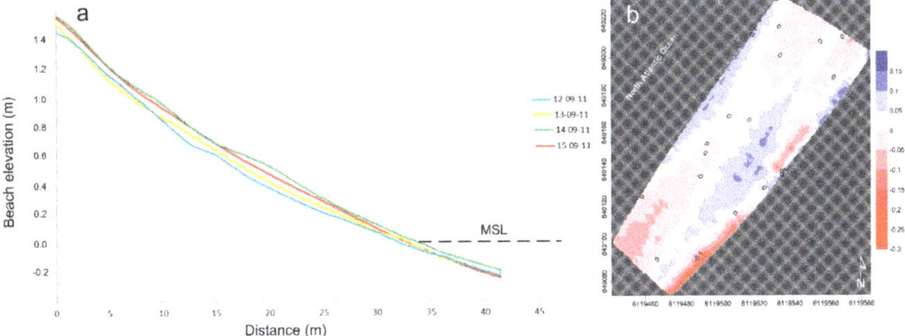

Figure 6. (**a**) Profile 2 evolution and (**b**) 3D morphological changes at Portrush Southern Sector.

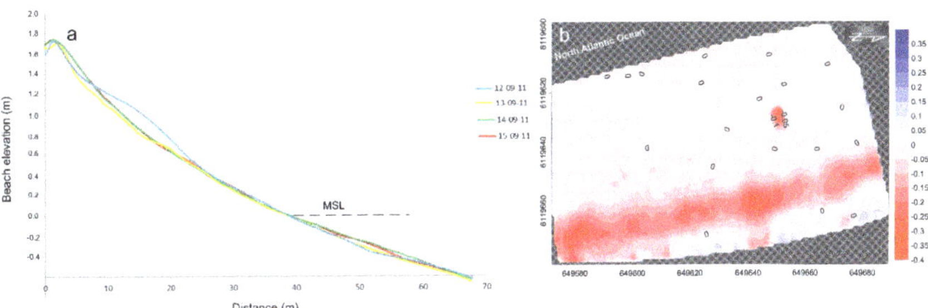

Figure 7. (**a**) Profile 2 evolution and (**b**) 3D morphological changes at Portrush Northern Sector.

5.2.3. Portrush, Northern Sector

The main morphological and volumetric changes were recorded in the first stages of the storm, i.e., on the 13th September (Figure 7a and Table 2). A well-developed berm, 20–30 cm in height present along the whole sector investigated, was completely eroded with a sediment loss of approximately 432.6 m^3 (Table 2). The berm was essentially transported southward to feed the southern sector of this pocket beach and, to a lesser extent, the middle and lower part of the profile (Figure 7a). This is the result of a relevant (2000 N/s) southward directed wave energy flux, forced by northerly approaching waves observed during that period (Figure 4e,f). In the following surveys only very small changes, of few centimetres, were observed. Overall, the beach recorded 30–35 cm of vertical erosion (Figure 7a,b) along the upper part, through the erosion of the berm, which corresponded to a volumetric change of −233.3 m^3 (Table 2).

5.2.4. Whiterocks

Prior to the storm, the beach presented a well-developed berm, ca. 30–40 cm in height (Figure 8a). During the first stages of the storm, on the 13 September, relevant morphological changes took place linked to the erosion of the berm and the landward and seaward transport of the eroded sand according to a process of beach flattening—but no net erosion was recorded (Table 2), this indicating a supply of sediments from the western part of Curran strand driven by eastward alongshore wave energy fluxes under low energy westerly approaching waves (Figure 4e,f). No relevant morphological changes took place in following days and accretion was recorded along all the profile (Figure 8a and Table 2). Overall, at the end of the storm, the beach recorded a volumetric accretion of 632.2 m^3 visible on the dry beach and at the lower foreshore (Figure 8b).

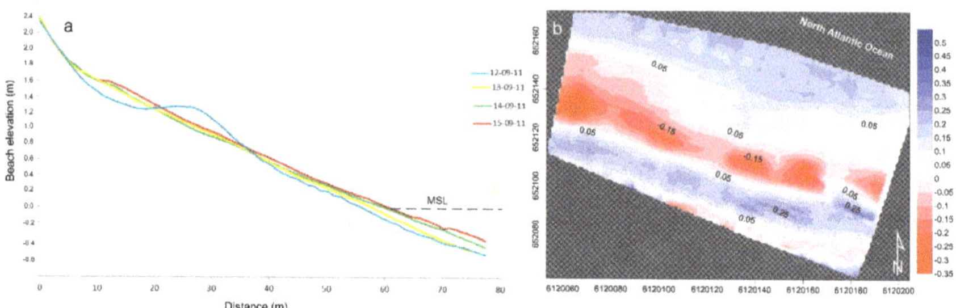

Figure 8. (**a**) Profile 2 evolution and (**b**) 3D morphological changes at Whiterocks.

6. Discussion

The Northern and Western coasts of Ireland are classified as high-energy coastlines and are often affected by energetic storms and the tail-end of a small number of Atlantic Hurricanes [32,33,55–57]. The effects of some of these events have been described by different authors [32,55–60], but the specific impacts of a recent post-tropical cyclone on Northern Irish beaches have not been investigated so far.

Coastal morphological changes investigated in this paper were related to the effects of the post-tropical cyclone Katia that impacted the coast of Northern Ireland from the 12 to the 15 September 2011 with sustained winds of ca. 95 km h^{-1} and offshore waves in excess of 5 m in height. The cyclone originated as a tropical depression over the eastern Atlantic on 29 August, strengthened into a tropical storm the following day and than developed into a hurricane by 1st September, becoming a Category 4 hurricane with winds of 225 km h^{-1} by 4 September moving eastward towards the east coast of USA and downgraded to a post-storm cyclone one day before reaching the British Isles on the 12–13 September 2011.

The earliest reported high-magnitude event in the Northwest of Ireland was the 'Night of the Big Wind', which was probably the tail-end of a hurricane that hit Ireland in January 1839. More recently, an analysis of instrumental storm records since the 1950s [60], identified Hurricane Debbie as the largest storm that impacted Ireland during the second half of the 20th Century [31,60,61]. Hurricane Debbie moved west across the Atlantic still maintaining hurricane-force winds and affecting Ireland in September 1961, with winds of up to 181 km h^{-1} for more than 5 h along the west of Ireland [60]. Furthermore, Guisado-Pintado and Jackson [32,33] mentioned the effects of Hurricane Charley (1986), which was considered a post-tropical storm when it reached the south coast of Ireland, and described in detail the effects of the extratropical storm Ophelia. This event, re-classified as a "post-tropical" storm but being previously a Category 3 hurricane [61], reached the western coast of Ireland on the 16 October 2017, triggering a nationwide severe weather warning and causing substantial coastal flooding.

During storm Katia, offshore wave buoys recorded values in excess of 5 m at the peak of the event but nearshore significant wave height ranged from 1 to 3 m, reflecting relevant wave energy reduction linked to a variety of dissipation processes, that is: although offshore conditions in the exposed Atlantic section of the coast were very energetic, when waves propagate and refract to reach the study areas (Figure 2), they dissipate significant amounts of energy due to bottom friction, refraction and diffraction along the wide, shallow and irregular shelf and shoreface, and indented coastline [56]. This was particularly noticeable at Magilligan, as this area is protected towards the west by a resistant rocky headland and the shoreline is situated on the lee side of the Innishowen Peninsula that prevents the arrival of large amounts of wave energy from the SE, SW or NW quadrants [57]. Finally, at Magilligan, because of the wide shoreface adjacent to it and its long-term association with a large ebb-delta (Tuns Bank), bottom friction dissipation and refraction are more significant than at other sites examined here and the waves arrived almost perfectly shore parallel. This leads to a very small angle of approach and minimal alongshore wave energy fluxes. As a result, at Magilligan, there was limited wave forcing over the whole period, both in terms of wave height and power, but also reduced wave steepness and almost insignificant alongshore fluxes. Specifically, with respect to other locations, wave power observed here was the lowest (i.e., 5×10^3 kW m^{-1} during the peak of the storm), much smaller than what was observed during the last phases of Ophelia storm (20×10^3 kW m^{-1}) [32]. As a result, only minor morphological changes were recorded at Magilligan with erosion of the upper beach and flattening of the foreshore, a trend similar—but with much smaller vertical morphological variations—to the one observed at Five Finger Strand in NW Ireland [32].

Less significant nearshore wave refraction favoured higher (compared to Magilligan) wave height and wave power values in Portrush and Whiterocks. Wave power ranged from 10 to 20×10^3 kW m^{-1} and presented a clear longshore component, predominantly southward at both Portrush sectors and eastward at Whiterocks. Hence, longshore transport was evident at Portrush sectors that are included within a morphological sediment cell [9] enclosed by a harbour, in the northern end, and a headland in the southern end. The northern sector presented erosion while the southern sector accreted, probably from a point of pivoting in beach planform located in between the two sectors as observed at other morphological cells [9,62], which is consistent with a short-term rotational response identified in small embayments in various other settings [63–65] and in other studies of large embayed beaches [66–68].

At Whiterocks, a trend similar to the one noted at Portrush was observed. Berm erosion took place by means of a general beach flattening and this resulted in an overall accretion due to an influx of sediment from the western section of the beach.

Overall, storm Katia produced relatively limited impacts on the beaches of the north coast of Northern Ireland. This was also noted by Guisado-Pintado and Jackson [32] and Cooper et al. [56] that indicated that, to have a relevant impact on dissipative beaches of Northern Ireland, storms need to be directed onshore and coincide with high tide, rendering storm duration and offshore intensity of less importance. Katia, which reached approximately 96 km h^{-1} presented much lower intensity that Debbie (181 km h^{-1}), but was marginally higher than Ophelia (gusts of 74 km h^{-1}) and had a longer duration than the previous events. This was not reflected in terms of morphological changes since during Katia the storm track and prevailing wind directions, which are relevant aspects in determining storm damages [23], were from the SW and W directions, while the beaches investigated are mainly exposed to N, NW and NE quadrants. Interestingly, this was also the case for previous hurricane/post-tropical cyclones Debbie and Ophelia. As observed by Cooper et al. [56], Northern Irish coasts facing NE and N are more susceptible to lower magnitude and longer duration storms, characterised by short sea waves, from a northerly direction. Further, as reported by Cooper et al. [56] and Guisado-Pintado and Jackson [32,33] and confirmed by this study, morphological changes produced on the coast were very localized and dependant on nearshore wave propagation driving cross and alongshore energy fluxes, since the volume and direction of transport during storm impact was highly site-specific. A similar trend, i.e., changes in the direction of longshore transport and morphological and volumetric (positive/negative) modalities of beach response to storms impacts, was recorded in SW England during the 2013–2014 storm winter season [69]. At Magilligan and Whiterocks, erosion processes did not greatly affect the dry beach and did not impact at all on the frontal dunes since this depends on storm peak coincidence with high tidal levels [32,56,65]. As observed by Cooper et al. [56] at Magilligan, the formation of the local dune escarpment is relatively rare and is typically associated with storms that occur at or close to high tide, with predicted tidal elevations of 2.1 m (high spring tide). During Katia storm, maximum water elevation was 1.4 m (high neap tide). The upper beach in Portrush, which is backed by a concrete concave seawall, experienced more relevant vertical changes, especially at the southern sector where the upper beach connects directly with the seawall and the backshore is inexistent.

7. Conclusions

Despite offshore wave buoys values in excess of 5 m wave height at the peak of the post-tropical cyclone Katia on the 13 September 2011, nearshore significant wave height ranged from 1 to 3 m. This was due to the dissipation processes experienced by waves as they propagate and refract along a shallow and irregular shelf and shoreface to reach the studied coastal sectors. Such propagation process produced a very small angle of approach and minimal alongshore wave energy fluxes at Magilligan and, as a result, this location only recorded minor morphological changes with erosion of the upper beach and flattening of the foreshore. More limited nearshore wave refraction favoured higher waves at Portrush and Whiterocks, leading to increased wave power with a clear longshore component, ranging from southward at both Portrush sectors to eastward at Whiterocks. Longshore transport at Portrush favoured erosion in the northern sector and accretion in the southern sector, which is consistent with a short-term rotational response. At Whiterocks, a trend similar to the one noted at Portrush was observed, i.e., an overall accretion due to an influx of sediment from the western section of the beach.

Katia post-tropical cyclone, as other similar events—e.g., hurricane Debbie and post-tropical cyclone Ophelia, produced moderate impacts in the beaches investigated because the storm track and prevailing wind directions were from the SW and W directions, while the monitored beaches are mainly exposed to N, NW and NE quadrants. Overall, it is not possible to predict a general and homogeneous response of the north coast of Northern Ireland to such kind of events because morphological changes produced are very site-specific and dependant on water level during the storm and, especially, wave transformation across the shelf and shoreface that controls the volume and direction of sediment transport and hence, beach morphological response.

Author Contributions: Conceptualization, G.A., C.L. and D.J.; methodology, T.S. and C.L.; software, M.T.; investigation, G.A. and T.S.; resources, D.J.; data curation, C.L. and M.T.; writing—original draft preparation, G.A.; writing—review and editing, C.L., D.J. and T.S.; All authors have read and agreed to the published version of the manuscript.

Funding: This research received no external funding.

Acknowledgments: This work is a contribution to the Andalusia (Spain) PAI Research Group RNM-328. Carlos Loureiro contribution is developed in the framework of H2020 MSCA NEARControl project, which received funding from the European Union's Horizon 2020 Research and Innovation programme under the Marie Skłodowska-Curie grant agreement No. 661342.

Conflicts of Interest: The authors declare no conflict of interest.

References

1. National Hurricane Center and Central Pacific Hurricane Center. Available online: https://www.nhc.noaa.gov (accessed on 18 April 2020).
2. Finkl, C.W.; Kruempfel, C. Threats, obstacles and barriers to coastal environmental conservation: Societal perceptions and managerial positionalities that defeat sustainable development. In Proceedings of the 1st International Conference on Coastal Conservation and Management in the Atlantic and Mediterranean Seas, Algarve, Portugal, 17–20 April 2005; Veloso-Gomez, F., Taveira Pinto, F., da Neves, L., Sena, A., Ferreira, O., Eds.; University of Porto: Porto, Portugal, 2005; pp. 3–28.
3. Molina, R.; Manno, G.; Lo Re, C.; Anfuso, G.; Ciraolo, G. A Methodological Approach to Determine Sound Response Modalities to Coastal Erosion Processes in Mediterranean Andalusia (Spain). *J. Mar. Sci. Eng.* **2020**, *8*, 154. [CrossRef]
4. Klein, Y.L.; Osleeb, J.P.; Viola, M.R. Tourism generated earnings in the coastal zone: A regional analysis. *J. Coast. Res.* **2004**, *20*, 1080–1088.
5. Jones, A.; Phillips, M. *Disappearing Destinations*; CABI: Wallingford, UK, 2011; p. 273.
6. EEA. *The Changing Face of Europe's Coastal Areas*; Report No 6. Luxembourg office for official publications of the European Communities; Breton, F., Meiner, A., Eds.; EEA: Copenhagen, Denmark, 2006.
7. Tourism Northern Ireland. 2019. Available online: https://tourismni.com (accessed on 2 February 2020).
8. Meyer-Arendt, K. Grand Isle, Louisiana: A historic US Gulf Coast Resort Adapts to Hurricanes, Subsidence and Sea Level Rise. In *Disappearing Destinations*; Jones, A., Phillips, M., Eds.; CABI: Wallingford, UK, 2011; pp. 203–217.
9. Carter, R.W.G. *Coastal Environments*; Academic Press: Cambridge, MA, USA, 1988; p. 617.
10. Dolan, R.; Davis, R.E. An intensity scale for Atlantic coast northeast storms. *J. Coast. Res.* **1992**, *8*, 352–364.

11. Morton, R.A.; Sallenger, A.H. Morphological impacts of extreme storms on sandy beaches and barriers. *J. Coast. Res.* **2003**, *19*, 560–573.

12. Sallenger, A. *Island in a Storm: A Rising Sea, a Vanishing Coast, and a Nineteenth-Century Disaster that Warns of a Warmer World*; Public Affairs: New York, NY, USA, 2009; p. 294.

13. Beudin, A.; Ganju, N.K.; Defne, Z.; Aretxabaleta, A.L. Physical response of a back-barrier estuary to a post-tropical cyclone. *JGR Oceans* **2017**, *122*, 5888–5904. [CrossRef]

14. Bonazzi, A.; Cusack, S.; Mitas, C.; Jewson, S. The spatial structure of European wind storms as characterized by bivariate extreme-value Copulas. *Nat. Hazards Earth Syst. Sci.* **2012**, *12*, 1769–1782. [CrossRef]

15. Anfuso, G.; Rangel-Buitrago, N.; Cortés-Useche, C.; Iglesias Castillo, B.; Gracia, F.J. Characterization of storm events along the Gulf of Cadiz (eastern central Atlantic Ocean). *Int. J. Climatol.* **2016**, *36*, 3690–3707. [CrossRef]

16. Rangel-Buitrago, N.; Anfuso, G. Winter wave climate, storms and regional cycles: The SW Spanish Atlantic coast. *Int. J. Climatol.* **2013**, *33*, 2142–2156. [CrossRef]

17. Ferreira, Ó. Storm groups versus extreme single storms: Predicted erosion and management consequences. *J. Coast. Res.* **2005**, *42*, 221–227.

18. Almeida, L.P.; Ferreira, O.; Vousdouskas, M.I.; Dodet, G. Historical variation and trends in storminess along the Portuguese South coast. *Nat. Hazards Earth Syst. Sci.* **2011**, *11*, 2407–2417. [CrossRef]

19. Phillips, M. Consequences of short-term changes in coastal processes: A case study. *Earth Surface Processe. Landf.* **2008**, *33*, 2094–2107. [CrossRef]

20. Phillips, M.; Crisp, S. Sea level trends and NAO influences: The Bristol Channel/Seven Estuary. *Glob. Planet. Chang.* **2010**, *73*, 211–218. [CrossRef]

21. Dodet, G.; Castelle, B.; Masselink, G.; Scott, T.; Davidson, M.; Floc'h, F.; Jackson, D.; Suanez, S. Beach recovery from extreme storm activity during the 2013–14 winter along the Atlantic coast of Europe. *Earth Surf. Process. Landf.* **2019**, *44*, 393–401. [CrossRef]

22. Masselink, G.; Scott, T.; Poate, T.; Russell, P.; Davidson, M.; Conley, D. The extreme 2013/2014 winter storms: Hydrodynamic forcing and coastal response along the southwest coast of England. *Earth Surf. Process. Landf.* **2016**, *41*, 378–391. [CrossRef]

23. Castelle, B.; Marieu, V.; Bujan, S.; Splinter, K.D.; Robinet, A.; Sénéchal, N.; Ferreira, S. Impact of the winter 2013–2014 series of severe Western Europe storms on a double-barred sandy coast: Beach and dune erosion and megacusp embayments. *Geomorphology* **2015**, *238*, 135–148. [CrossRef]

24. Blaise, E.; Suanez, S.; Stephan, P.; Fichaut, B.; David, L.; Cuq, V.; Autret, R.; Houron, J.; Rouan, M.; Floc'h, F.; et al. Bilan des tempêtes de l'hiver 2013–2014 sur la dynamique du recul de trait de côte en Bretagne. *Geomorphol. Relief Process. Environ.* **2015**, *21*, 267–292. [CrossRef]

25. Autret, R.; Dodet, G.; Fichaut, B.; Suanez, S.; David, L.; Leckler, F.; Ardhuin, F.; Ammann, J.; Grandjean, P.; Allemand, P.; et al. A comprehensive hydro-geomorphic study of cliff-top storm deposits on Banneg Island during winter 2013–2014. *Mar. Geol.* **2016**, *382*, 37–55. [CrossRef]

26. Santos, V.M.; Haigh, I.D.; Wahl, T. Spatial and Temporal Clustering Analysis of Extreme Wave Events around the UK Coastline. *J. Mar. Sci. Eng.* **2017**, *5*, 28. [CrossRef]

27. Van Loon, H.; Rodgers, J.C. The seesaw in winter temperatures between Greenland and Northern Europe: Part I. *Gen. Descr. Mon. Weather Rev.* **1978**, *106*, 296–310. [CrossRef]

28. Grams, C.M.; Blumer, S.R. European high-impact weather caused by the downstream response to the extratropical transition of North Atlantic Hurricane Katia (2011). *Geophys. Res. Lett.* **2015**, *42*, 8738–8748. [CrossRef]

29. Hart, R.E.; Evans, J.L. A Climatology of the Extratropical Transition of Atlantic Tropical Cyclones. *J. Clim.* **2001**, *14*, 546–564. [CrossRef]

30. Cooper, J.A.G.; Orford, J.D. Hurricanes as agents of mesoscale coastal change in Western Britain and Ireland. *J. Coast. Res.* **1998**, *26*, 123–128.

31. MacClenahan, P.; McKenna, J.; Cooper, J.A.G.; O'Kane, B. Identification of highest magnitude coastal storm events over western Ireland on the basis of wind speed and duration thresholds. *Int. J. Climatol.* **2001**, *21*, 829–842. [CrossRef]

32. Guisado-Pintado, E.; Jackson, D.W.T. Multi-scale variability of storm Ophelia 2017: The importance of synchronised environmental variables in coastal impact. *Sci. Total Environ.* **2018**, *630*, 287–301. [CrossRef] [PubMed]

33. Guisado-Pintado, E.; Jackson, D.W.T. Coastal Impact from High-Energy Events and the Importance of Concurrent Forcing Parameters: The Cases of Storm Ophelia (2017) and Storm Hector (2018) in NW Ireland. *Front. Earth Sci.* **2019**. [CrossRef]

34. Kossin, J.P.; Emanuel, K.A.; Vecchi, G.A. The poleward migration of the location of tropical cyclone maximum intensity. *Nature* **2014**, *509*, 349–352. [CrossRef]

35. Goldenberg1, S.; Landsea, C.; Mestas-Nuñez, A.; Gray, W. The Recent Increase in Atlantic Hurricane Activity: Causes and Implications. *Science* **2001**, *293*, 474–479. [CrossRef]

36. Lozano, I.; Devoy, R.J.N.; May, W.; Andersen, U. Storminess and vulnerability along the Atlantic coastlines of Europe: Analysis of storm records and greenhouse gases induced climate scenario. *Mar. Geol.* **2004**, *210*, 205–225. [CrossRef]

37. Jackson, D.W.T.; Cooper, J.A.G.; del Rio, L. Geological control of beach morphodynamic state. *Mar. Geol.* **2005**, *216*, 297–314. [CrossRef]

38. Jackson, D.W.T.; Cooper, J.A.G. Application of the equilibrium planform concept to natural beaches in Northern Ireland. *Coast. Eng. J.* **2009**, *57*, 112–123. [CrossRef]

39. Lynch, K.; Jackson, D.W.T.; Cooper, J.A.G. The fetch effect on aeolian sediment transport on a sandy beach: A case study from Magilligan Strand, Northern Ireland. *Earth Surf. Process. Landf.* **2016**, *41*, 1129–1135. [CrossRef]

40. Backstrom, J.T.; Jackson, D.W.T.; Cooper, J.A.G. Shoreface Dynamics of two High-Energy Beaches in Northern Ireland. *J. Coast. Res.* **2007**, *50*, 594–598.

41. Carter, R.W.G.; Bartlett, D.J. Coastal Erosion in Northeast Ireland—Part I: Sand beaches, dunes and river mouths. *Ir. Geogr.* **1990**, *23*, 1–6. [CrossRef]

42. Plets, R.; Clements, A.; Quinn, L.; Strong, J.; Breen, J.; Edwards, H. Marine substratum map of the Causeway Coast, Northern Ireland. *J. Maps.* **2012**, *8*, 1–13. [CrossRef]

43. Rangel-Buitrago, N.; Anfuso, G. Coastal storm characterization and morphological impacts on sandy coasts. *Earth Surf. Process. Landf.* **2011**, *36*, 1997–2010. [CrossRef]

44. Lee, G.; Nicholls, R.J.; Birkemeier, W.A. Storm-induced profile variability of the beach-nearshore profile at Duck, North Carolina, USA, 1981–1991. *Mar. Geol.* **1998**, *148*, 163–177. [CrossRef]

45. Loureiro, C.; Ferreira, Ó.; Cooper, J.A.G. Non-uniformity of storm impacts on three high-energy embayed beaches. *J. Coast. Res.* **2014**, *70*, 326–331. [CrossRef]

46. Holthuijsen, L.; Booij, N.; Ris, R.C. A Spectral Wave Model for the Coastal Zone. In Proceedings of the 2nd International Symposium on Ocean Wave Measurement and Analysis, New Orleans, LA, USA, 25–28 July 1993; pp. 630–641.

47. Booij, N.; Ris, R.C.; Holthuijsen, L.H. A third-generation wave model for coastal regions: 1. Model description and validation. *J. Geophys. Res.* **1999**, *104*, 7649–7666. [CrossRef]

48. EMODnet Bathymetry Consortium. EMODnet Digital Bathymetry (DTM). *EMODnet Bathymetry Consort.* **2018**. [CrossRef]

49. Loureiro, C.; Ferreira, O.; Cooper, J.A.G. Geologically constrained morphological variability and boundary effects on embayed beaches. *Mar. Geol.* **2012**, *339–331*, 1–15. [CrossRef]

50. Matias, A.; Carrasco, A.R.; Loureiro, C.; Masselink, G.; Andriolo, U.; McCall, R.; Ferreira, O.; Plomaritis, T.A.; Pacheco, A.; Guerreiro, M. Field measurements and hydrodynamic modelling to evaluate the importance of factors controlling overwash. *Coast. Eng.* **2019**, *152*, 103523. [CrossRef]

51. Salmon, S.; Holthuijsen, L. Modeling depth-induced wave breaking over complex coastal bathymetries. *Coast. Eng.* **2015**, *105*, 21–35. [CrossRef]

52. Komar, P.D. *Beach Processes and Sedimentation*; Prentice Hall: Upper Saddle River, NJ, USA, 1998; p. 544.

53. Harley, M.D.; Turner, I.L.; Short, A.D.; Ranasinghe, R. A reevaluation of coastal embayment rotation: The dominance of cross-shore versus alongshore sediment transport processes, Collaroy-Narrabeen Beach, southeast Australia. *J. Geophys. Res.* **2011**, *166*, F04033. [CrossRef]

54. Aagaard, T.; Masselink, G. The Surf Zone. In *Handbook of Beach and Shoreface Morphodynamics*; Short, A.D., Ed.; Wiley: Hoboken, NJ, USA, 1999; pp. 72–118.

55. Backstrom, J.; Jackson, D.; Cooper, A.; Loureiro, C. Contrasting geomorphological storm response from two adjacent shorefaces. *Earth Surf. Process. Landf.* **2015**, *40*, 2112–2120. [CrossRef]

56. Cooper, J.A.G.; Jackson, D.W.T.; Navas, F.; McKenna, J.; Malvarez, G. Identifying storm impacts on an embayed, high-energy coastline: Examples from western Ireland. *Mar. Geol.* **2004**, *210*, 261–280. [CrossRef]

57. Loureiro, C.; Cooper, J.A.G. Temporal variability in winter wave conditions and storminess in the Northwest of Ireland. *Ir. Geogr.* **2018**, *51*, 155–170.
58. Scheffers, A.; Scheffers, S.; Kelletat, D.; Browne, T. Wave-emplaced coarse debris and Megaclasts in Ireland and Scotland: Boulder transport in a high energy littoral environment. *J. Geol.* **2009**, *117*, 553–573. [CrossRef]
59. Erdmann, W.; Kelletat, D.; Kuckuck, M. Boulder ridges and Washover features in Galway Bay, western Ireland. *J. Coast. Res.* **2017**, *33*, 997–1021. [CrossRef]
60. Hickey, K.R.; Connolly-Johnston, C. The Impact of Hurricane Debbie (1961) and Hurricane Charley (1986) on Ireland. In *Advances in Hurricane Research—Modelling, Meteorology, Preparedness and Impacts*; Hickey, K., Ed.; InTech: Rijeka, Croatia, 2012; Chapter 9.
61. NOAA. *National Hurricane Center: Post-Tropical Cyclone Ophelia Discussion Number 28 NWS*; Miami FL AL172017 1100 PM AST Sun; NOAA: Silver Spring, MD, USA, 2017.
62. Anfuso, G.; Pranzini, E.; Vitale, G. An integrated approach to coastal erosion problems in northern Tuscany (Italy): Littoral morphological evolution and cell distribution. *Geomorphology* **2011**, *129*, 204–214. [CrossRef]
63. Loureiro, C.; Ferreira, O. Mechanisms and timescales of beach rotation. In *Sandy Beach Morphodynamics*; Jackson, D.W.T., Short, A.D., Eds.; Elsevier: Dordrecht, The Netherlands, 2020; in press.
64. Coco, G.; Senechal, N.; Rejas, A.; Bryan, K.R.; Capo, S.; Parisot, J.P.; Brown, J.A.; MacMahan, J.H.M. Beach response to a sequence of extreme storms. *Geomorphology* **2014**, *204*, 493–501. [CrossRef]
65. Dissanayake, P.; Brown, J.; Wisse, P.; Karunarathna, H. Effects of storm clustering on beach/dune evolution. *Mar. Geol.* **2015**, *370*, 63–75. [CrossRef]
66. Ojeda, E.; Guillen, J. Shoreline dynamics and beach rotation of artificial embayed beaches. *Mar. Geol.* **2008**, *253*, 51–62. [CrossRef]
67. Turki, I.; Medina, R.; Coco, G.; Gonzalez, M. An equilibrium model to predict shoreline rotation of pocket beaches. *Mar. Geol.* **2013**, *346*, 220–232. [CrossRef]
68. Thomas, T.; Phillips, M.R.; Lock, G. An analysis of subaerial beach rotation and influences of environmental forcing adjacent to the proposed Swansea Bay tidal lagoon. *Appl. Geogr.* **2015**, *62*, 276–293. [CrossRef]
69. Burvingt, O.; Masselink, G.; Russell, P.; Scott, T. Classification of beach response to extreme storms. *Geomorphology* **2017**, *295*, 722–737. [CrossRef]

Article

Tsunami Propagation and Flooding in Sicilian Coastal Areas by Means of a Weakly Dispersive Boussinesq Model

Carlo Lo Re *, Giorgio Manno and Giuseppe Ciraolo

Department of Engineering (DI), University of Palermo, Viale delle Scienze, Bd. 8, 90128 Palermo, Italy;
giorgio.manno@unipa.it (G.M.); giuseppe.ciraolo@unipa.it (G.C.)
* Correspondence: carlo.lore@unipa.it; Tel.: +39-238-965-24

Received: 27 April 2020; Accepted: 15 May 2020; Published: 19 May 2020

Abstract: This paper addresses the tsunami propagation and subsequent coastal areas flooding by means of a depth-integrated numerical model. Such an approach is fundamental in order to assess the inundation hazard in coastal areas generated by seismogenic tsunami. In this study we adopted, an interdisciplinary approach, in order to consider the tsunami propagation, relates both to geomorphological characteristics of the coast and the bathymetry. In order to validate the numerical model, comparisons with results of other studies were performed. This manuscript presents first applicative results achieved using the weakly dispersive Boussinesq model in the field of tsunami propagation and coastal inundation. Ionic coast of Sicily (Italy) was chosen as a case study due to its high level of exposure to tsunamis. Indeed, the tsunami could be generated by an earthquake in the external Calabrian arc or in the Hellenic arc, both active seismic zones. Finally, in order to demonstrate the possibility to give indications to local authorities, an inundation map, over a small area, was produced by means of the numerical model.

Keywords: tsunami propagation; tsunami flooding; sicilian coast; coastal hazard

1. Introduction

In the study of the tsunami propagation phenomenon is very important to model the wave frequency dispersion because of its significant role during wave transformation from deep to intermediate waters. During the propagation to the shore, dispersive waves refract, shoal, due to coastal bathymetry morphology. Long waves have an high impact on surf-zone dynamics, sediment transport and beach erosion. In the Mediterranean sea, the effects of tsunamis on the coasts could be similar to the effects of large storms [1,2] and the detailed modeling of the shoreline movement is important in order to avoid big uncertainties [3]. Tsunamis are huge waves generated by earthquakes, submarine volcanic eruptions or landslides. In very deep oceanic waters tsunami do not dramatically increase in height. But as the waves travel onshore, they increase in height as the bathymetry gradually decrease, becoming potentially destructive (e.g., the tsunami in Indian Ocean in December 2004, or in Japan in March 2011). Usually the tsunami were modelled as solitary waves and obliviously the shoaling, breaking, and run-up are phenoma of major interest for researcher [4–9]. The high computational power of modern computers and parallel computing make it possible to solve more and more complex fluid dynamics problems. Indeed, it is possible to solve 3D Reynolds Averaged Navier–Stokes (RANS) equations and use methods like Smoothed Particle Hydrodynamics (SPH) or Volume Of Fluids (VOF) [10–12]. Unfortunately, 3D tsunami modeling requires high computational efforts that are not consistent with practical purposes. To overcome this problem we used the weakly dispersive model described by [4,13]. This kind of approach, also called the non-hydrostatic Non Linear Shallow Water Equations (NLSWE) method, usually solves, in the horizontal coordinates and time, free surface motion with a single value function. This requires a much lower vertical resolution than 3D methods. Moreover, the used model has an accurate modeling technique of wetting/drying processes [4,13]. The Western Ionian area, due to the clash between the African and Eurasian tectonic plates, is exposed to a high seismic risk causing possible tsunamis whose origin is directly related to earthquakes. In particular, the Greek and Italian (Calabrian and Sicilian ones) coastal areas are among the most exposed sites to such a hazard. These coastal areas, have a strong anthropization which makes them more vulnerable [14]. In order to plan actions useful for risk mitigation, it is necessary to produce flooding and exposure maps that can be used for prevention and protection purposes. In the last decades, due to the recent tsunami disasters, researchers developed numerical models of increasingly quality. Ref Samaras et al. [15], to simulate the effects of a tsunami striking the Greek and Sicilian coasts, used two-dimensional Boussinesq equations with a high order of approximation. Ref Schambach et al. [16] used a non-hydrostatic three-dimensional model coupled with a non-linear and dispersive two-dimensional model to simulate the propagation of a tsunami on the coast, generated by the earthquake struck the city of Messina on 28th December 1908. Ref Mueller et al. [17] used, instead, the well-known Cornell Multi-grid Coupled Tsunami model (COMCOT), which solves the NLSWE in spherical and Cartesian coordinates, to analyze the effects of scenarios of a flood near the Maltese coasts. The same two-dimensional model was adopted by [18] to examine the characteristics of an earthquake-induced tsunami in the south of the province of Bali (Indonesia). In this paper, we present preliminary results regarding hypothetical strike of an earthquake induced tsunami on Sicilian coast. Furthermore, an inundation map, over a very small area, was produced by means of the weakly dispersive Boussinesq model [4,13].

2. Materials and Methods

Many Mediterranean coastal areas are potentially exposed to the tsunami risk [15]. Specifically, the Sicilian coasts are highly exposed, because they have morphological characteristics able to enhance flooding effects and because they are densely populated and plenty of infrastructure. One of the most exposed Sicilian coastal area, to probable tsunami events, is the Ionian Mediterranean area [19]. In fact, two important tectonic structures are located in this area, the external Calabrian Peloritano Arch and the Hellenic Arch; both originated due to the clash between the Eurasian Plate and the African Plate, Figure 1).

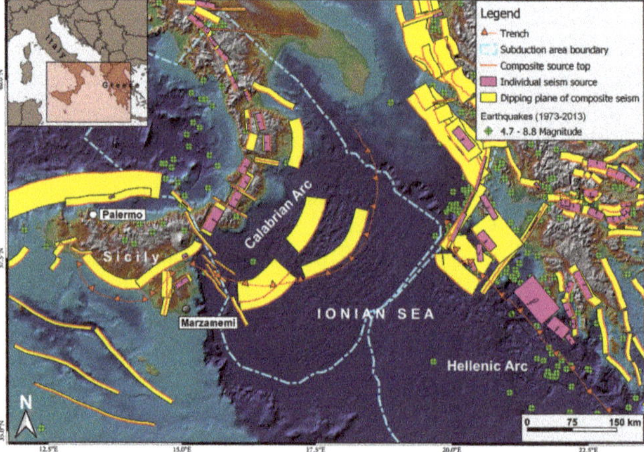

Figure 1. Possible seismic sources in the Ionian Sea. The data-set is taken from the DISS Working Group [20], Caputo and Pavlides [21].

The tectonic structure of this area includes also several smaller plates [19] making more difficult the analysis of earthquake-induced tsunami (Figure 1). The outer Calabrian Arc has impressive deep reverse fault systems (Figure 1), with a predominantly NW-SE direction [14,22], which could originate earthquakes with significant magnitudes [23]. The Hellenic Arc (about 1000 km long) is also one of the most seismically active areas near Greece (Figure 1). This structure consists of three main elements: an outer area (South) consisting of three ocean tranches, an intermediate area with an island arc and a North area characterized by a volcanic islands arc [24]. The coastal area of study was the South of the Ionian Sicilian coast. The coast is articulated with low and rocky coastlines and with sandy beaches, with a slight slope and variable width, delimited by small promontories. This area, depicted in Figure 2 was used for numerical test adopting a weakly dispersive numerical model (see Section 2.1). The location chosen as a case study is Marzamemi (from the Arabic marsa for port and memi for small), a little coastal village on the Sicilian Ionian area (Italy). This village was selected because it falls into specific typologies: (a) it has a big exposure to seismic areas that can cause tsunamis; (b) the coast has a flat topography (at about 300 m from the coastline altitudes ranging between 1 and 6 m above sea level); (c) in this coastal sector, the continental shelf is tight (about 17 km) and it is engraved by little canyons; (d) despite being a fishing village, Marzamemi is densely populated both during the summer and during the international frontier film festival; (e) it is a site of archaeological-industrial interest because, in its main square, an old tuna factory is still present. Figure 2 shows an overview of studied coastal area, a magnification of the promontory area of Marzamemi and the boundaries of the numerical domain. The village develops on the promontory northward the small fishery port, this promontory is partially exposed to the wave action. In particular, the shoreline of the promontory northern part is preceded by the rocky shelf that has small water depths (about 0.2 m).

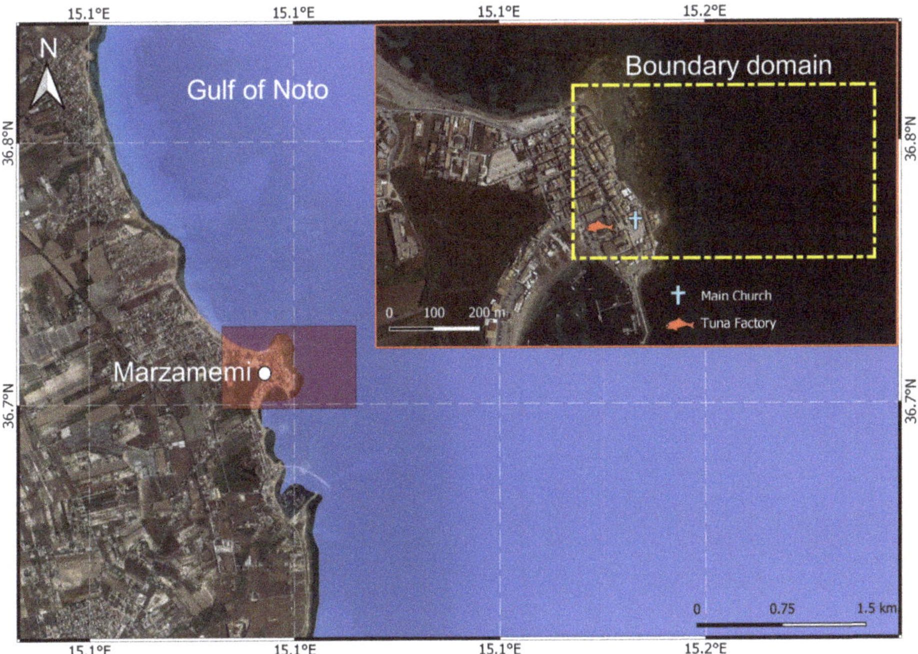

Figure 2. The case study area. The red rectangle shows the Marzamemi promontory, the yellow dash-dot line highlights the boundaries of the numerical model.

2.1. The Numerical Model

The numerical model here adopted is a weakly dispersive Boussinesq type of model ([4,13]), which is a depth integrated model derived from the incompressible continuity and averaged Navier-Stokes Reynolds momentum equations. Generally, a good numerical model for water waves should guarantee a balance between the frequency dispersion and nonlinearity and the Boussinesq type of models are ones of the most suitable. Basically, the governing equations include a non-hydrostatic pressure term in order to better reproduce the frequency dispersion than the classical hydrostatic model (NLSWE). The model dispersive properties were achieved by adding the non-hydrostatic pressure component in the governing equations. In the z momentum equation, both the vertical local and convective acceleration terms were kept. The numerical solver has shock capturing capabilities and easily addresses wetting/drying problems. The governing equations are written in a conservative form, this property guarantees that the models can properly simulate discontinuous flows (e.g., breaking, hydraulic jumps, and bores) [25–27]. In the following are listed the non-hydrostatic depth-integrated continuity and momentum equations:

$$\frac{\partial h}{\partial t} + \frac{\partial(Uh)}{\partial x} + \frac{\partial(Vh)}{\partial y} = 0 \tag{1}$$

$$\frac{\partial(Uh)}{\partial t} + \frac{\partial(U^2h)}{\partial x} + \frac{\partial(UVh)}{\partial y} = -gh\frac{\partial h}{\partial x} - gh\frac{\partial z_b}{\partial x} - \frac{1}{2}\frac{\partial(q_bh)}{\partial x} - q_b\frac{\partial z_b}{\partial x} - \frac{gn^2(Uh)\sqrt{(Uh)^2+(Vh)^2}}{h^{7/3}} \tag{2}$$

$$\frac{\partial(Vh)}{\partial t} + \frac{\partial(UVh)}{\partial x} + \frac{\partial(V^2h)}{\partial y} = -gh\frac{\partial h}{\partial y} - gh\frac{\partial z_b}{\partial y} - \frac{1}{2}\frac{\partial(q_bh)}{\partial y} - q_b\frac{\partial z_b}{\partial y} - \frac{gn^2(Vh)\sqrt{(Uh)^2+(Vh)^2}}{h^{7/3}} \tag{3}$$

$$\frac{\partial(Wh)}{\partial t} + \frac{\partial(UWh)}{\partial x} + \frac{\partial(VWh)}{\partial y} = q_b \tag{4}$$

where U, V and W are the depth-integrated velocity components in the x, y and z coordinates. Uh, Vh and Wh are the specific flow rate components. $q_b = \hat{q}_b/\rho$, $\hat{q}_b$ is the dynamic pressure and n is the Manning coefficient. Indeed, the total pressure was decomposed by means of:

$$p = \rho g(H - z) + \hat{q} \tag{5}$$

Figure 3 shows the definition scheme of the adopted variables, the subscript b refers to bottom.

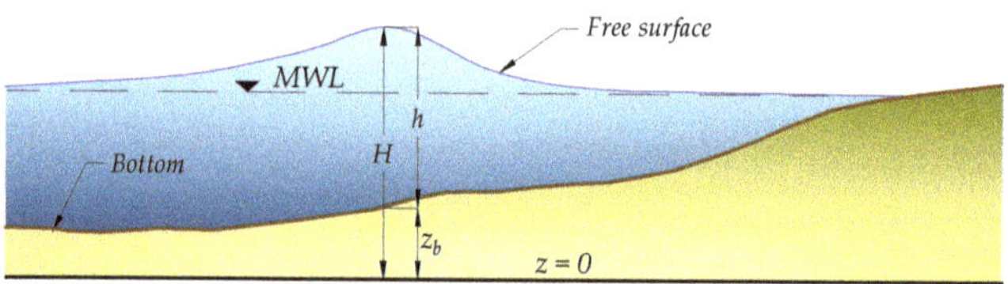

Figure 3. Definition scheme of the adopted variables.

The governing equations Equations (1)–(5) are a system of Partial Differential Equations in the unknown variables h, Uh, Vh, Wh and q_b. The solution of the system of equation was performed using the fractional time step procedure [13]. The governing equations were solved using a fractional time step procedure, where a hydrostatic problem and a non-hydrostatic problem are sequentially solved. The dynamic pressure terms in the momentum equations are neglected when solving the hydrostatic problem and were kept in the non-hydrostatic problem. Furthermore the hydrostatic problem was solved by a prediction-correction scheme, in the corrector step of the hydrostatic problem, a large linear system for the unknown water levels and dynamic pressures is solved.

3. Validation of the Numerical Model

3.1. The Carrier and Greenspan Numerical Solution

In order to validate the numerical model two comparisons with the results of other authors were performed. A test to propagates a sinusoidal wave train incident an inclined plane was performed. Ref Carrier and Greenspan [28] proposed an analytical solution derived by the Airy's approximation of the NLSWE. This analytical solution became a standard test for run-up and run-down modelling. A sinusoidal wave, 0.006 m height and a period of 10 s, was used to force the weakly dispersive Boussinesq model. The wave train propagates in a numerical flume with a water depth of 0.5 m and a slope of 1:25. The envelope of the free surface, computed by the weakly dispersive Boussinesq model at different time steps, was plotted in Figure 4 superimposed with the analytical solution of [28]. Figure 4b shows a magnification of an intermediate time step identified with a black circle in Figure 4a.

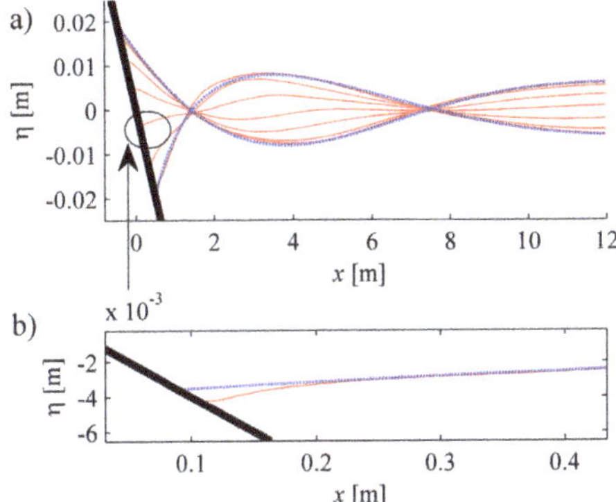

Figure 4. (a) Envelope of free surface of sine wave run-up on a planar beach. Comparison between the weakly dispersive model (blue dotted lines) and Carrier and Greenspan [28] analytical solution (continuous red lines). (b) A zoom of the surface elevation near the planar beach at an intermediate time step.

Figure 5 shows the oscillations of the run-up (R), compared to the analytical solution of Carrier and Greenspan [28], nevertheless, the slight underestimation of the maxima of R the result of the non-hydrostatic weakly dispersive model are very good. The proposed model shoreline horizontal velocity was also compared with the analytical solution Figure 6. The horizontal shoreline velocity is almost the same both in the numerical and in the analytical solution.

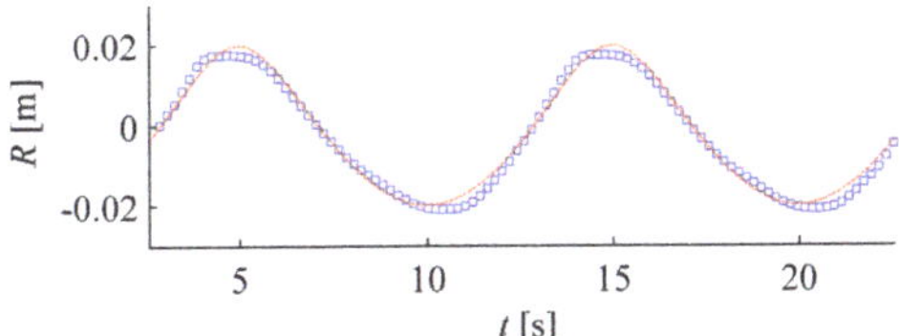

Figure 5. Shoreline vertical motion R of sine wave run-up on a planar beach. Comparison between adopted model (blue dotted lines) and the analytical solution by Carrier and Greenspan [28] (continuous red lines).

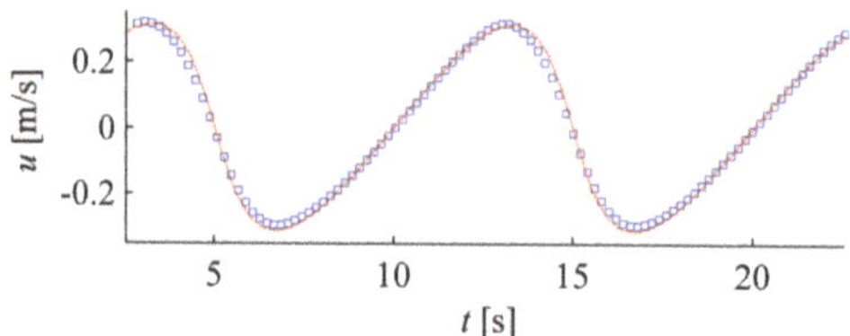

Figure 6. Shoreline velocity of monochromatic wave run-up on a planar beach. Comparison between adopted model (blue dotted lines) and the analytical solution by Carrier and Greenspan [28] (continuous red lines).

3.2. The Fringing Reef Experiment

The second test case was the solitary wave propagation over a reef. The wave transformation over an idealized fringing reef highlights the model's capability in resolving nonlinear dispersive solitary waves, considering also wave breaking. The experiments results used for the comparison were carried out at the O.H. Hinsdale Wave Research Laboratory of Oregon State University where a model of a flat dry reef was used to represent a real fringing reef [5,6]. The numerical model replicated the real flume that was 48.8 m long, 2.16 m wide, and 2.1 m high. The computational grid was simply built with equilateral triangles of edge length equal to 0.08 m. The total number of triangles was 55,550 and the nodes were equal to 28,813 and the time step was $dt = 0.02$ s. Figure 7 shows the numerical domain and the surface elevation at $t^* = t \cdot \sqrt{g/h_0} = 55.1$ and the red line shows a local zoom of the triangular mesh.

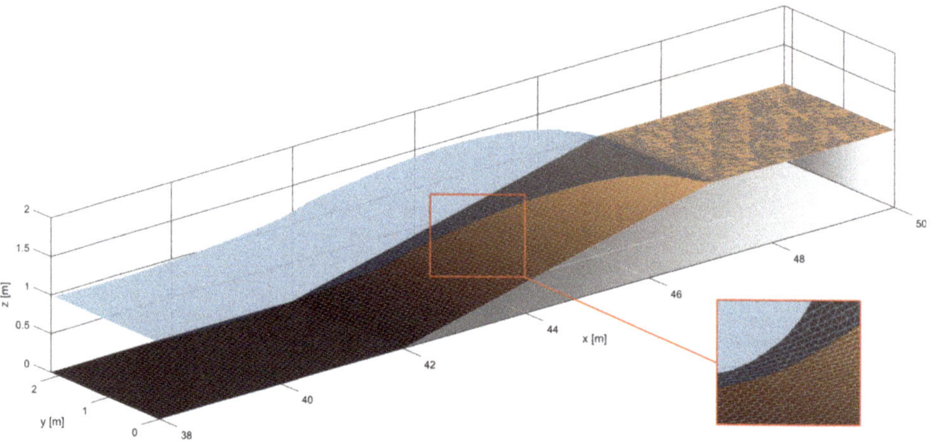

Figure 7. The numerical domain and the triangular equilateral mesh used in the flat reef run-up test. The red box highit a magnified area of the mesh. The surface elevation correspond to $t^* = 55.1$.

A solitary wave with a dimensionaless wave height of $H/h_0 = 0.5$ was generated at the inlet of numerical flume and a Manning coefficient $n = 0.012$ was adopted to reproduce the roughness of the bottom of the flume. In Figure 8 are shown the model results compared to the measured data [5] at 13 dimensionaless time steps $t^* = t \cdot \sqrt{g/h_0}$.

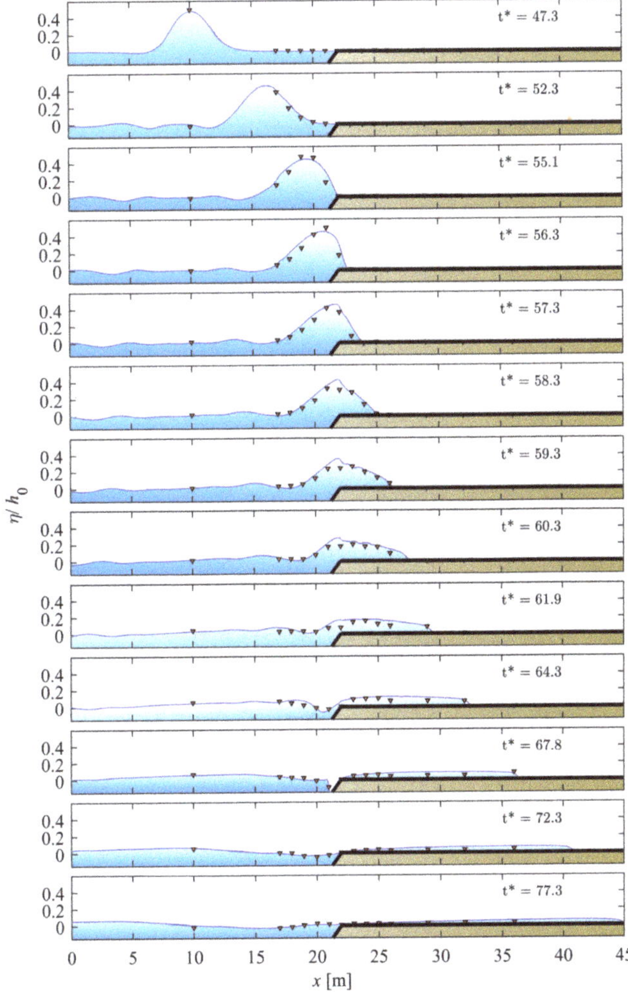

Figure 8. Surface elevations of solitary wave over a flat reef with $H/h_0 = 0.5$ and 1:5 slope. Solid blue lines are the weakly dispersive model results and the red triangles are measured data. Each subplot shows the results at a dimensionaless time step.

The solitary wave becomes steeper as it propagates over the slope and it rises over the coral reef shoals without breaking into a typical plunge.

In the dimensionless time step $t^* = 64.3$, the numerical model shows a draw-down in front of the flat reef, that produces a back-reflected wave (Figure 8). Instead, the high-speed water sheet over the reef quickly runs up. The numerical model accurately replicated the physics of the phenomenon at each time step (Figure 8). Moreover, the numerical model well reproduces the solitary wave propagation over the edge reef as shown in Figure 9. The time series refers to a point far 22 m from the wavemaker as described in [29], once more the agreement with the numerical model is good.

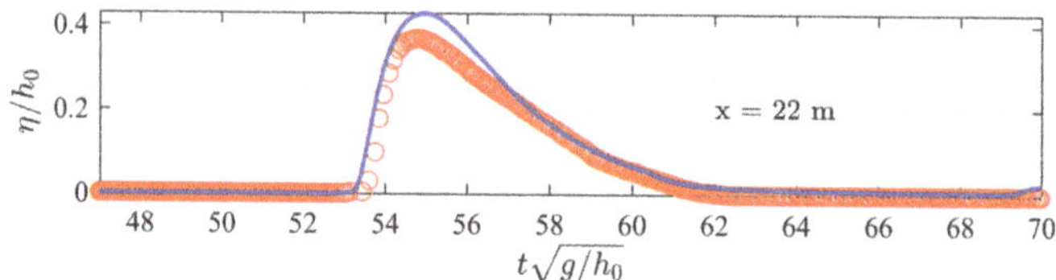

Figure 9. Time series of the surface elevation at edge of the reef. The blue line is the numerical model results the red circles are the measurements by [29].

4. Results and Discussions of a Real Case of Tsunami Propagation

In order to make an inundation map of a small interest area a numerical model was used in the case study of Marzamemi (see Section 2).

A triangular mesh was built using the code proposed by Engwirda [30]. The unstructured mesh has the following characteristics: 11,714 triangles, 5990 nodes (see Figure 10a). The triangles size was determined by means of a density function related to bathymetry, making larger triangles in deeper waters.

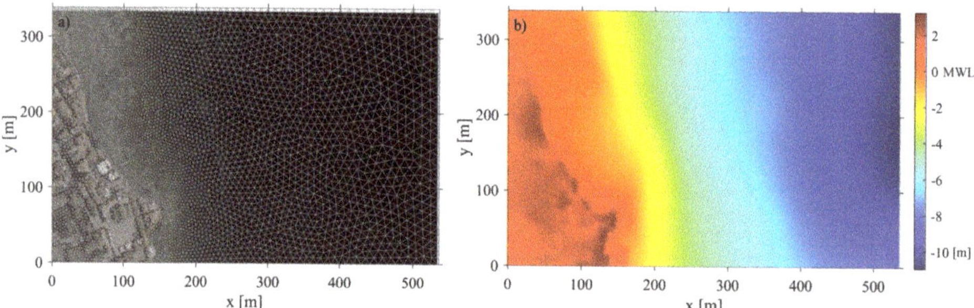

Figure 10. The Marzamemi numerical domain. (**a**) The used triangular mesh, (**b**) the bathymethry of the studied area, the color bar shows the elevation in meters above mean water level. In the subplots the coordinate origin are E = 510,569 m; N = 4,066,207 m; WGS84-UTM33N reference system.

The bathymetry of the model was obtained from regional digital bathymetric charts whereas digital elevation model was built by the regional topographic maps at 1:10,000 scale (see Figure 10b). The two lateral sides of the domain were walls with a free slip condition, in the bottom was calculated a friction term using the Manning's roughness coefficient equal to $0.012 \ s/m^{1/3}$ see Equations (2) and (3). The adopted time step was $\Delta t = 0.2$ s and the incoming tsunami wave was simulated as 2 m height solitary wave. The propagation of a tsunami wave and the subsequent flooding areas are shown in Figure 11 in which each subplot shows the surface elevation (above Mean Water Level (MWL)) in a specific time after the start of the simulation. At the initial time step a solitary wave, 2 m height, was generated in the eastern domain side, about 500 m far from the coastline. This wave is linked to a possible earthquake with a return period of about 2000 years [31], neglecting a statistical study regarding catastrophic events. The earthquake-induced tsunami is related to a hypocentral point near to seafloor and the fault mechanism is reverse. In particular, it was taken into account, as a potential tsunami source, an earthquake about 200 km far from the coast. It is important to point out that in this manuscript we are presenting a preliminary study. Analysis of tsunami propagation in the Ionian sea area including statistical studies about earthquake-induced tsunami hazard are ongoing.

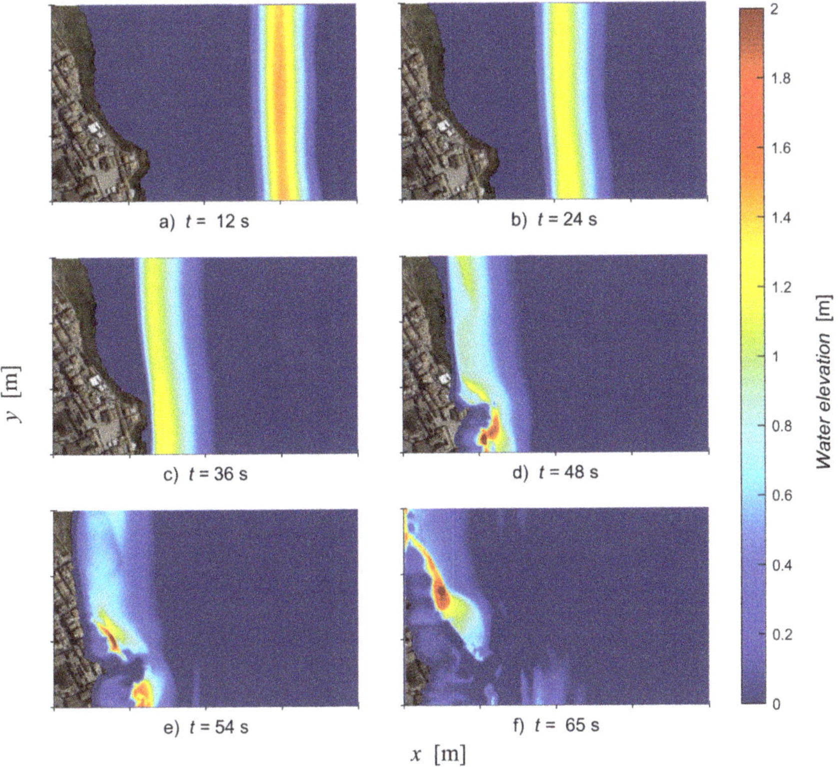

Figure 11. Water surface elevation. The origin of coordinate axis is E = 510,569 m; N = 4,066,207 m; WGS84 UTM33N reference system.

Figure 11 reports the water surface elevation at several time steps. The water elevation was calculated regarding the initial condition (still water level), thus it coincides with the water depth where the points originally were dry. In the subplot (a) of Figure 11, the tsunami propagation at time $t = 12$ s is shown. In this time the wave is about 300 m far from the coastline and its shape begins to change due to the frequency dispersion. Simultaneously, the wavefront starts to rotate as a result of the change of bathymetry, at the $t = 24$ s, in Figure 11b, it is clearly distinguishable the refraction process. At $t = 36$ s Figure 11c, the wave reaches the coast near the most exposed stretch in which there is the main square of the village of Marzamemi. In Figure 11d, $t = 48$ s, it is described the shoaling of the tsunami and the initial flooding inside the main streets of the village reaching the wave an elevation of 2 m. In this time the northern part of the village is not yet flooded. In the next subplot, (e) $t = 54$ s, a complete inundation of the main square and the littoral promenade is shown although with small water depth (20–40 cm). At same time step, it is possible to see the wave breaking over the seawall that should protect the habitations and the road infrastructure. The last time step, (f) shows the flooding of the whole studied coastal area. A magnification of the last time step of the simulation is shown in Figure 12 with the x and y coordinates in the local reference system and the water height measured above the MWL. At $t = 65$ s, the church and the ancient tuna factory, XV century, were flooded.

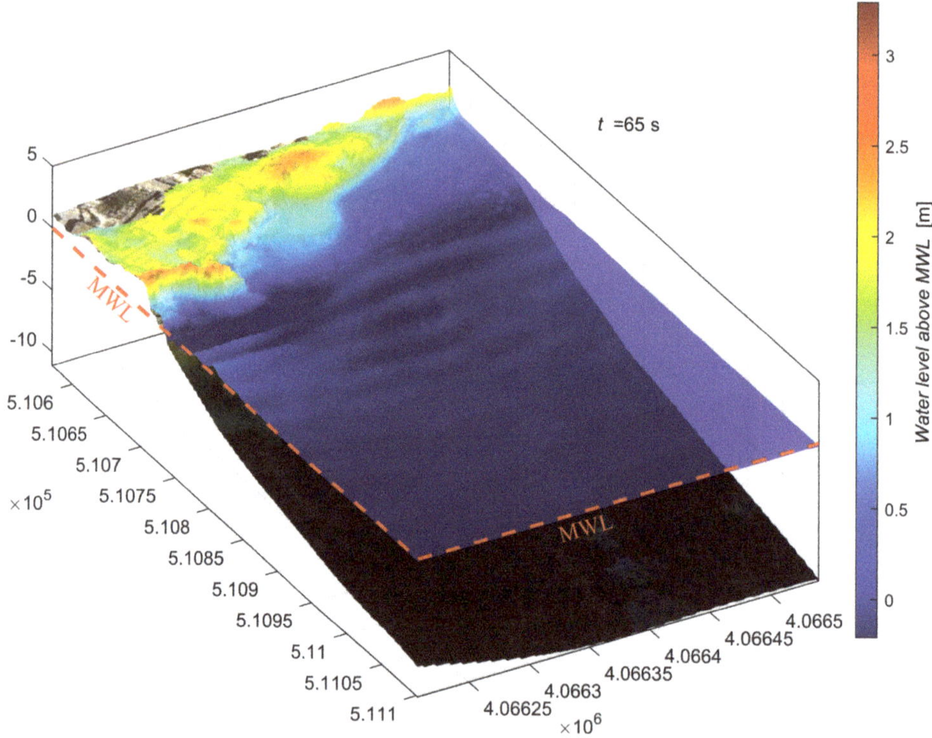

Figure 12. Magnification of subplot (f) of Figure 11. The red lines shows the MWL the color bar shows the water level above the MWL.

5. Concluding Remarks

In this paper, preliminary studies regarding the tsunami flooding hazards were presented. The tsunami is an highly nonlinear and dispersive wave and it must be modeled using appropriated numerical model. Indeed, the adopted Boussinesq type of model, in its category of depth integrated, guarantees a balance between the frequency dispersion and non-linearity. Moreover, it was possible, by means of the Delaunay mesh grid, to have very detailed results with a minor computational cost. As a consequence of this, the model better assess the coastal flooding hazard. The numerical model was validated through the analytical solution of Carrier and Greenspan [28] and with the experimental study presented by Roeber et al. [5]. The comparisons with the numerical model show excellent agreements. Finally, it was applied the numerical modeling procedure to a real case in order to perform the propagation of the tsunami and to evaluate its impact on the coast and the subsequent coastal flooding. To assess both the possibility of coastal flooding and their extent, an earthquake, that cause a tsunami, was generated off the Ionian coast (return period c.a. 2000 years).The results of the tsunami propagation show that the extent of the flooded areas is about 100 m inward the shoreline. All roads near the shoreline, including the historic village square, are flooded. The water levels, although not extremely high, could cause dangers mainly in the periods of the year when the population density grows considerably. These results, although preliminary, highlight the extreme fragility of this coastal site. For this reason, further numerical modeling is ongoing, taking into account also the structural response of the buildings inside the village of Marzamemi. This preliminary investigation is the basis for further studies that are already underway, their results will be useful for civil protection agencies in order to project emergency management plans.

Author Contributions: Data curation, C.L.R. and G.M.; Formal analysis, C.L.R. and G.M.; Methodology, C.L.R., G.M., and G.C.; Software, C.L.R. and G.M.; Supervision, G.C.; Geological and geomorphological supervision, G.M.; Hydarulic modelling C.L.R.; Writing—original draft, C.L.R., G.M. and G.C.; Writing—review & editing, C.L.R., G.M. and G.C. All authors have read and agreed to the published version of the manuscript.

Funding: This research was funded by SIMIT THARSY Tsunami Hazard Reduction System C1-3.2-5-INTERREG V-A Italia-Malta.The APC was funded by SIMIT THARSY.

Acknowledgments: We thank our scientific responsible, Goffredo La Loggia, for assistance to tasks development during the project SIMIT-THARSY.

Conflicts of Interest: The authors declare no conflict of interest.

References

1. Molina, R.; Manno, G.; Lo Re, C.; Anfuso, G.; Ciraolo, G. Storm Energy Flux Characterization along the Mediterranean Coast of Andalusia (Spain). *Water* **2019**, *11*, 509. [CrossRef]

2. Molina, R.; Manno, G.; Lo Re, C.; Anfuso, G.; Ciraolo, G. A Methodological Approach to Determine Sound Response Modalities to Coastal Erosion Processes in Mediterranean Andalusia (Spain). *J. Mar. Sci. Eng.* **2020**, *8*, 154. [CrossRef]

3. Manno, G.; Lo Re, C.; Ciraolo, G. Uncertainties in shoreline position analysis: the role of run-up and tide in a gentle slope beach. *Ocean Sci.* **2017**, *13*, 661–671. [CrossRef]

4. Lo Re, C.; Musumeci, R.E.; Foti, E. A shoreline boundary condition for a highly nonlinear Boussinesq model for breaking waves. *Coast. Eng.* **2012**, *60*, 41–52. [CrossRef]

5. Roeber, V.; Cheung, K.F.; Kobayashi, M.H. Shock-capturing Boussinesq-type model for nearshore wave processes. *Coast. Eng.* **2010**, *57*, 407–423. [CrossRef]

6. Roeber, V.; Cheung, K.F. Boussinesq-type model for energetic breaking waves in fringing reef environments. *Coast. Eng.* **2012**, *70*, 1–20. [CrossRef]

7. Zhang, H.; Zhang, M.; Ji, Y.; Wang, Y.; Xu, T. Numerical study of tsunami wave run-up and land inundation on coastal vegetated beaches. *Comput. Geosci.* **2019**, *132*, 9–22. [CrossRef]

8. Marivela, R.; Weiss, R.; Synolakis, C. The Temporal and Spatial Evolution of Momentum, Kinetic Energy and Force in Tsunami Waves during Breaking and Inundation. *arXiv* **2016**, arXiv:1611.04514.

9. Manoj Kumar, G.; Sriram, V.; Didenkulova, I. A hybrid numerical model based on FNPT-NS for the estimation of long wave run-up. *Ocean Eng.* **2020**, *202*, 107181. [CrossRef]

10. Wei, Z.; Dalrymple, R.A.; Rustico, E.; Hérault, A.; Bilotta, G. Simulation of Nearshore Tsunami Breaking by Smoothed Particle Hydrodynamics Method. *J. Waterway Port Coast. Ocean Eng.* **2016**, *142*, 05016001. [CrossRef]

11. Qin, X.; Motley, M.; LeVeque, R.; Gonzalez, F.; Mueller, K. A comparison of a two-dimensional depth-averaged flow model and a three-dimensional RANS model for predicting tsunami inundation and fluid forces. *Nat. Hazards Earth Syst. Sci.* **2018**, *18*, 2489–2506. [CrossRef]

12. Qu, K.; Ren, X.; Kraatz, S. Numerical investigation of tsunami-like wave hydrodynamic characteristics and its comparison with solitary wave. *Appl. Ocean Res.* **2017**, *63*, 36–48. [CrossRef]

13. Arico, C.; Lo Re, C. A non-hydrostatic pressure distribution solver for the nonlinear shallow water equations over irregular topography. *Adv. Water Resour.* **2016**, *98*, 47–69. [CrossRef]

14. Presti, V.L.; Antonioli, F.; Auriemma, R.; Ronchitelli, A.; Scicchitano, G.; Spampinato, C.; Anzidei, M.; Agizza, S.; Benini, A.; Ferranti, L.; et al. Millstone coastal quarries of the Mediterranean: A new class of sea level indicator. *Quat. Int.* **2014**, *332*, 126–142. [CrossRef]

15. Samaras, A.; Karambas, T.V.; Archetti, R. Simulation of tsunami generation, propagation and coastal inundation in the Eastern Mediterranean. *Ocean Sci.* **2015**, *11*, 643–655. [CrossRef]

16. Schambach, L.; Grilli, S.T.; Kirby, J.T.; Shi, F. Landslide Tsunami Hazard Along the Upper US East Coast: Effects of Slide Deformation, Bottom Friction, and Frequency Dispersion. *Pure Appl. Geophys.* **2019**, *176*, 3059–3098. [CrossRef]

17. Mueller, C.; Micallef, A.; Spatola, D.; Wang, X. The Tsunami Inundation Hazard of the Maltese Islands (Central Mediterranean Sea): A Submarine Landslide and Earthquake Tsunami Scenario Study. *Pure Appl. Geophys.* **2020**, *177* 1617–1638. [CrossRef]

18. Suardana, A.M.A.P.; Sugianto, D.N.; Helmi, M. Study of Characteristics and the Coverage of Tsunami Wave Using 2D Numerical Modeling in the South Coast of Bali, Indonesia. *Int. J. Oceans Oceanogr.* **2019**, *13*, 237–250.

19. Papadopoulos, G.A.; Fokaefs, A. Strong tsunamis in the Mediterranean Sea: A re-evaluation. *ISET J. Earthq. Technol.* **2005**, *42*, 159–170.

20. DISS Working Group. Database of Individual Seismogenic Sources (DISS). In *A Compilation of Potential Sources for Earthquakes Larger than M 5.5 in Italy and Surrounding Areas*; Istituto Nazionale di Geofisica e Vulcanologia: Rome, Italy, 2018.[CrossRef]

21. Caputo, R.; Pavlides, S. *The Greek Database of Seismogenic Sources (GreDaSS), version 2.0.0: A Compilation of Potential Seismogenic Sources (Mw > 5.5) in the Aegean Region*; University of Ferrara: Ferrara, Italy, 2013. [CrossRef]

22. Gutscher, M.A.; Roger, J.; Baptista, M.A.; Miranda, J.M.; Tinti, S. Source of the 1693 Catania earthquake and tsunami (southern Italy): New evidence from tsunami modeling of a locked subduction fault plane. *Geophys. Res. Lett.* **2006**, *33*, L08309. [CrossRef]

23. Catalano, R.; Doglioni, C.; Merlini, S. On the mesozoic Ionian basin. *Geophys. J. Int.* **2001**, *144*, 49–64. [CrossRef]

24. Papadopoulos, G.A.; Daskalaki, E.; Fokaefs, A.; Giraleas, N. Tsunami hazard in the Eastern Mediterranean sea: Strong earthquakes and tsunamis in the west Hellenic arc and trench system. *J. Earthq. Tsunami* **2010**, *04*, 145–179. [CrossRef]

25. LeVeque, R.J. *Numerical Methods for Conservation Laws*; Birkhäuser Boston: Basel, Switzerland, 2013.

26. Toro, E.F. *Riemann Solvers and Numerical Methods for Fluid Dynamics: A Practical Introduction*; Springer Science & Business Media: New York, NY, USA, 2013.

27. Stelling, G.S.; Duinmeijer, S.P.A. A staggered conservative scheme for every Froude number in rapidly varied shallow water flows. *Int. J. Numer. Methods Fluids* **2003**, *43*, 1329–1354. [CrossRef]

28. Carrier, G.; Greenspan, H. Water waves of finite amplitude on a sloping beach. *J. Fluid Mech.* **1958**, *4*, 97–109. [CrossRef]

29. Roeber, V. Boussinesq-Type Model for Nearshore Wave Processes in Fringing Reef Environment. Ph.D. Thesis, University of Hawaii at Manoa, Honolulu, HI, USA, December 2010.

30. Engwirda, D. Locally-Optimal Delaunay-Refinement and Optimisation-Based Mesh Generation. Ph.D. Thesis, The University of Sydney, Sydney, Australia, 2014.

31. Basili, R.; Brizuela, B.; Herrero, A.; Iqbal, S.; Lorito, S.; Maesano, F.E.; Murphy, S.; Perfetti, P.; Romano, F.; Scala, A.; et al. *NEAM Tsunami Hazard Model 2018 (NEAMTHM18): Online Data of the Probabilistic Tsunami Hazard Model for the NEAM Region from the TSUMAPS-NEAM Project*; Istituto Nazionale di Geofisica e Vulcanologia (INGV): Roma, Italy, 2018. [CrossRef]

 *water*

Technical Note

Influence of Different Sieving Methods on Estimation of Sand Size Parameters

Patricio Poullet *, Juan J. Muñoz-Perez *, Gerard Poortvliet, Javier Mera, Antonio Contreras and Patricia Lopez

CASEM (Centro Andaluz Superior de Estudios Marítimos), Universidad de Cádiz, 11510 Cádiz, Spain; poor0023@hz.nl (G.P.); javier.merabaston@hotmail.com (J.M.); antonio.contreras@uca.es (A.C.); patricia.lopezgarcia@uca.es (P.L.)
* Correspondence: ppoullet@miteco.es (P.P.); juanjose.munoz@uca.es (J.J.M.-P.); Tel.: +34-956-546356 (J.J.M.-P.)

Received: 27 March 2019; Accepted: 23 April 2019; Published: 26 April 2019

Abstract: Sieving is one of the most used operational methods to determine sand size parameters which are essential to analyze coastal dynamics. However, the influence of hand versus mechanical shaking methods has not yet been studied. Herein, samples were taken from inside the hopper of a trailing suction dredger and sieved by hand with sieves of 10 and 20 cm diameters on board the dredger. Afterwards, these same samples were sieved with a mechanical shaker in the laboratory on land. The results showed differences for the main size parameters D_{50}, standard deviation, skewness, and kurtosis. Amongst the main results, it should be noted that the highest values for D_{50} and kurtosis were given by the small sieves method. On the other hand, the lowest values were given by the mechanical shaker method in the laboratory. Furthermore, standard deviation and skewness did not seem to be affected by the sieving method which means that all the grainsize distribution was shifted but the shape remained unchanged. The few samples that do not follow these patterns have a higher percentage of shells. Finally and definitely, the small sieves should be rejected as a sieving method aboard.

Keywords: D_{50}; sieving; sand size; sand parameters; coastal dynamics

---

1. Introduction

Sand-size parameters are essential to study the coastal dynamics and other geomorphological behaviours of beaches [1–3]. Main parameters must include measure of: average grain size (D_{50}), spread of the size around the average (Standard Deviation, σ, or Sorting), degree of asymmetry (Skewness) and degree of peakedness (Kurtosis). Parameters such as D_{50} and sorting (σ) are necessary for calculating equilibrium profiles or estimating sediment transport [4,5], to check if the borrowed sand is suitable to substitute the native sand eroded from the beach [6,7] or to calculate the required amount of sand for beach nourishment [8]. Analysis of sand-size distribution also gives essential hints to the origin, depositional environment and movement history [9]. Moreover, sand-size analysis is an indispensable mechanism to subdivide facies and environments [10] and it also makes possible to see how a beach reacts to storms [11]. Different methods to analyse sand-size particles and some comparisons between them were studied by some researchers, such as; laser diffraction [12], laser granulometer and sedigraph [13], microtac [14], comparison of laser grain size with pipette and sieve [15], image analysis [16] and sieving [17]. Due to the simplicity use and economy, the sieving method was chosen to be the method to perform the analysis of sand-size inside a dredger because of the special circumstances on board a ship.

In particular, in order to make decisions about the landfill area, the coastal manager needs to know the size parameters of the dredged sand before it gets dumped onto the beach. This means that the analysis must be done on board the dredger. Therefore, due to the usual shortage of space inside the dredger and the consequent absence of an adequate laboratory, sieving is usually done by hand. However different sieving methods can influence not only the D_{50} but also all other parameters.

Thus, due to the already established importance of sand size determination, the aim of this paper is to find out the influence of using different sieving methods on the results obtained for sand size parameters. The sieving methods were by hand (shaking manually) with 10-cm diameter sieves versus 20-cm sieves, and mechanical sieving (shaking machine) with 20-cm sieves.

2. Area of Study

The samples were taken from inside the hopper of a trailing suction dredger used for two beach nourishments (Costa Ballena and Punta Candor). Sand had been borrowed from the Meca sandbank, located in the Gulf of Cadiz close to the Strait of Gibraltar (Figure 1a). It has a depth of approximately 15–20 m and contains up to 25 millions of m^3 of sand that can be used for beach nourishments (further data can be found in reference [18]). A previous study demonstrated that there was no serious impact to the environment of the area as a result of the sand removal [19].

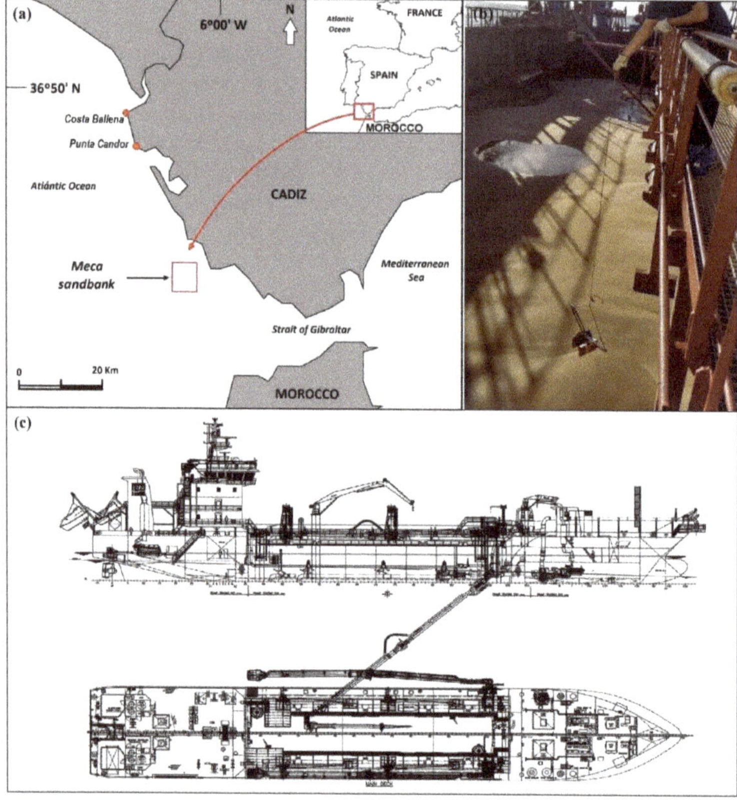

Figure 1. (**a**) Location of Costa Ballena and Punta Candor nourished beaches and the borrow site named "Placer de Meca" (Meca sandbank) (SW Spain); (**b**) sampling method aboard the dredger; (**c**) scheme of Njord dredger (https://rohde-nielsen.com).

Moreover, the effects of the turbidity are also negligible due to the limited percentage of fines [10]. Tidal range is mesotidal, varying between 1.10 m and 3.22 m, and has a semidiurnal periodicity. The beaches considered in this study were composed of fine-medium sand, very similar to the borrowed sediment. The average D_{50} is about 0.25 mm, consisting of 90% quartz and 10% calcium carbonate [20].

3. Materials and Methods

Nineteen samples were taken (Figure 1b) from the hopper of the Njord dredger (Figure 1c) on different days once the dredging operations were finished (for security measures). The same sample was sieved on board, by hand, with two different kinds of sieves—small (10 cm diameter) and large (20 cm diameter)—with the same sieving time (10 min in all cases) and saved. Afterwards, the samples were sieved at the laboratory inland with a mechanical shaker and the big sieves. The last method of sieving, the results of which will be taken as a reference, cannot always be performed on the dredger due to several reasons. First of all, a stable energy supply is not easy to get onto the dredger. Moreover, the machine performing the sieving has to deal with the instability and the vertical acceleration induced by waves.

The mesh sizes of the eight sieves required by the Spanish Coastal Administration were: 2 mm, 1 mm, 710 µm, 500 µm, 355 µm, 250 µm, 125 µm and 62.5 µm. Thus, the same sample, about 100 g, was sieved by the three different methods. The amount of sediment is important because, obviously, a large volume of sediment means a lower chance for the grain to pass through the net.

After the initial weighting, the samples were put in the upper sieve and the sieves were shaken by hand. It should be noted that finger pressing was not allowed. After weighting the sand accumulated in each sieve, required values (D_{16}, D_{50} and D_{84}) were obtained (Table 1) as well as the other main parameters [21] as sorting, skewness and kurtosis (though these last two parameters are not used for beach nourishment projects) (Figure 2). They were calculated by using the corresponding equations (see Table 2, [21,22]) which are based on the phi unit scale (Equation (1)), and where to convert from phi units to millimetres, the inverse equation (Equation (2)) is used:

$$\varphi = -\log_2 D \tag{1}$$

$$D = 2^{-\varphi} \tag{2}$$

Table 1. Values of the parameters D_{16}, D_{50} and D_{84}, obtained by three different methods, for the 19 samples taken from the dredger hopper.

| | Manual Shaking | | | | | | Mechanical Shaking | | |
| | Small Sieves | | | Large Sieves | | | Laboratory | | |
Sample	D_{16}	D_{50}	D_{84}	D_{16}	D_{50}	D_{84}	D_{16}	D_{50}	D_{84}
1	0.217	0.300	0.379	0.201	0.291	0.35	0.177	0.274	0.346
2	0.218	0.300	0.375	0.203	0.297	0.384	0.143	0.281	0.347
3	0.261	0.314	0.418	0.229	0.307	0.422	0.149	0.296	0.383
4	0.220	0.301	0.383	0.179	0.278	0.349	0.138	0.256	0.343
5	0.194	0.290	0.354	0.174	0.269	0.339	0.139	0.263	0.335
6	0.204	0.297	0.385	0.192	0.29	0.377	0.178	0.277	0.35
7	0.179	0.280	0.358	0.171	0.267	0.349	0.171	0.265	0.346
8	0.177	0.280	0.407	0.168	0.261	0.38	0.177	0.277	0.354
9	0.186	0.284	0.351	0.182	0.278	0.344	0.17	0.261	0.335
10	0.193	0.301	0.539	0.177	0.285	0.571	0.141	0.281	0.549
11	0.257	0.314	0.440	0.206	0.298	0.409	0.164	0.307	0.417
12	0.220	0.295	0.349	0.179	0.278	0.348	0.14	0.272	0.346
13	0.255	0.315	0.497	0.205	0.305	0.499	0.151	0.306	0.491
14	0.181	0.284	0.385	0.211	0.297	0.363	0.187	0.286	0.354
15	0.199	0.292	0.354	0.187	0.28	0.339	0.25	0.307	0.417
16	0.182	0.280	0.350	0.163	0.244	0.326	0.137	0.247	0.327
17	0.186	0.288	0.395	0.165	0.251	0.339	0.16	0.235	0.337
18	0.240	0.306	0.448	0.24	0.306	0.448	0.239	0.306	0.45
19	0.226	0.303	0.394	0.141	0.272	0.342	0.17	0.263	0.339

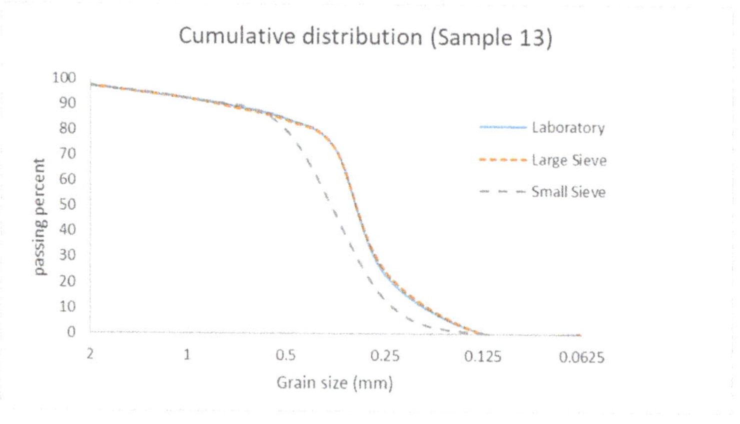

Figure 2. Cumulative distribution of the passing percentage of samples 13 and 17 for the three sieving methods.

Table 2. Formulae used for calculation of the main granulometric parameters, according to Folk and Ward graphical measures [21,22].

Mean	Standard Deviation or Sorting	Skewness	Kurtosis
$M_\varphi = \frac{\varphi_{16}+\varphi_{50}+\varphi_{84}}{3}$	$\sigma_\varphi = \frac{\varphi_{84}-\varphi_{16}}{4} + \frac{\varphi_{95}-\varphi_5}{6}$	$Sk_\varphi = \frac{\varphi_{16}+\varphi_{84}-2\varphi_{50}}{2(\varphi_{84}-\varphi_{16})} + \frac{\varphi_5+\varphi_{95}-2\varphi_{50}}{2(\varphi_{95}-\varphi_5)}$	$K_\varphi = \frac{\varphi_{95}-\varphi_5}{2.44(\varphi_{75}-\varphi_{25})}$

4. Results and Discussion

4.1. D_{50} (Median Grain Diameter)

A comparison of the cumulative distribution of the passing percentage for the three methods is shown in Figure 2. Only two samples (13 and 17) have been presented to show an example of the results. Though the 19 samples are not shown, this figure gives a real image of the differences among the methods.

On the other hand, Figure 3a shows how D_{50} ranges from approximately 0.24 mm to 0.32 mm. The values obtained with the automatic sieving, carried out in the laboratory, are always the lowest number. On the other side, the values obtained by hand with the small sieves are always the biggest. Thus, the value of the large sieve is usually between the small sieving and the automatic shaking, values being closer to the latter. A possible explanation is that the surface in the 20 cm sieves is four times bigger than the 10 cm sieves. More holes for the grains to pass through mean more grains passing during the same sieving time (10 min in all cases). This increases the amount of sand in the lower sieves, and thus decreases the D_{50}.

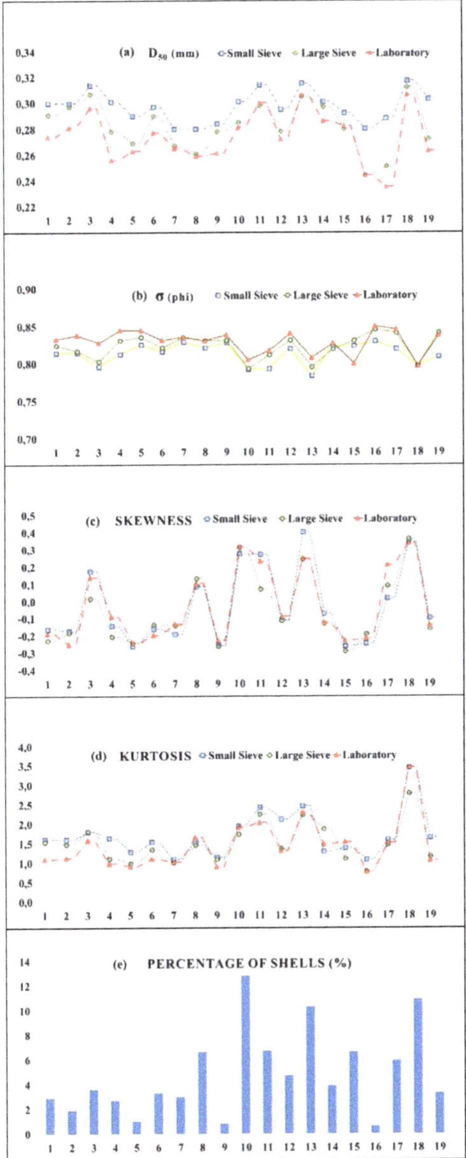

Figure 3. Values of the granulometric parameters obtained with three different types of sieving for the samples taken on board of the dredger: (**a**) D_{50}; (**b**) sorting; (**c**) skewness; (**d**) kurtosis; (**e**) percentage of shells.

Further on, we consider as reference those values obtained in the laboratory by the mechanical shaker since it is the standard procedure. Probably, the higher and constant intensity of the mechanical shaker in comparison to the manual sieving is the reason why the values of D_{50} of the former are smaller than the latter. Furthermore, though the sieving time did remain constant in all cases, the operator on board may get tired during the shaking of the sieves, which would influence the intensity of the sieving. Though the average difference was about 10%, it can be seen how the highest differences between laboratory D_{50} and small sieve D_{50}, which coincide with the finest sands (samples 4, 16, 17), can be as high as 20–25%. The manual and consequently less intensity sieving, plus the small diameter of the small sieve, could be the cause of the accumulation of fine sands in the upper sieves.

4.2. Sorting or Standard Deviation

Looking at the sorting graph in Figure 3b, the results for all the sieving methods were really close to each other, values ranging from 0.79 to 0.84 phi and averaging 0.82. This information shows us that the method of sieving did not affect the sorting. Moreover, according to Folk [23], this sediment is moderately sorted, being aware that the limit for poor sorted sediment (1 phi) is not close.

4.3. Skewness

In general, the same pattern for all three sieving methods can be obtained (Figure 3c). The values of skewness for every individual sample, as in the case of the sorting, did not seem to depend on the sieving method (this means that all the grainsize distribution is shifted but the shape did not change). Most of them were between near-symmetrical (−0.1 to +0.1) and coarse-skewed (−0.3 to +0.3). Again, samples 10, 13 and 18 were the only ones that come out of the pattern, with values larger than 0.3 (very coarse-skewed). Thus, a high percentage of shells may also increase the skewness.

4.4. Kurtosis

Looking at Figure 3d, no platykurtic coefficient was observed at all. All the values ranged from mesokurtic (0.90–1.10) to leptokurtic (<1.50) and very leptokurtic (>1.50). No clear pattern could be determined.

4.5. Percentage of Shells

The amount of shells per individual sample was determined with the help of a microscope and is shown in Figure 3e. Since the particles of shells are bigger than the particles of sand, the biggest amount of shells were found in the upper sieves. The three samples (10, 13 and 18) with the highest value of shells (above 10%) were also the samples with the highest standard deviation.

4.6. Summary Table

Percent Relative Error (%) has been calculated for the main sediment parameters (mean size, sorting, skewness and kurtosis) and results are shown in Table 3. Laboratory sieve results (standard procedure) were taken as a reference to compare the results. It could be verified that the error was greater in all cases for the small sieves as discussed previously. Skewness was the only parameter without a significant difference. Thus, we can assume that the large sieve is the best way to sieve aboard and small sieves should be avoided.

Table 3. Percent Relative Error (%) calculated for granulometric parameters obtained with small and large sieves taking as a reference the laboratory sieve results (the standard procedure). σ: Standard deviation.

		D_{50}	σ	Skewness	Kurtosis
Percent Relative Error (%)	Small Sieve	8.8	2.4	28.2	26.5
	Large Sieve	3.2	1.3	29.3	14.6

5. Conclusions

The importance of the D_{50} (as well as sorting, skewness and kurtosis) in understanding the coastal dynamics and other geomorphological behaviours of beaches is well known. And, nevertheless, the influence of the method of sieving had not yet been considered in the determination of these parameters. For this reason, 19 samples were taken on board while dredging the borrow site named Placer de Meca (Meca sandbank) for a beach nourishment in the Gulf of Cadiz (SW of Spain). They were analysed with three different sieving methods (small vs. big sieves and mechanical shaker vs. manual shaking procedure).

The results showed a pattern for the D_{50} value. The biggest values were always obtained with the small sieves. The values with the machine in the laboratory always gave the lowest values.

Variance of sorting values was negligible. Thus the method of sieving does not influence the sorting and, by the way, nor the skewness either. This fact means that all the grainsize distribution was shifted but the shape remained unchanged.

The kurtosis gives a similar pattern as the D_{50} parameter: the values for the small sieves were the highest. The bigger sieves gave lower values whereas the sieving with the automatic shaker gave the lowest values in general.

The laboratory method is the standard procedure, the results of which were taken as a reference. But this sieving method cannot always be performed on the dredger. Therefore, looking at the results, we can assume that a large sieve is the best way to manually sieve aboard. Definitely, a small sieve should be rejected as a sieving method during dredging.

Author Contributions: Investigation and Conceptualization, P.P., J.J.M.-P., J.M., A.C., P.L.; Data collection, P.P., J.M., G.P., P.L.; methodology, P.P., J.J.M.-P., J.M., G.P.; writing, P.P., J.J.M.-P., G.P., J.M., A.C., P.L.; Review and editing, P.P., J.J.M.-P., A.C., P.L.

Funding: This research received no external funding.

Acknowledgments: The authors acknowledge the valuable comments and suggestions given by Enzo Pranzini and two anonymous reviewers, which greatly improved the manuscript.

Conflicts of Interest: The authors declare no conflict of interest.

References

1. Guillen, J.; Hoekstra, P. The "equilibrium" distribution of grain size fractions and its implications for cross-shore sediment transport: A conceptual model. *Mar. Geol.* **1996**, *135*, 15–33. [CrossRef]
2. Medellin, G.; Torres-Freyermuth, A.; Tomasicchio, G.R.; Francone, A.; Tereszkiewicz, P.A.; Lusito, L.; Palemon-Arcos, L.; Lopez, J. Field and numerical study of resistance and resilience on a sea breeze dominated beach in Yucatan (Mexico). *Water* **2018**, *10*, 1806. [CrossRef]
3. Pranzini, E.; Anfuso, G.; Cinelli, I.; Piccardi, M.; Vitale, G. Shore protection structures increase and evolution on the Northern Tuscany Coast (Italy): Influence of tourism industry. *Water* **2018**, *10*, 1647. [CrossRef]
4. Gravens, M.B.; Emersole, B.A.; Walton, T.L.; Wise, R.A. Beach Fill Design. In *Coastal Engineering Manual*; Warp, D.L., Ed.; U.S. Army Corps of Engineers: Washington, DC, USA, 2002.
5. Giardino, A.; Diamantidou, E.; Pearson, S.; Santinelli, G.; den Heijer, K. A regional application of bayesian modeling for coastal erosion and sand nourishment management. *Water* **2019**, *11*, 61. [CrossRef]
6. Pranzini, E.; Anfuso, G.; Muñoz-Perez, J.J. A probalistic approach to borrow sediment selection in beach nourishment projects. *Coastal Eng.* **2018**, *139*, 32–35. [CrossRef]
7. Saponieri, A.; Valentini, N.; Di Risio, M.; Pasquali, D.; Damiani, L. Laboratory investigation on the evolution of a sandy beach nourishment protected by mixed soft-hard system. *Water* **2018**, *10*, 1171. [CrossRef]
8. USACE. Coastal Engineering Manual, 2002. V-4-24 EM 1110-2-1100. Part III Coastal Sediment Processes. Available online: https://www.publications.usace.army.mil/USACE-Publications/Engineer-Manuals (accessed on 22 December 2018).
9. Bernabeu, A.M.; Medina, R.; Vidal, C. An equilibrium profile model for tidal environments. *Sci. Mar.* **2002**, *66*, 325–335. [CrossRef]

10. Roman-Sierra, J.; Navarro, M.; Muñoz-Perez, J.J.; Gomez-Pina, G. Turbidity and other effects resulting from Trafalgar Sandbank Dredging and Palmar Beach Nourishment. *J. Waterw. Port Coastal Ocean Eng.* **2011**, *137*, 332–343. [CrossRef]

11. Larson, M.; Kraus, N.C. Mathematical modelling of the fate of beach fill. *Coastal Eng.* **1991**, *16*, 83–114. [CrossRef]

12. Blott, S.J.; Pye, K. Particle size distribution analysis of sand-sized particles by laser diffraction: an experimental investigation of instrument sensitivity and the effects of particle shape. *Sedimentology* **2006**, *53*, 671–685. [CrossRef]

13. Magno, M.C.; Venti, F.; Bergamin, L.; Galianone, G.; Pierfranceschi, G.; Romano, E. A comparison between Laser Granulometer and Sedigraph in grain size analysis of marine sediments. *Measurement* **2018**, *128*, 231–236. [CrossRef]

14. Austin, L.G.; Shah, I. A method for inter-conversion of microtrac and sieve size distributions. *Powder Technol.* **1983**, *35*, 271–278. [CrossRef]

15. Konert, M.; Vandenberghe, J. Comparison of laser grain size analysis with pipette and sieve analysis: a solution for the underestimation of the clay fraction. *Sedimentology* **1997**, *44*, 523–535. [CrossRef]

16. Orru, C.; Chavarrias, V.; Uijttewaal, W.S.J.; Blom, A. Image analysis for measuring the size stratification in sand-gravel laboratory experiments. *Earth Surf. Dyn.* **2014**, *2*, 217–232. [CrossRef]

17. Pope, L.R.; Ward, C.W. *Manual on Test Sieving Methods*; ASTM: West Conshohocken, PA, USA, 1998; Volume 56.

18. Muñoz-Perez, J.J.; Roman-Sierra, J.; Navarro-Pons, M.; Neves, M.G.; del Campo, J.M. Comments on "Confirmation of Beach Accretion by Grain-Size Trend Analysis: Camposoto Beach, Cádiz, SW Spain" by Pizot, E., Anfuso, G., Méar, Y., Bellido, C.; 2013. *Geo-Mar. Lett.* **2014**, *34*, 75–78. [CrossRef]

19. Ministry of Environment. Environmental impact study: Use of Meca sandbank to Cadiz beach nourishment. In *Resolution of the Secretariat of the Environment*; Boletin Oficial del Estado (BOE): Madrid, Spain, 2003; pp. 17266–17270.

20. Muñoz-Perez, J.J.; Gutierrez, J.M.; Naranjo, J.M.; Torres, E.; Fages, L. Position and monitoring of anti-trawling reefs in the Cape of Trafalgar (Gulf of Cadiz, SW Spain). *Bull. Mar. Sci.* **2000**, *67*, 761–772.

21. Folk, R.L.; Ward, W.C. A study in the significance of grain size parameters. *J. Sediment. Petrol.* **1957**, *27*, 3–26. [CrossRef]

22. Blott, S.J.; Pye, K. Gradistat: A grain size distribution and statistics package for the analysis of unconsolidated sediments. *Earth Surf. Proc. Land* **2001**, *26*, 1237–1248. [CrossRef]

23. Folk, R.L. *Petrology of Sedimentary Rocks*; Hemphill Publishing Company: Austin, TX, USA, 1974.

 *water*

Article

The Origin of Sand and Its Colour on the South-Eastern Coast of Spain: Implications for Erosion Management

Francisco Asensio-Montesinos [1], Enzo Pranzini [2], Javier Martínez-Martínez [3], Irene Cinelli [2], Giorgio Anfuso [1] and Hugo Corbí [4,*]

[1] Department of Earth Sciences, Faculty of Marine and Environmental Sciences, University of Cádiz, CASEM, 11510 Puerto Real, Cádiz, Spain; francisco.asensio@uca.es (F.A.-M.); giorgio.anfuso@uca.es (G.A.)
[2] Department of Earth Science, University of Florence, Via Micheli 4, 50121 Florence, Italy; enzo.pranzini@unifi.it (E.P.); irene.cinelli@hotmail.it (I.C.)
[3] Spanish Geological Survey (IGME), Calle Ríos Rosas, 23, 28003 Madrid, Spain; javier.martinez@igme.es
[4] Department of Earth Sciences and the Environment, University of Alicante, Apdo. Correos 99, San Vicente del Raspeig, 03080 Alicante, Spain
* Correspondence: hugo.corbi@ua.es

Received: 21 December 2019; Accepted: 25 January 2020; Published: 30 January 2020

Abstract: Sand colour can give important information about mineral composition and, consequently, sediment source areas and input systems. Beach appearance, which is mostly linked to sand colour, has a relevant economic function in tourist areas. In this paper, the colour of 66 sand samples, collected along both natural and nourished beaches in the western Mediterranean coast of Spain, were assessed in CIEL*a*b* 1976 colour space. The obtained results showed relevant differences between natural and artificially nourished beaches. The colour of many nourished beaches generally differs from the native one because the origin of the injected sand is different. The native sand colour coordinates' range is: L* (40.16–63.71); a* (−1.47–6.40); b* (7.48–18.06). On the contrary, for nourished beaches' the colour range is: L* (47.66–70.75); a*(0.72–5.16); b* (5.82–18.82). Impacts of beach nourishment on the native sand colour were studied at San Juan beach, the most popular one along the study area. Nourishment works were performed after severe erosion, usually linked to anthropic activities/structures and storm events, but also to increase beach width and hence benefit tourism.

Keywords: Alicante; beach nourishment; CIEL*a*b*; coastal management; Costa Blanca; mineralogical characterization; sediment colour; tourism; western Mediterranean

---

1. Introduction

Coastal sedimentological studies allow the understanding of both natural processes and anthropogenic interventions, e.g., coastal dynamics and required management actions. Beach sediments characteristics play an essential role in coastal ecology but also in the choice of a pleasant site by beach users [1]. The most important characteristics of the sediment are texture (grain size and sorting), mineralogy, colour and morphology of grains (roundness and sphericity). These characteristics should be preserved during anthropogenic actions since their modifications can cause undesired impacts. First, sediment colour and texture directly influence the perception of beach users [2] and coastal scenery characteristics [3]: beach sediment colour is one out of the 18 natural parameters considered in the probably most commonly used landscape assessment method, i.e., the Coastal Scenic Evaluation [4]. This methodology, applied in >1000 coastal sites around the world [5], including the study area of this paper [6], is based on enquires that revealed that golden or white sand is better valued than dark sand. In this way, beach users also prefer sandy sediments to gravel or pebble ones [4]. Second, beach nourishment can cause ecological and physical impacts on the coast. For example, nourishment works can degrade different beach habitats and affect several marine species [7]. Addition of new sand volumes can involve a change in sand colour, which can modify sand temperature with consequences on biological processes of different species, e.g., the sex of marine turtles [1]. Finally, beach nourishment can also modify beach characteristics such as beach profile and/or water colour. Consequently, the modification of beach colour due to sediment injection can cause discontent between local beach users and the economic impacts to the local economy [8,9]. Sardinia (Italy) constitutes an example of this socioeconomic problem, where a court case was opened stimulated by stakeholders because a natural white beach was replaced by a black one [10]. On the contrary, nourishment works can sometimes improve the natural sand colour making the beach more attractive to tourists, this way incrementing the economic value [11].

Previous studies carried out along the study area of this paper, were focused on biological, mineral, and textural characteristics [12–17]. However, no attention was focused on the chromatic analysis of native sand and its modification after beach nourishment. Such studies on beach colour are relatively recent and they have important implications for coastal conservation and beach management. Sand colour characteristics can be used to create a catalogue of sand compatibility and improve nourishment works [3,9,18]. This catalogue should be useful to identify the origin of sediments, as well as to detect the occurrence and level of oil contamination, e.g., after the beach oiling in NW Spain linked to the sinking of the Prestige [19]. In Tuscany (Italy), compatibility studies in sand colour [20] have been carried out to avoid future management issues. Further, the chromatic analysis of sand in combination with the study of other sedimentological parameters such as the granulometry or mineralogical composition can be used for didactic activities or as an interesting geo-sedimentary resource to outreach geoscience among the general public [21–23].

The colour of solid objects mainly depends on three factors: the light source, the observer, and especially the characteristics of the material surface [24]. The colour of a stimulus can be assessed in the CIEL*a*b* colour space, recommended by the Commission Internationale de l'Eclairage [25] and is appropriate for sediment colour comparison [26]. On the one hand, the coordinate L* is the Lightness component, and it ranges between 0 and 100 (from black to white, respectively). It is directly associated with the visual sensation of luminosity. On the other hand, the coordinates a* and b* are denominated "Chromaticity". The a* coordinate defines the deviation of the achromatic point corresponding to Clarity, to red if a* >0 and to green if a* <0. Similarly, the coordinate b* defines the deviation to yellow if b* >0, and to blue if b* is <0 [24]; both can theoretically shift from −200 to +200, with values external to ±100 being unusual [3,26]. Several authors have quantified the colour of beach sand from different countries (e.g., Italy, Belgium, Cuba, Portugal, New Zealand and Japan) in the CIEL*a*b* colour space [3,9,26–29].

The most important characteristics of coastal sediments are particle size distribution, mineralogy, and colour (which is generally the least studied). For this reason, the main objective of this study was to carry out the first chromatic characterization of beach sands from the south-eastern coast of Spain. Samples from natural and altered (nourished) beaches were taken in order to analyse the influence of the origin of sediments on the actual beach colour and sediment chromatic characteristics were compared with the mineralogical composition of the studied sands. Colour studies in sandy beaches are very simple to perform compared to complex mineralogical studies, so it is mandatory to advance in this type of research in order to establish other objective criteria to adequately choose borrowed sand for nourishment projects. This study is the starting point for future research on sand colour characteristics in the study area in order to establish objective criteria for designing future anthropogenic actions on the coast attempting to maintain original beach sediment characteristics.

Finally, the data shows the current colour of beach sand in different locations, which can be very interesting in applied environmental studies (e.g., in terms of the beach quality assessment, tourism or engineering activities). For instance, this study can be particularly interesting to identify the type of sand (natural or nourished), to nourish beaches attempting to employ natural-like sand colour, and even to detect future changes in the sand colour or the presence of contaminants.

2. Materials and Methods

2.1. Study Area

The studied area covers about 200 kilometres of coastline of the western Mediterranean (SE Spain) (Figure 1). It includes 60 beaches from San Pedro del Pinatar (Murcia Province) to Dénia (Alicante Province, Figure 1, Table 1). The studied coast has very diverse habitats: dunes in the south [30], high cliffs of the Betic Cordillera in the north coast of Alicante [31], as well as beaches located on the Nueva Tabarca Island [23]. From a geological standpoint, this coastline is divided into three different domains: the southern sector mainly belongs to the Bajo Segura basin [32,33], the northern one pertains to the Prebetic zone of the External Zones of the Betic Range [34], while the Island of Tabarca is related to the Internal Zones of the Betic Range [35]. The different geological contexts in which the beaches are included establish broadly three different source areas of sediments, which constitute an intrinsic factor that determines natural sand composition and hence the native beach colour.

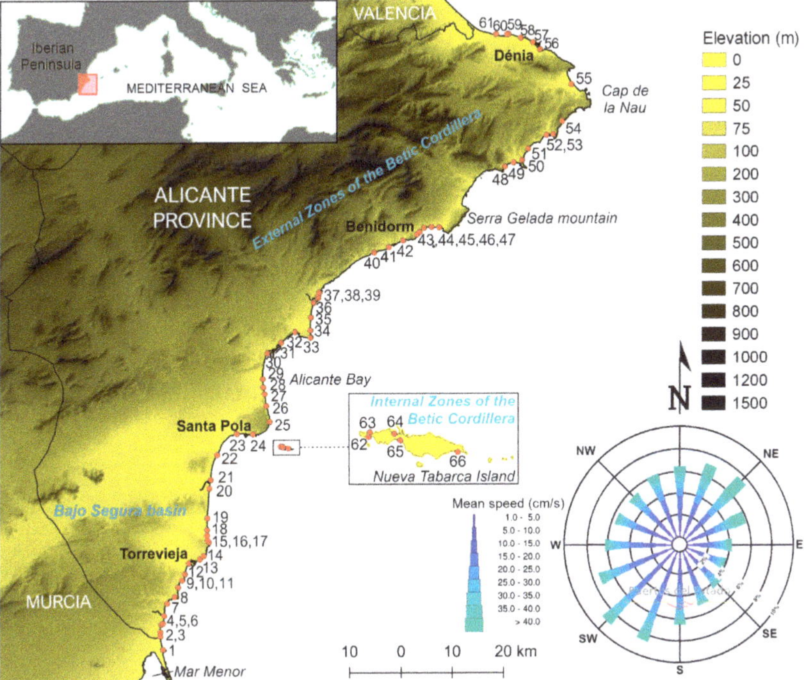

Figure 1. Location map showing the position of the 66 studied samples. Detailed information according to site numbering is showed in Table 1. Currents in Alicante area, observation period: 2006–2020 (source: www.puertos.es, accessed in January 2020).

Different natural elements reinforce the tourist attraction of the studied coastal scenery, such as cliffs, rocky shores, golden sand, dunes, turquoise water colour, and other coastal landscape features such as islands, lagoons, caves, rock ridges, and irregular headlands. However, most of the studied sites have been affected by infrastructures such as groins, seawalls, houses/buildings, parking, and nourishment works among others. Such human actuations/structures are mostly linked to the tourism demand and strong urban growth in the area and they have been especially frequent during recent years [36]. These anthropic structures were recently observed [6] and took into consideration 26 parameters—18 physical and 8 human—that delineate coastal scenery, according to Coastal Scenic Evaluation (CSE) Methodology [4,37], applied in different countries [5,8,38,39]. This semi-objective method utilizes weighting parameters and fuzzy logic mathematics to obtain the scenery Decision Value parameter (*D*). This value allows the coastal areas to be classified into 5 classes, from Class I (extremely attractive/natural sites) to Class V (very unattractive/urbanized sites).

Table 1. Location and main characteristics of studied beaches: Map number (Figure 1), beach name, municipality, beach typology and sand condition with their information source.

No.	Beach	Municipality	Typology	Sand Condition	Source
1	Torre Derribada	San Pedro del Pinatar	Rural	Nourished	[40]
2	El Mojón	Pilar de la Horadada	Village	Nourished	[41]; Environmental services [a]
3	Las Higuericas	Pilar de la Horadada	Village	Native	[42]
4	El Puerto	Pilar de la Horadada	Urban	Nourished	[41]; Environmental services [a]
5	El Conde	Pilar de la Horadada	Urban	Native	[42]
6	Mil Palmeras	Pilar de la Horadada	Urban	Nourished	Environmental services [a]
7	La Glea	Orihuela	Village	Native	[42]
8	Cabo Roig	Orihuela	Urban	Nourished	[41]; Coastscape 1956 [b]
9	La Estaca	Orihuela	Urban	Native	[42]
10	Punta Prima	Orihuela	Urban	Nourished	[41,42]
11	Cala Ferri	Torrevieja	Rural	Native	[41]
12	Los Náufragos	Torrevieja	Urban	Nourished	[41]; Coastscape 1956 [b]
13	Playa del Cura	Torrevieja	Urban	Nourished	[41]
14	Los Locos	Torrevieja	Urban	Native	[41]
15	Cala Cervera	Torrevieja	Urban	Native	[41]
16, 17	La Mata	Torrevieja	Village	Native	[41]
18	Torrelamata	Torrevieja	Urban	Native	[41]
19	Les Ortigues	Guardamar del Segura	Remote	Native	[41]
20	Els Vivers	Guardamar del Segura	Rural	Native	[41]
21	Els Tossals	Guardamar del Segura	Remote	Native	[41,42]
22	El Pinet	Elx	Rural	Native	[41,42]
23	Playa Lisa	Santa Pola	Urban	Nourished	[41]
24	Bernabeu II	Santa Pola	Urban	Nourished	[41]; Coastscape 1956 [b]
25	Calas del Cuartel	Santa Pola	Rural	Nourished	[43,44]
26	El Carabassí	Elx	Remote	Native	[41,42]
27	Arenales del Sol	Elx	Urban	Native	[41,42]
28	El Altet	Elx	Remote	Native	[41,42]
29	El Saladar	Alicante	Urban	Native	[41,42]
30	San Gabriel	Alicante	Urban	Native	Coastscape 1956 [b]
31	El Postiguet	Alicante	Urban	Nourished	[41]
32	La Albufereta	Alicante	Urban	Nourished	[41,42]
33	Cala Palmera	Alicante	Rural	Native	[41]
34, 35	San Juan	Alicante	Urban	Nourished	[41]
36	Muchavista	El Campello	Urban	Nourished	[41]
37–39	Carrer La Mar	El Campello	Urban	Nourished	[41,45]
40	Bon-Nou	La Vila Joiosa	Rural	Native	[41]
41	Playa Centro	La Vila Joiosa	Urban	Nourished	Coastscape 1956 [b]; [45]
42	El Torres	La Vila Joiosa	Rural	Native	[41]; Coastscape 1956 [b]
43	Cala Finestrat	Finestrat	Urban	Nourished	[46]
44–46	Ponent	Benidorm	Urban	Nourished	[47]
47	Llevant	Benidorm	Urban	Nourished	[48]
48	Port Blanc	Calp	Rural	Native	-
49	Arenal Bol	Calp	Urban	Nourished	[42]
50	La Fossa	Calp	Urban	Native	[41]
51	Cala Fustera	Benissa	Village	Nourished	[49]
52	L'Ampolla	Teulada	Village	Native	[41]
53	El Portet	Teulada	Village	Nourished	[50]
54	El Moraig	Benitatxell	Rural	Native	[34]
55	L'Arenal	Xàvia	Urban	Nourished	[51]
56	Marineta Cassiana	Dénia	Village	Nourished	[15,41]
57	Punta del Raset	Dénia	Village	Nourished	[41,42]
58	Les Marines	Dénia	Village	Nourished	[42]
59	Les Bovetes	Dénia	Village	Native	[42]
60	Molins i Palmeres	Dénia	Village	Native	[42]
61	Racó de L'Alberca	Dénia	Village	Native	[42]
62	Porta O (Sud)	Alicante	Rural	Native	[34,52]
63	Porta O (Nord)	Alicante	Rural	Native	[34,52]
64	Port Vell	Alicante	Rural	Native	[34,52]
65	Tabarca	Alicante	Rural	Native	[34,52]
66	Platja Gran	Alicante	Remote	Native	[34,52]

[a] Environmental services of the municipality (Pilar de la Horadada); [b] Coastscape 1956: Comparison among images of the American Flight, B (1956–1957 years.) and current images.

2.2. Sampling Design

A total of 66 sand samples were gathered in 16–19 February 2019, at sixty Mediterranean beaches (Table 1). Each sample was collected in the middle of the backshore (i.e., in the dry beach) and contained approximately 300 g of sand. First, the surface sand was removed and then, the sample was gathered in the first 20 cm of depth. On the longest beaches such as "La Mata", "San Juan", "Carrer la Mar", and "Ponent", several samples (2–3) were collected (Table 1). In this paper, only sandy beaches were investigated. For this reason, the north-eastern region has some coastal sectors not sampled due to the presence of cliffs or beaches mainly composed of gravel and boulders (Figure 1). Complementary information about the existence of anthropic actions on the sampled beaches was collected from both press news and coastal management actions by involved municipalities (Table 1).

2.3. Sand Colour Measurement and Analysis

The colour of sand samples was determined in the laboratory of the University of Florence (Italy) using the CIEL*a*b* 1976 colour space [25]. It required about 100 g of sand per sample to measure the colour. Moreover, mineralogical composition was quantified under a binocular microscope (Nikon SMZ 1500, Nikon Corp., Japan) from the Earth Sciences laboratory of the Alicante University (Spain). For this analysis a representative sample (about 200 sand grains) was analysed and classified taking into account three main groups of components: quartz, (i.e., leucocratic component), rock fragments, and shell fragments (i.e., melanocratic components), then the values were recalculated to 100% in order to synthesize and examine the results.

Principal Component Analysis (PCA) was carried out in order to establish multivariate relationships between samples as well as connections among them considering the CIEL*a*b* coordinates as input data.

Equations (1)–(3) were applied to quantify the chromatic change in a representative nourished beach (sample No. 35, Table 1). For this, a native sample was required and measured in the CIEL*a*b* colour space with the rest of the 66 samples.

$$\text{Hue (h)} = \arctan(b^*/a^*) \tag{1}$$

$$\text{Chroma (C*)} = (a^{*2} + b^{*2})^{1/2} \tag{2}$$

The distance between colours is given by the Euclidean distance:

$$\Delta E^*ab = [(\Delta L^*)^2 + (\Delta a^*)^2 + (\Delta b^*)^2]^{1/2} \tag{3}$$

Sedimentological analysis (particle size distribution) was performed via dry sieving at 1 phi interval in order to have a more detailed database. For this characterization, initially, each sample contained 200 g of sand. The software used for data analysis was "Gradistat" [53] and granulometric parameters were calculated according to Folk and Ward (1957) [54].

3. Results and Discussion

3.1. Typology of Beaches

Table 1 shows the main beaches under study, and their typology according to the Bathing Area Registration and Evaluation (BARE) classification system [55]. This classification is based on the level of urbanization, accessibility, etc., and establishes four basic types: remote, rural, village, and urban. Remote areas are mainly defined by difficulty of access. They may be adjacent to rural areas and, occasionally, village environments. Usually, remote beaches are not supported by public transport and have very limited (0–5) or no temporary summer housing [55]. Rural areas are located outside the urbanized environment. They are not accessible by public transport and present virtually no facilities. Housing is limited in number, generally ranging from 0 to 10. Rural beaches do not have permanent community focal centres (e.g., schools, shops, cafes and bars) and they have little or no beachfront development [55]. Village areas are located outside the focal urban environment and related to a small and permanent population. Village beaches may be reached by public or private transport [55]. Urban areas have large populations with well-established public services and a well-marked central business district. In the proximity of urban areas can be found commercial activities such as fishing/boating harbours and marinas. Urban beaches are located within or adjacent to the urban area and are generally freely open to the public [55].

The selected study case is an excellent example of coastal development under high touristic pressure, with many coastal sites showing very urbanized conditions (34 studied beaches classified as Urban in Table 2) and the rest of the cases are located in urbanized environments (13 Village beaches) or they show the presence of houses (14 Rural beaches). Only a very few coastal sites (5) can be classified as natural beaches, i.e., beaches with few anthropic interventions. They were classified as Remote.

Table 2. Sector analysis approach (Sand condition and beach typology). Consider that the Chi-square Test was performed to see the level of significance among them.

	Remote	Rural	Village	Urban	Total
Nourished	0	2	6	24	32
Native	5	12	7	10	34
Total	5	14	13	34	66

Chi-square Test: $\alpha = 0.05$; df = 3; $X^2 = 17.94$; p-value = 0.00045; $X^2_{3,\,0.05} = 7.814$; sig = yes.

As a consequence of the urban and touristic pressure that affects the study area, some beaches have been modified by anthropogenic activities, but others have preserved their original state, i.e., only receive sand through natural processes. For this reason, a new classification was made according to sand condition: "Nourished" and "Native" (Table 1).

Nourished beaches are common in the study area [41,49–51,56–58], causing anthropogenic modifications in the original appearance of the coast [59]. Construction of dams on land that retain sediments, massive construction of buildings near the coast, creating new harbours and breakwaters, beach nourishment, as well as sand movements from beach to beach are some examples. These coastal modifications that have been carried out since approximately the middle of the 20th century [60], have changed the native sand colour among other sediment characteristics. Figures A1–A5 show the current state (February 2019) of the sampled sites. Sand colour observed in these images is a result of multiple factors acting simultaneously: the camera used, hour, luminosity of the moment (e.g., sunny or shaded), beach surface characteristics, etc. All these extrinsic determinants modify the perception of the real intrinsic sand colour, which is mainly a consequence of the mineralogical composition. For these reasons it is necessary to apply an objective methodology to quantify the chromatic appearance of samples, i.e., in a laboratory with a quasi-uniform colour space method (CIEL*a*b*).

3.2. Sand Colour Analysis

An interesting diversity of colours emerged along the study area (Figure 2, Table A1) mainly due to two factors: the different origin of sediments (intrinsic factor) and the anthropic modifications that have been carried out since approximately the middle of the 20th century (extrinsic factor).

Figure 2. Sand samples from the 66 coastal sites. One shoot in sunlight including X-rite colour table and centimetric scale.

Regarding the colour parameters (the L*, a* and b* values) expressed in Figure 3a,b, the following considerations can be argued:

3.2.1. The L* Value

Four groups are differentiated according to the measured Lightness value: (i) samples with L* >60 (seven samples); (ii) samples with intermediate values (50 < L* < 60); (iii) samples with L* ≈ 48 (four samples); and (iv) samples with L* <46 (five samples) (Figure 3a).

These results highlight the relevant underlying influence of the geological context on the final appearance of the coastal deposit. This geological frame defines the natural source area of beach grains. In this sense, the above first group includes the southern beaches. Their geological context is mainly related to the Bajo Segura basin. This source area is mostly constituted of Neogene-Quaternary sediments mainly related to continental and coastal deposits [61]. Also, this area is characterized by unaltered coastal stretches. The northern beaches related to the External Zones of the Betic range are grouped in the second group. They are the most modified beaches and only a few of them are natural (Figure 3a). The dark colour of the northernmost beaches (included in the third group) is related to the contributions of different rivers and streams (e.g., Girona River and "Barranc de L'Alberca"). However, the darkest sands correspond to those included in the fourth group. These samples are included in the geological context of the Tabarca Island (Figure 3a). In this island the outcrops are mainly materials from Internal Zones of the Betic range. The low Lightness of these natural beaches is justified by the high content of dark rock fragments that comes from the black dolomites and diabases, which outcrop along the island [23,52].

Two samples differ significantly from the general trend defined above. On the one hand, sample No. 62 differs from the rest of the samples located in Tabarca Island due to the influence of the yellowish calcarenites (Miocene age) that outcrop at this point of the island [23,34]. On the other hand, sample number 51 shows the highest Lightness value (i.e., 70.75, Table 3, Figure 3a) because this sand is from an inland quarry.

3.2.2. The a* and *b Plane

The parameters a* and b* (Table A1) are positively correlated among them ($R^2 = 0.89$; p-value <0.05; Figure 3b and Table A2). The found values were more uniform than the Lightness ones. On the one hand, the parameter a* ranges from −1.47 (sample No. 66, Platja Gran) to 6.4 (sample No. 61, Racó de L'Alberca; Figures 2 and A5). On the other hand, parameter b* varies from 5.82 (sample No. 25, Calas del Cuartel; Figures 2 and A2) to 18.82 (sample No. 51, Fustera Beach; Figures 2 and A4). Those samples with negative a* and low b* values have a characteristic green colour and are from the eastern side of Tabarca Island where the beaches are dominated by metagabbro clasts proceeding from eroded cliffs [23]. On the opposite side, several native samples (No. 11, 14, 15, 17, 19, 28, 40, 54, 59, 60, 61, Table 1) stand out, with their values towards the yellow-red colour (Figure 2) resulting in an attractive golden sand.

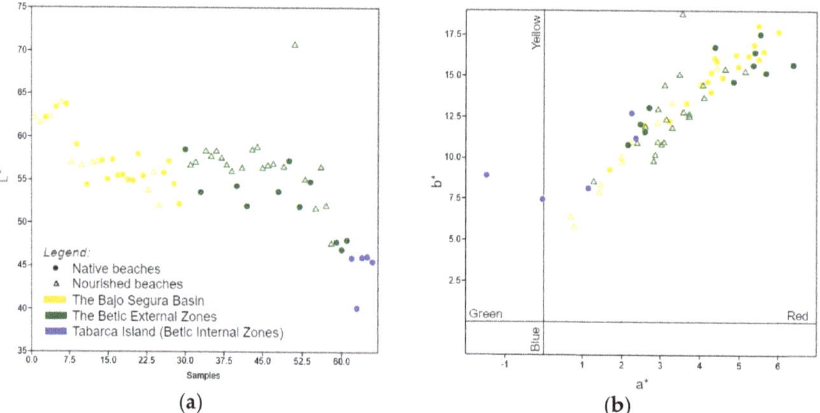

Figure 3. (**a**) The Lightness (L*) from sample number 1 to sample number 66, the beaches of Tabarca Island were enumerated, the last ones from west to east (Figure 1 and Table A1). (**b**) Sample position in the a*, b* plane (CIEL*a*b* colour space; illuminant D65, Table A1). Colour (yellow, green and blue) refers to the geological domains of the study area (respectively, The Bajo Segura basin, The Betic External Zones, and the Tabarca Island which belongs to the Betic Internal Zones).

3.3. Mineralogical Characterization

Mineralogical composition of the studied sands is very diverse, showing variations in the type of minerals and also in the relative proportion in which they are represented. The colour of each grain is directly related to its mineralogical composition and, therefore, the global colour of the deposit is the result of the presence of certain grains and their abundance.

Leucocratic components (mainly transparent grains with angular morphology) were identified, and they correspond mainly to quartz (SiO_2). They are very well represented in all samples in variable proportions (from 8.08% to 72.26%, Table 3). Melanocratic components correspond mainly to both rock fragments (basically calcarenites and dark carbonate rocks) and other minerals such as pyroxenes, amphiboles, and biotites. In general terms, melanocratic components are more abundant than leucocratic ones, especially at Tabarca Island (Figure 4). Their abundance varies along the study area from 27.74% to 91.92% (Table 3). Finally, bioclasts (i.e., small fragments of shells, sea urchins, calcareous algae, corals, etc.) were also counted and their proportion in the study area is very low, almost insignificant. Due to the dark aspect of bioclasts and their low abundance, they were included in the melanocratic fraction.

Concerning the abundance of leucocratic and melanocratic components, 14 samples contained less than 50% of leucocratic components, while the rest (52) contained ≥50% of melanocratic components (Figure 4). Table 3 shows median, mean, as well as minimum and maximum values of the colour data (CIEL*a*b*) and mineralogical composition, i.e., percentage of leucocratic and melanocratic components.

Table 3. Summary of colour and mineralogical parameters.

	L*	a*	b*	Leucocratic Components	Melanocratic Components
Min.	40.16	−1.470	5.82	8.08	27.74
Median	56.44	3.400	13.04	39.30	60.70
Mean	55.48	3.452	13.04	38.23	61.77
Max.	70.75	6.400	18.82	72.26	91.92

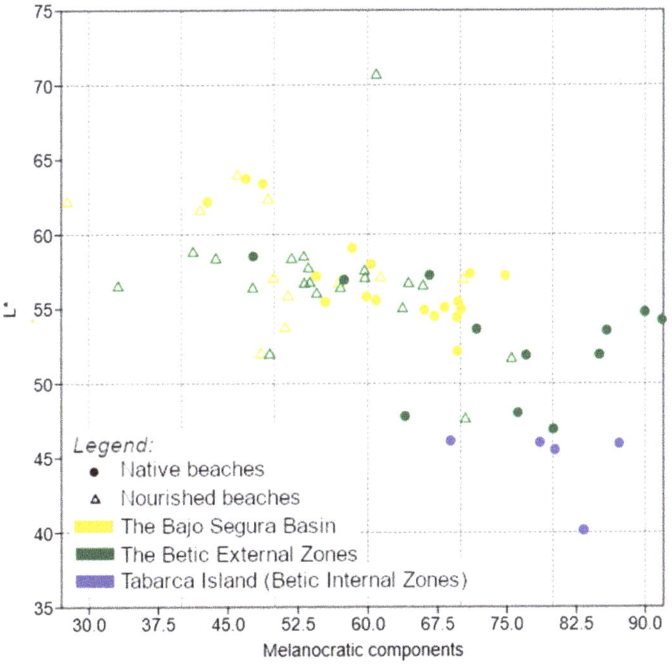

Figure 4. Sample position according to the Lightness (L*) and percentage of melanocratic components. Colour (yellow, green, and blue) indicates the geological domains to which the studied beaches belong (respectively, The Bajo Segura basin, The Betic External Zones, and the Tabarca Island which belongs to the Betic Internal Zones).

3.4. Statistical Analysis

In order to find correlations among the different colour and mineralogical parameters, the Pearson correlation statistic coefficient was obtained (Table A2). On the one hand, the highest correlation was found among a* and b* parameters ($R^2 = 0.89$; p-value = 2.2×10^{-16}). On the other hand, L* parameter shows a good correlation with the leucocratic component ($R^2 = 0.62$; p-value = 2.187×10^{-8}).

Quartz grains normally present light colours and their high proportion in samples determines sand clarity. In this way, the content of quartz grains is inversely proportional to the a* and b* colour parameters as was previously concluded for the Belgian coast [9].

Principal Component Analysis (PCA) is a multivariate-statistical technique used for reducing the dimensionality of data. The resulting analysis is shown in a two-dimensional plot composed of two axes or components, PC1 and PC2 (Figure 5). PC1 explained 63.57% of the total variance and PC2 32.83%. The cumulative proportion of the second principal component was 96.39% (the data were previously normalized). In the graph, the three colour variables (L*, a*, and b*) are represented with arrows and each one of the 66 investigated sites is located in the graph according to its relation with the variables, a technique applied previously in another sand colour study with important implications for beach nourishment and coastal management [26]. The resulting distribution of the samples in the graph is due to both their geographical location mainly related to the geological context and anthropic actions carried out on the deposit (i.e., nourishment works). Four groups of beaches were identified in the graph of Figure 5 according to their geographical location ("neighbouring beaches"): (i) the southernmost beaches (n = 7); (ii) the northernmost beaches (n = 4); (iii) beaches of Santa Pola (n = 3); and (iv) beaches of Tabarca Island (n = 5).

Moreover, very interesting groupings can be observed according to sand condition consulted in the bibliography (Table 1). A group of natural samples (+10) from different municipalities is shown towards the arrows of the colour variables (a* and b*, Figure 5). On the left side of the graph (Figure 5) there are generally those beaches identified as natural or without alterations in colour of sand. The sand composition of the northern beaches was also considered as native (Table 1) where sediment supplies mainly arrive from the river and nearby streams. The sand of Tabarca beaches is also native and its source area is the island itself and seabed. In the central sector of Figure 5 there are beaches located in the central-northern part of the study area. Some of them correspond to modified beaches by nourishment works and emplacement structures, e.g., harbours (No. 38, 39, 41, 46, 48, 53, 57).

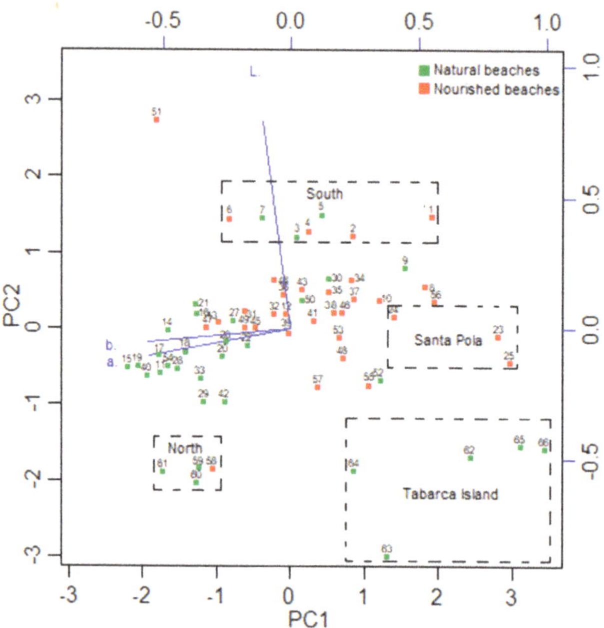

Figure 5. Principal Component Analysis (PCA) graph: representation of sand samples according to colour variables (L*, a*, and b*). See Table A1 for numerical information of each variable. Green colour: native samples; red colour: altered samples (information obtained from bibliography data, Table 1).

3.5. Grain Size Parameters

The sediment studied from the 66 samples was classified mainly as coarse sand (78.8%), very coarse sand (12.6%), and medium sand (6.5%). Beaches in the southernmost part of the study area were composed essentially of fine sand, while from the central part to the north, different types of beaches can be observed with rocky shores, gravel, pebbles, with fine, medium and coarse sand. Table A1 shows some of the grain size parameters of each sample.

3.6. Colour Modification due to Human Activities

As a consequence of the urban and touristic pressure, artificial beaches were created and deep modifications in the native coastal deposits were carried out. The sand used to nourish the beaches came from different sources: (i) submerged sandbanks (e.g., San Juan beach, No. 34–35), sometimes near to the mouth of harbours; (ii) from other beaches (e.g., El Puerto beach, No. 4); and even (iii) from a quarry inland (Fustera Beach, No. 51).

At the beginning of the 1990s one of the largest beach modifications in the province of Alicante was carried out in San Juan beach (Figure 6). Thanks to a sample of sand from the late 1980s it was possible to compare the native colour of the sand with the current one of the nourished beach, whose sand grains come from the littoral bottoms in front of the Serra Gelada Mountain (Benidorm, Alicante) [57]. Table 4 shows the colorimetric parameters of two samples from different years (1989/2019) and gathered in the same coastal sector of the beach. Difference of colour values between these two samples was estimated [62].

Table 4. Coordinates in CIEL*a*b* colour space and their differences at the San Juan beach.

Sample	L*	a*	b*	ΔL*	Δa*	Δb*
San Juan II (2019 year.)	57.76	3.03	10.86	3.85	1.15	5.84
San Juan (1989 year.)	61.61	4.53	16.70			

Equations (1) and (2) show the mathematical expression for quantifying the chromatic change in terms of Hue differences (Δh) and Chroma differences (ΔC*). The parameter Total Colour Difference (ΔE*, Equation (3)) is also used. The above calculations show that the colour of the beach after nourishment works can change even when sediments come from near submerged areas. The colour change is registered in all the chromatic parameters L*, a*, and b* (Table 4), and differences of $-0.4°$ (h), -6 (C*), and 7.15 (ΔE*) are evidenced by different equations. All these values reflect a significant aesthetical change of the coastal deposit after the nourishment. On the one hand, the negative values of Δh and ΔC* mean that the new appearance of the altered San Juan beach is duller and more brown than the native aspect. Therefore, the new beach colour is more distant from the ideal golden or light sand appreciated by beach users [6]. On the other hand, the high ΔE* value indicates that the chromatic change is clearly perceptible by local beach users, this occurs when the Total Colour Difference parameter exceeds three units (ΔE* > 3) [63]. Another similar example is the Varadero beach (Cuba) where beach nourishment works performed several times in the past decades with nearshore sands changed the original beach colour that was the second lightest one out of 100 in Cuba, now occupying only the fiftieth position in rank [26].

San Juan beach is a clear example showing how the construction on the coast imposes a barrier to wind dynamics that can even cause the disappearance of beaches with the consequent requirement of nourishment works (Figure 6). In addition, river channelling and sediment retention in upstream reservoirs, which contribute to the erosion of beaches [64,65], have played a significant role in the reduction of the sediment input to this representative beach of south-eastern Spain. Historical bathymetric measures in reservoirs of the study area confirm this issue (Centre for Hydrographic Studies). Indeed, San Juan beach has been the subject of geological outreach activities to promote the Geoheritage of the area, which has also very didactic geological and sedimentary outcrops [66]. In the didactic activities undertaken [67], the sedimentological analyses of the sand (colour, sorting and presence of bioclasts) has been used to understand and reflect on the differences between the natural and the nourishment sediments of the beach, and how important it is to attempt to archive equivalent sands to the original in beach nourishment works.

Figure 6. San Juan beach. From the original condition to the present one [57,68–70].

The study case of "Cala Fustera" (No. 51, Table A1, Figures 7 and A4) constitutes another example of significant coastal alteration after nourishment works. In this case, the previous sand deposit was replaced in 2017 by artificial quarry sand. Figure 7 shows visually differences in the colour of the coast of the Fustera Beach (now composed of quarry sand) with respect to the native colour in this coastal area, which can be seen at the Pinets beach. This original deposit is mainly composed of rocky shore with *Posidonia oceanica* banquettes (that are the characteristics that Fustera Beach should have). Unfortunately, no sand sample from the native Fustera Beach deposit was available. However, colour differences have been assessed assuming that the original aspect of the beach was similar to the sand colour of adjacent beaches. In this case, the chromatic change parameters take values of $\Delta h = 0.01$, $\Delta C^* = 8.13$ and $\Delta E = 20.55$, indicating that the new beach appearance is much whiter and bright than the original aspect.

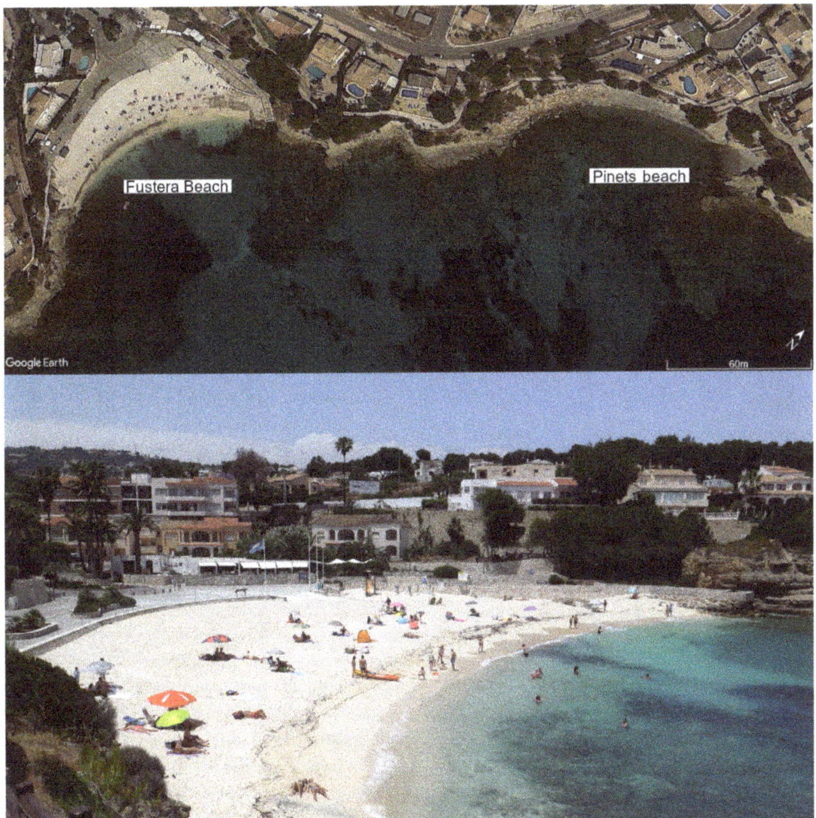

Figure 7. Fustera Beach nourished in 2017 with quarry sand (No. 51, Table A1), and Pinets beach, non-altered site. The contrast between the artificial beach vs the natural one is obvious (2018 year.). The bottom of the figure shows Fustera Beach in summer 2017.

A third example of sand colour modification due to management actions is the case of El Puerto beach (sample No. 4) located in Pilar de la Horadada, Alicante. In this case, sand from the El Puerto beach (a site next to a harbour from where sand is usually dredged, Figure 8) was transferred to other beaches such as "El Mojón" or "Mil Palmeras". Figure 8 shows the contact between the old and new coastal deposit. The colour difference of the added sand can be justified because the southern beaches are grouped at the top of Figure 5 with important differences in a* and b* parameters.

Figure 8. El Puerto beach. One of the places where sand is dredged, mixed and distributed to other beaches [71,72]. Image source: Google Earth Pro (18 March 2017).

4. Conclusions

This research is a new contribution to the diverse geological heritage of the southeast of the Iberian Peninsula. Sand colour of natural beaches is influenced by the geological context in which they are located: External and Internal zones of the Betic Cordillera, and the Bajo Segura basin. These three main source areas of sediments determine natural sand composition and thus, the native beach colour. Because of this, the resulting colour of each beach is the consequence of their mineralogical composition linked to the source and hydrographic basin. In this sense, natural beaches related to the geological context of the Bajo Segura basin are clearer than the rest due to the high content in quartz grains, especially the southernmost ones. On the contrary, the darkest beaches are located in Tabarca Island, whose rocky outcrops mainly belong to the Internal zones of the Betic Cordillera. Therefore, although geology is the basis of the native beach colour, human factors can modify sand characteristics, as in the case of several studied beaches, especially those related to the geological context of the External Zones of the Betic Cordillera.

To measure the colour in sandy sediments it is adequate to apply CIEL*a*b* methodology to objectively quantify the chromatic appearance of each sample. In the study area, generally nourished/altered beaches differ from the native ones. The alteration of the sand colour by anthropic actions transformed the natural colour increasing green and blue tones (a* and b* colour components) and Lightness (positive values in L*). Despite this, a big group of natural beaches was identified with the highest a* and b* values (towards red and yellow tones). Colour and mineralogical parameters are related among them, such as the Lightness and the content of leucocratic components. Many beaches with native sand present a high content of melanocratic components (>60%), while most of the nourished beaches normally show less than 60% of such components. Principal Component Analysis shows good groupings associated with the mineralogical composition (related to the geographical location) and anthropic vs natural beaches. The position of the samples in the PCA graph is very useful to compare the sand colour from different beaches and also to interpret the limited information about nourishment works carried out in the past.

The 66 investigated samples differ in colour parameters and the beaches that receive more tourists have been nourished and tend to be clearer (e.g., "Ponent", "El Postiguet", "San Juan"). Results show that a few nourished beaches, e.g., "Playa del Cura", "Llevant", "Arenal Bol", "El Postiguet", and "Ponent", have a similar colour to the nearby native ones. The rest of the nourished beaches have lower values of a* and b* colour parameters, moving towards green and blue tones, e.g., "Punta Prima", "Marineta Cassiana", "Playa Lisa", "Bernabeu II", "Calas del Cuartel", etc. Despite this, native clear samples generally presented higher values in the a* and b* parameters, this giving a more golden visual appearance, e.g., Ferri Beach, Cervera Beach, "Los Locos", "Les Ortigues", "La Mata" or "Bon-nou" among others. Today, the beach colour is not taken into consideration in some nourishment works, at least in the study area where "El Puerto" in the south or Fustera Beach in the north are clear examples.

This study showed with analytical data a post-nourishment change in sand colour, a common consequence also produced in other coastal sites such as Italy, Belgium or Cuba. Such changes may not be appreciated by national and international tourists as well as by old local beachgoers. The CIEL*a*b* colour space used to characterise sediment colour and Lightness is a useful tool to avoid significant changes in sand colour that can affect coastal scenery, beachgoers preferences, and the biological processes of some species. Native sediment colour is a natural characteristic of the coastal landscape and for this reason it should be maintained, at least in rural beaches and natural parks. It is essential to preserve the most natural coastal areas such as the dunes of Guardamar or the Island of Nueva Tabarca, considered by the Spanish Geological Survey as places of geological interest, mainly due to their geomorphological, sedimentological and stratigraphic particularities. In urban beaches, where all is artificial, a change of sand colour can be accepted in order to improve the beach attractiveness. Regarding future research prospects on the colour of beach sand, it would be interesting to analyse the sand properties against time in order to detect possible changes in the coastal environment.

Sand is a limited and increasingly valuable resource. Its use and relocation generates strong environmental, economic, and social impacts, especially on the coast. For this reason, it is preferable to opt for dune restoration and the elimination of unnecessary structures/constructions (on the coast or inland) that promote beach erosion instead of proposing nourishment projects that involve large economic investments and that may affect different coastal and marine habitats as well as associated species. Finally, the method used in this paper is easy to apply and can be very useful to carry out appropriate erosion management actions by means of beach nourishment works in the future. Consequently, the current conclusions are of particular interest to future actions to preserve the natural environments and original characteristics of native deposits.

Author Contributions: Conceptualization, F.A.-M. and H.C.; methodology, E.P., I.C. and H.C.; software, F.A.-M.; formal analysis, E.P.; investigation, F.A.-M. and H.C.; resources, E.P.; data curation, F.A.-M.; writing—original draft, F.A.-M., J.M.-M. and H.C.; writing—review and editing, E.P. and G.A.; visualization, J.M.-M.; supervision, H.C.; funding acquisition, H.C. and G.A. All authors have read and agreed to the published version of the manuscript.

Funding: This research received no external funding.

Acknowledgments: Special thanks go to Laura Piqueras for her technical support in the laboratory. The authors would also like to thank Jesús Soria for providing us with a 30-year-old sand sample (native) from San Juan beach.

Conflicts of Interest: The authors declare no conflict of interest.

Appendix A

As complementary material, photographs of each of 66 sampling points (Figures A1–A5) and other relevant numerical data (Tables A1 and A2) are attached.

Figure A1. Beaches investigated: from "Torre Derribada" (No. 1) to "Cala Cervera" (No.15).

Figure A2. Beaches investigated: from "La Mata I" (No. 16) to "San Gabriel" (No.30).

Figure A3. Beaches investigated: from "El Postiguet" (No. 31) to "Ponent II" (No.45).

Figure A4. Beaches investigated: from "Ponent III" (No. 46) to "Molins i Palmeres" (No.60).

Figure A5. Beaches investigated: from "Racó de L'Alberca" (No. 61) to "Platja Gran" (No.66).

Table A1. Coordinates in CIEL*a*b* colour space of the 66 investigated samples and percentage of sand grains by quartz (leucocratic component), rock fragments and shell fragments (melanocratic components). The three granulometry parameters most commonly used are included. See Figure 1 for number location.

No.	Colour Parameters			Mineralogy		Grain Size Parameters (µm)		
	L*(D65)	a*(D65)	b*(D65)	Leucocratic Comp. (%)	Melanocratic Comp. (%)	Mean	Mode	Median D50
1	62.2	1.43	7.94	72.26	27.74	151.1	152.5	151.1
2	61.65	2.26	10.95	57.92	42.08	185.2	152.5	161.2
3	62.16	3.25	12.24	57.14	42.86	190.8	152.5	166.7
4	62.41	2.95	12.17	50.59	49.41	152.5	152.5	152.5
5	63.38	2.6	12	51.14	48.86	153.6	152.5	153.6
6	64	4.03	14.52	53.91	46.09	193.2	152.5	169.8
7	63.71	3.68	13.32	52.97	47.03	194.7	152.5	170.6
8	57.06	1.48	8.37	50.00	50.00	149.2	152.5	149.2
9	59.06	1.71	9.27	41.63	58.37	182.9	152.5	159.3
10	56.7	2.01	10.08	43.10	56.90	153.8	152.5	153.8
11	54.4	5.66	16.49	30.35	69.65	194.3	152.5	171.0
12	57.02	3.32	13.36	29.62	70.38	150.9	152.5	150.9
13	57.14	4.45	15.51	38.56	61.44	177.8	152.5	155.7
14	57.19	5.53	16.01	25.10	74.90	233.8	302.5	269.0
15	55.07	6.04	17.7	31.66	68.34	239.5	302.5	276.3
16	57.33	4.39	16.07	28.90	71.10	150.1	152.5	150.1
17	55.44	5.41	16.9	30.21	69.79	198.7	152.5	176.0
18	55.54	4.95	16.29	39.02	60.98	189.6	152.5	162.3
19	54.95	5.53	18.06	29.87	70.13	154.5	152.5	154.5
20	54.9	4.59	14.89	33.84	66.16	224.7	302.5	253.3
21	57.95	4.44	15.9	39.57	60.43	149.7	152.5	149.7
22	55.43	4.3	14.02	44.49	55.51	199.9	302.5	180.0
23	53.78	0.72	6.41	48.78	51.22	149.3	152.5	149.3
24	55.9	2.02	9.76	48.47	51.53	151.3	152.5	151.3
25	52	0.81	5.82	51.38	48.62	151.4	152.5	151.4
26	55.78	4.3	15.19	40.08	59.92	152.6	152.5	152.6
27	57.15	4.21	14.65	45.45	54.55	188.7	152.5	164.4
28	54.48	5.27	16.21	32.74	67.26	236.2	302.5	272.4
29	52.14	5	15.57	30.29	69.71	296.0	302.5	296.0
30	58.52	2.6	11.6	52.22	47.78	181.8	152.5	157.8
31	56.76	3.48	15.12	35.58	64.42	174.9	152.5	155.3
32	57.11	3.11	14.46	40.27	59.73	232.5	152.5	258.9
33	53.53	4.38	16.76	14.22	85.78	469.5	605.0	537.2
34	58.41	2.83	9.85	48.12	51.88	193.0	152.5	166.2
35	57.76	3.03	10.86	46.32	53.68	187.6	152.5	159.4
36	58.41	3.73	12.57	56.28	43.72	234.0	302.5	269.8
37	57.58	2.87	10.23	40.33	59.67	199.2	152.5	177.3
38	56.79	3.09	11	46.15	53.85	232.7	302.5	266.0
39	56.07	3.73	12.72	45.43	54.57	239.4	302.5	276.6
40	54.26	5.55	17.54	8.08	91.92	904.3	605.0	682.0
41	56.43	3.3	11.88	52.26	47.74	232.2	302.5	265.3
42	51.92	4.86	14.65	15.00	85.00	386.1	302.5	331.9
43	58.55	2.94	13	46.82	53.18	183.6	152.5	159.8
44	58.85	3.58	12.84	58.72	41.28	176.9	152.5	156.3
45	56.44	4.1	13.71	42.91	57.09	258.2	302.5	291.0
46	56.75	2.93	11.02	46.75	53.25	191.0	152.5	163.7
47	56.93	4.64	15.44	42.53	57.47	196.5	152.5	174.4
48	53.62	2.48	12.06	28.23	71.77	151.5	152.5	151.5
49	56.59	4.07	14.47	34.01	65.99	182.5	152.5	158.8
50	57.23	2.71	13.07	33.33	66.67	180.5	152.5	157.5
51	70.75	3.55	18.82	39.04	60.96	275.2	302.5	280.9
52	51.88	2.16	10.81	22.85	77.15	176.0	152.5	155.8
53	55.07	2.61	11.94	36.22	63.78	290.1	302.5	292.3
54	54.77	5.41	16.44	10.00	90.00	624.0	605.0	615.2
55	51.72	2.4	10.95	24.42	75.58	237.1	302.5	273.7
56	56.55	1.29	8.57	66.79	33.21	189.6	152.5	167.6
57	52.02	3.15	12.39	50.43	49.57	185.1	152.5	161.4
58	47.66	5.16	15.32	29.44	70.56	292.6	302.5	292.6
59	47.78	5.37	15.66	35.96	64.04	299.4	302.5	299.4
60	46.91	5.69	15.19	19.90	80.10	572.2	605.0	559.0
61	48.01	6.4	15.7	23.75	76.25	470.3	605.0	538.1
62	45.95	1.16	8.14	12.81	87.19	380.3	302.5	330.7
63	40.16	2.37	11.21	16.67	83.33	593.4	605.0	593.4
64	46.03	2.26	12.73	21.33	78.67	464.1	605.0	534.1
65	46.13	−0.03	7.48	31.05	68.95	348.9	302.5	304.4
66	45.54	−1.47	8.93	19.70	80.30	605.0	605.0	605.0

Table A2. Bivariate Pearson correlation matrix for the analyzed parameters. Relevant correlations marked in bold.

	L*	a*	b*	Leucocratic Components	Melanocratic Components
L*	1	0.058	0.119	0.624	−0.624
a*		1	0.890	−0.302	0.302
b*			1	−0.402	0.402
Leucocratic components				1	−1
Melanocratic components					1

References

1. Pranzini, E.; Vitale, G. Beach Sand Colour: The Need for a Standardised Assessment Procedure. *J. Coast. Res.* **2011**, *61*, 66–69. [CrossRef]
2. Roca, E.; Riera, C.; Villares, M.; Fragell, R.; Junyent, R. A combined assessment of beach occupancy and public perceptions of beach quality: A case study in the Costa Brava, Spain. *Ocean. Coast. Manag.* **2008**, *51*, 839–846. [CrossRef]
3. Pranzini, E.; Simonetti, D.; Vitale, G. Sand colour rating and chromatic compatibility of borrow sediments. *J. Coast. Res.* **2010**, *26*, 798–808. [CrossRef]
4. Ergin, A.; Karaesmen, E.; Micallef, A.; Williams, A.T. A new methodology for evaluating coastal scenery: Fuzzy logic systems. *Area* **2004**, *36*, 367–386. [CrossRef]
5. Rangel-Buitrago, N. *Coastal Scenery: Evaluation and Management*; Springer: Cham, Switzerland, 2018.
6. Asensio-Montesinos, F.; Anfuso, G.; Corbí, H. Coastal scenery and litter impacts at Alicante (SE Spain): Management issues. *J. Coast. Cons.* **2019**, *23*, 185–201. [CrossRef]
7. Peterson, C.H.; Bishop, M.J. Assessing the environmental impacts of beach nourishment. *Bioscience* **2005**, *55*, 887–896. [CrossRef]
8. Anfuso, G.; Williams, A.T.; Hernández, J.C.; Pranzini, E. Coastal scenic assessment and tourism management in western Cuba. *Tour. Manag.* **2014**, *42*, 307–320. [CrossRef]
9. Cárdenes, V.; Rubio, A. Measure of the color of beach nourishment sands: A case study from the Belgium coast. *Trabajos de Geología* **2015**, *35*, 7–18. [CrossRef]
10. Pranzini, E. Protection studies at two recreational beaches: Poetto and Cala Gonone beaches, Sardinia, Italy. In *Beach Management*; Williams, A.T., Micallef, A., Eds.; Earthscan publishers: London, UK, 2009; pp. 287–306.
11. Houston, J.R. The economic value of beaches-a 2013 update. *Shore Beach* **2013**, *81*, 3–11.
12. Aragonés, L.; López, I.; Villacampa, Y.; Serra, J.C.; Saval, J.M. New methodology for the classification of gravel beaches: Adjusted on Alicante (Spain). *J. Coast. Res.* **2014**, *3*, 1023–1034. [CrossRef]
13. Corbí, H.; Asensio-Montesinos, F.; Ramos-Esplá, A.A. The littoral bottoms of Benidorm Island (western Mediterranean Sea): Eco-sedimentological characterization through benthic foraminifera. *Thalassas* **2016**, *32*, 105–115. [CrossRef]
14. López, I.; López, M.; Aragonés, L.; García-Barba, J.; López, M.P.; Sánchez, I. The erosion of the beaches on the coast of Alicante: Study of the mechanisms of weathering by accelerated laboratory tests. *Sci. Total Environ.* **2016**, *566*, 191–204. [CrossRef] [PubMed]
15. Pagán, J.I.; Aragonés, L.; Tenza-Abril, A.J.; Pallarés, P. The influence of anthropic actions on the evolution of an urban beach: Case study of Marineta Cassiana beach, Spain. *Sci. Total Environ.* **2016**, *559*, 242–255. [CrossRef] [PubMed]
16. López, M.; Baeza-Brotons, F.; López, I.; Tenza-Abril, A.J.; Aragonés, L. Mineralogy and morphology of sand: Key parameters in the durability for its use in artificial beach nourishment. *Sci. Total Environ.* **2018**, *639*, 186–194. [CrossRef] [PubMed]
17. Pardo-Pascual, J.E.; Sanjaume, E. Beaches in Valencian Coast. In *The Spanish Coastal Systems*; Springer: Cham, Switzerland, 2019; pp. 209–236.
18. Pranzini, E. Il colore della sabbia: Percezione, caratterizzazione e compatibilità nel ripascimento artificiale delle spiagge. *Studi Costieri.* **2008**, *15*, 101–122.
19. Fernández-Fernández, S.; Bernabeu, A.M.; Rey, D.; Rubio, B.; Vilas, F. Determinación del color como herramienta de detección de contaminación por fuel en playas arenosas. *Geogaceta* **2011**, *50*, 165–168.

20. Bigongiari, N.; Cipriani, L.E.; Pranzini, E.; Renzi, M.; Vitale, G. Assessing shelf aggregate environmental compatibility and suitability for beach nourishment: A case study for Tuscany (Italy). *Mar. Pollut. Bull.* **2015**, *93*, 183–193. [CrossRef]

21. Cárdenes, V.; Pedrosa, E.S. Propuestas de actividad didáctica para el estudio del, tamaño y composición de arenas de playa. Un caso práctico en la costa de Bélgica. *Enseñanza de las Ciencias de la Tierra* **2015**, *23*, 315.

22. Corbí, H.; Martínez-Martínez, J. Interpretando ambientes sedimentarios: Taller de sedimentología con arenas como actividad didáctica de Ciencias de la Tierra. *Enseñanza de las Ciencias de la Tierra* **2015**, *23*, 242.

23. Corbí, H.; Martínez-Martínez, J.; Martin-Rojas, I. Linking Geological and Architectural Heritage in a Singular Geosite: Nueva Tabarca Island (SE Spain). *Geoheritage* **2019**, *11*, 703–716. [CrossRef]

24. Gilabert, E.J. *Medida del color*; Universidad Politécnica de Valencia: Valencia, Spain, 1992; p. 177.

25. Comission Internationale de l'Eclairage. *Recommendations on Uniform Color Spaces: Color-Difference Equations; Psychometric Color Terms*; CIE Publication 15 Bureau Central de la CIE Paris: Paris, France, 1976.

26. Pranzini, E.; Anfuso, G.; Botero, C.M.; Cabrera, A.; Campos, Y.A.; Martinez, G.C.; Williams, A.T. Sand colour at Cuba and its influence on beach nourishment and management. *Ocean. Coast. Manag.* **2016**, *126*, 51–60.

27. Guedes, A.; Ribeiro, H.; Valentim, B.; Noronha, F. Quantitative colour analysis of beach and dune sediments for forensic applications: A Portuguese example. *Forensic Sci. Int.* **2009**, *190*, 42–51. [CrossRef] [PubMed]

28. Harris, A.C.; Weatherall, I.L. Geographic variation for colour in the sandburrowing beetle *Chaerodes trachyscelides* White (Coleoptera: Tenebrionidae) on New Zealand beaches analysed using CIELAB L* values. *Biol. J. Linn. Soc.* **1991**, *44*, 93–104. [CrossRef]

29. Tsujimoto, G.; Tamai, M. Analysis of beach sand colour and its application to sedimentation. In Proceedings of the 7th International Conference on Asian and Pacific Coasts (APAC 2013), Bali, Indonesia, 24–26 September 2013; pp. 141–146.

30. Estévez, A.; Vera, J.A.; Alfaro, P.; Andreu, J.M.; Tent-Manclús, J.E.; Yébenes, A. Geología de la provincia de Alicante. *Enseñanza de las Ciencias de la Tierra* **2004**, *12*, 2–15.

31. Alfaro, P.; Andreu, J.M.; Estévez, A.; Tent-Manclús, J.E.; Yébenes, A. *Geología de Alicante*; AEPECT—University of Alicante: Alicante, Spain, 2004.

32. Corbí, H.; Soria, J.M.; Lancis, C.; Giannetti, A.; Tent-Manclús, J.E.; Dinarès-Turell, J. Sedimentological and paleoenvironmental scenario before, during, and after the Messinian Salinity Crisis: The San Miguel de Salinas composite section (western Mediterranean). *Mar. Geol.* **2016**, *379*, 246–266.

33. Corbí, H.; Soria, J.M. Late Miocene–early Pliocene planktonic foraminifer event-stratigraphy of the Bajo Segura basin: A complete record of the western Mediterranean. *Mar. Petrol. Geol.* **2016**, *77*, 1010–1027.

34. Alfaro, P.; Andreu, J.M.; Estévez, A.; Pina, J.A.; Yébenes, A. *Itinerarios Geológicos por la Provincia de Alicante*; Publicaciones de la Universidad de Alicante: Alicante, Spain, 2008; p. 317.

35. Martínez-Martínez, J.; Corbí, H.; Martin-Rojas, I.; Baeza-Carratalá, J.F.; Giannetti, A. Stratigraphy, petrophysical characterization and 3D geological modelling of the historical quarry of Nueva Tabarca island (western Mediterranean): Implications on heritage conservation. *Eng Geol.* **2017**, *231*, 88–99. [CrossRef]

36. Díaz Orueta, F.; Lourés Seoane, M.L. La globalización de los mercados inmobiliarios: Su impacto sobre la Costa Blanca. *CyTET* **2008**, *XL*, 77–92.

37. Ergin, A.; Williams, A.T.; Micallef, A. Coastal scenery: Appreciation and evaluation. *J. Coast. Res.* **2006**, *22*, 958–964. [CrossRef]

38. Botero, C.; Anfuso, G.; Williams, A.T.; Palacios, A. Perception of coastal scenery along the Caribbean littoral of Colombia. *J. Coast. Res.* **2013**, *65*, 1733–1739. [CrossRef]

39. Williams, A.T.; Rangel-Buitrago, N.G.; Anfuso, G.; Cervantes, O.; Botero, C.M. Litter impacts on scenery and tourism on the Colombian north Caribbean coast. *Tour. Manag.* **2016**, *55*, 209–224. [CrossRef]

40. El Naturalista Digital. Las playas del Parque Regional de los "Arenales" de San Pedro, en regresión por las actuaciones sobre los arribazones que las protegen. Available online: elnaturalistadigital.blogspot.com/2006/07/las-playas-del-parque-regional-de-los.html (accessed on 13 January 2020).

41. Diputación Provincial de Alicante. *Senderos de la Arena. Guía de Playas de la Provincia de Alicante*; Diputación de Alicante, Área de Medio Ambiente: Alicante, Spain, 2010; p. 248.

42. Gobierno de España. Ministerio para la Transición Ecológica. Guía de Playas. Available online: https://www.miteco.gob.es/es/costas/servicios/guia-playas/ (accessed on 13 January 2020).

43. Instituto Geográfico Nacional. Plano del fondeadero de Lugar Nuevo. Dirección Hidrográfica, Cádiz, 1813. Available online: http://www.ign.es/web/catalogo-cartoteca/resources/pdfcards/card009746.pdf (accessed on 13 January 2020).

44. Corbí, H.; Riquelme, A.; Megías-Baños, C.; Abellan, A. 3-D morphological change analysis of a beach with seagrass berm using a terrestrial laser scanner. *ISPRS Int. Geo-Inf.* **2018**, *7*, 234.

45. Pagán, J.I.; López, M.; López, I.; Tenza-Abril, A.J.; Aragonés, L. Study of the evolution of gravel beaches nourished with sand. *Sci. Total Environ.* **2018**, *626*, 87–95.

46. Bajoelagua. Bruselas denuncia a España por las regeneraciones de playas en Baleares y Alicante. Available online: www.bajoelagua.com/articulos/noticias-buceo/1385.htm (accessed on 13 January 2020).

47. Aragonés, L.; García-Barba, J.; García-Bleda, E.; López, I.; Serra, J.C. Beach nourishment impact on *Posidonia oceanica*: Case study of Poniente Beach (Benidorm, Spain). *Ocean Eng.* **2015**, *107*, 1–12. [CrossRef]

48. Diario Información. Benidorm desiste de cubrir "el llosar" de Levante con arena. Available online: www.diarioinformacion.com/benidorm/2012/02/05/benidorm-desiste-cubrir-llosar-levante-arena/1219915.html (accessed on 13 January 2020).

49. Diario Información. 8.000 toneladas de arena para paliar los efectos del temporal en Cala Fustera de Benissa. Available online: www.diarioinformacion.com/marina-alta/2017/03/21/8000-toneladas-arena-paliar-efectos/1874226.html (accessed on 13 January 2020).

50. Chiva, L.; Pagán, J.I.; López, I.; Tenza-Abril, A.J.; Aragonés, L.; Sánchez, I. The effects of sediment used in beach nourishment: Study case El Portet de Moraira beach. *Sci. Total Environ.* **2018**, *628*, 64–73. [CrossRef] [PubMed]

51. Alicante Plaza. Xàbia retira el dique de arena de la playa que se alzó para frenar el temporal. Available online: alicanteplaza.es/Xbiaretiraeldiquedearenadelaplayaquesealzparafrenareltemporal (accessed on 13 January 2020).

52. Aberasturi, A.; Acosta, J.; Aguilera, J.C.; Alfaro, P.; Andreu, J.M.; Antón, I.; Zaragozí, A. Geolodía 17. *Alicante: Isla de Nueva Tabarca* **2017**, *7*, 43.

53. Blott, S.J.; Pye, K. GRADISTAT: A grain size distribution and statistics package for the analysis of unconsolidated sediments. *Earth Surf. Process. Landf.* **2001**, *26*, 1237–1248. [CrossRef]

54. Folk, R.L.; Ward, W.C. Brazos River bar: A study in the significance of grain size parameters. *J. Sed. Petrol.* **1957**, *27*, 3–26. [CrossRef]

55. Williams, A.T.; Micallef, A. *Beach Management. Principles and Practice*; Earthscan: London, UK, 2009; p. 480.

56. González-Correa, J.M.; Fernández-Torquemada, Y.; Sánchez-Lizaso, J.L. Short-term effect of beach replenishment on a shallow Posidonia oceanica meadow. *Mar. Environ. Res.* **2009**, *68*, 143–150.

57. López-Andrés, L. La Regeneración de las playas de San Juan y Muchavista en 1991. Available online: http://www.alicantevivo.org/2008/06/la-regeneracin-de-las-playas-de-san.html (accessed on 22 November 2019).

58. Tenza-Abril, A.J.; Pagán, J.I.; Aragonés, L.; Saval, J.M.; Serra, J.C.; López, I. 60 years of urban development in Denia and its influence on the Marineta Cassiana beach. *Int. J. of Sust. Dev. Plan* **2017**, *12*, 678–686. [CrossRef]

59. Bricio Garberí, L.; Negro Valdecantos, V.; Diez Gonzalez, J.J.; López Gutiérrez, J.S. Diseño funcional y ambiental de diques exentos de baja cota de coronación. *Ing. Civ.* **2010**, *158*, 53–61.

60. Martínez-Medina, A. Arquitectura del boom turístico (1953–1979). *In El turismo en Alicante y la Costa Blanca Canelobre* **2016**, *66*, 166–185.

61. Delgado, J.; Casado, C.L.; Estevez, A.; Giner, J.; Cuenca, A.; Molina, S. Mapping soft soils in the Segura river valley (SE Spain): A case study of microtremors as an exploration tool. *J. Appl. Geophys* **2000**, *45*, 19–32. [CrossRef]

62. Gilabert, E.; Martínez-Verdú, F. *Medida de la luz y el color. Tomo 2: Aplicaciones*; Valencia: Servicio de Publicaciones de la Universidad Politécnica de Valencia: Valencia, Spain, 2007; p. 251.

63. Martínez-Martínez, J.; Pola, A.; García-Sánchez, L.; Reyes Agustín, G.; Osorio Ocampo, L.S.; Macías Vázquez, J.L.; Robles-Camacho, J. Building stones used in the architectural heritage of Morelia (México): Quarries location, rock durability and stone compatibility in the monument. *Environ. Earth Sci.* **2018**, *77*, 167–182. [CrossRef]

64. Batalla, R.J. Sediment deficit in rivers caused by dams and instream gravel mining. Are view with examples from NE Spain. *Rev C. & G.* **2003**, *17*, 79–91.

65. Liquete, C.; Canals, M.; Arnau, P.; Urgeles, R.; Durrieu de Madron, X. The impact of humans on strata formation along Mediterranean margins. *Oceanography* **2004**, *17*, 70–79. [CrossRef]

66. Giannetti, A.; Monaco, P.; Falces-Delgado, S.; La Iacona, F.G.; Corbí, H. Taphonomy, ichnology, and palaeoecology to distinguish event beds in varied shallow-water settings (Betic Cordillera, SE Spain). *J. Iber Geol.* **2019**, *45*, 47–61. [CrossRef]

67. Cuevas-González, J.; Díez-Canseco, D.; Alfaro, P.; Andreu, J.M.; Baeza-Carratalá, J.M.; Benavente, D.; Blanco-Quintero, I.; Cañaveras, J.C.; Corbí, H.; Delgado, J.; et al. Geogymkhana-Alicante (Spain): Geoheritage through Education. *Geoheritage* **2020**, (in press).

68. Ayuntamiento de Alicante. Alicante en los años 30. Exposición en el archivo municipal. Available online: https://www.alicante.es/es/agenda/alicante-anos-30-exposicion-archivo-municipal (accessed on 13 January 2020).

69. Pinterest. Playa de San Juan en los años 60, Alicante, España. Available online: https://www.pinterest.es/pin/450782243935888731/?lp=true (accessed on 9 January 2020).

70. Biroupeplaja. Available online: http://www.biroupeplaja.ro/plaja-san-juan-din-alicante-spania-villa-royal/ (accessed on 9 January 2020).

71. Diario Información. Pilar de la Horadada regenera sus playas con más de 30000m3 de arena. Available online: https://www.diarioinformacion.com/vega-baja/2015/07/07/playas-pilar-han-sido-regeneradas/1653097.html (accessed on 9 January 2020).

72. La crónica independiente. Las playas del Pilar han sido regeneradas con más de 30.000 m^3 de arena, "el doble que otros años". Available online: http://lacronicaindependiente.com/2015/07/las-playas-del-pilar-han-sido-regeneradas-con-mas-de-30-000-m3-de-arena-el-doble-que-otros-anos/ (accessed on 9 January 2020).

MDPI

St. Alban-Anlage 66

4052 Basel

Switzerland

Tel. +41 61 683 77 34

Fax +41 61 302 89 18

www.mdpi.com

Water Editorial Office

E-mail: water@mdpi.com

www.mdpi.com/journal/water